Studies in Systems, Decision and Control

Volume 344

The series "Studies in Systems, Decision and Control" (SSDC) covers both new developments and advances, as well as the state of the art, in the various areas of broadly perceived systems, decision making and control–quickly, up to date and with a high quality. The intent is to cover the theory, applications, and perspectives on the state of the art and future developments relevant to systems, decision making, control, complex processes and related areas, as embedded in the fields of engineering, computer science, physics, economics, social and life sciences, as well as the paradigms and methodologies behind them. The series contains monographs, textbooks, lecture notes and edited volumes in systems, decision making and control spanning the areas of Cyber-Physical Systems, Autonomous Systems, Sensor Networks, Control Systems, Energy Systems, Automotive Systems, Biological Systems, Vehicular Networking and Connected Vehicles, Aerospace Systems, Automation, Manufacturing, Smart Grids, Nonlinear Systems, Power Systems, Robotics, Social Systems, Economic Systems and other. Of particular value to both the contributors and the readership are the short publication timeframe and the world-wide distribution and exposure which enable both a wide and rapid dissemination of research output.

Indexed by SCOPUS, DBLP, WTI Frankfurt eG, zbMATH, SCImago.

All books published in the series are submitted for consideration in Web of Science.

More information about this series at http://www.springer.com/series/13304

Cengiz Kahraman · Eda Bolturk
Editors

Toward Humanoid Robots: The Role of Fuzzy Sets

A Handbook on Theory and Applications

Springer

Editors
Cengiz Kahraman
Department of Industrial Engineering
Management Faculty
Istanbul Technical University
Maçka, Istanbul, Turkey

Eda Bolturk
Department of Industrial Engineering
Management Faculty
Istanbul Technical University
Maçka, Istanbul, Turkey

ISSN 2198-4182 ISSN 2198-4190 (electronic)
Studies in Systems, Decision and Control
ISBN 978-3-030-67162-4 ISBN 978-3-030-67163-1 (eBook)
https://doi.org/10.1007/978-3-030-67163-1

This Springer imprint is published by the registered company Springer Nature Switzerland AG
The registered company address is: Gewerbestrasse 11, 6330 Cham, Switzerland

Cengiz Kahraman · Eda Bolturk
Editors

Toward Humanoid Robots: The Role of Fuzzy Sets

A Handbook on Theory and Applications

Springer

Editors
Cengiz Kahraman
Department of Industrial Engineering
Management Faculty
Istanbul Technical University
Maçka, Istanbul, Turkey

Eda Bolturk
Department of Industrial Engineering
Management Faculty
Istanbul Technical University
Maçka, Istanbul, Turkey

ISSN 2198-4182 ISSN 2198-4190 (electronic)
Studies in Systems, Decision and Control
ISBN 978-3-030-67162-4 ISBN 978-3-030-67163-1 (eBook)
https://doi.org/10.1007/978-3-030-67163-1

This Springer imprint is published by the registered company Springer Nature Switzerland AG
The registered company address is: Gewerbestrasse 11, 6330 Cham, Switzerland

I dedicate this book to the medical doctors over the world who are struggling with the corona virus.

Prof. Cengiz Kahraman

I dedicate this book to my beloved parents, Müzeyyen and Nejdet and my brother Taha.

Dr. Eda Bolturk

Preface

Humanoid robots with their body shapes having a torso, a head, two arms, and two legs are the robots built to resemble all kinds of the movements of human body. Presently, humanoid robots are produced for functional purposes such as interacting with human tools and environments. For instance, a humanoid robot can work at a tourist information center. Humanoid robots may be also produced to resemble only a certain part of the human body. Some humanoid robots have heads designed to replicate human facial features such as eyes and mouths, whereas some humanoid robots have arms designed to do complex and toxic operations.

It cannot be stated that today's humanoid robots are very successful in imitating the movements of the human body. Real human emotions cannot be modeled by opening or closing their eyes while talking, moving them up or down or left and right. Humanoid robots must be able to express their emotions appropriately to what they see and/or hear. This can be possible using continuous logic, but never possible by discrete logic.

Fuzzy set theory based on a continuous logic has a great potential in modeling the movements and emotions of humanoid robots. Everything is a matter of degree in fuzzy logic. The facial expression of a person who becomes happy by hearing that her friend is married will be different from the facial expression of a person who is happy by hearing that her friend has successfully passed her Ph.D. exam. Hence, the way of smiling and its degree is different in each of these events. The vagueness and impreciseness in human thoughts can be captured by the fuzzy sets and their extensions.

The aim of this book is to present the theory and applications of fuzzy sets in modeling the movements and emotions of humanoid robots. This book involves 11 chapters, each written by the experts of that research area. The first chapter summarizes the present status of humanoid robots in the world. Besides, a literature review on humanoid robot publications is given and books, conferences, institutes, and journals on humanoid robots are presented.

The second chapter presents a comprehensive literature review on the recent developments and theories associated with fuzzy set extensions. The recently developed fuzzy set extensions from type-2 fuzzy sets to t-spherical fuzzy sets are classified and presented by their main definitions and operations.

The third chapter presents some fuzzy modeling techniques for facial expressions of a humanoid robot depending on the degrees of the emotions. Larger degree of emotion causes a stronger indicator of the facial mimic. Intuitionistic fuzzy sets and Pythagorean fuzzy sets are employed for modeling facial expressions.

The fourth chapter argues that to make sure that human-like robots exhibit human-like behavior, it is needed to use fuzzy techniques. The authors provide details of this usage. The chapter is intended for both researchers and practitioners who are very familiar with fuzzy techniques and also for researchers and practitioners who do not know these techniques—but who are interested in designing human-like robots.

The fifth chapter presents the metaheuristics methods used in the control of robots. A literature review and graphical analyses are given on this research area. The chapter shows that metaheuristics can be used as important building blocks in humanoid robots together with fuzzy set theory.

The sixth chapter tries to find the best-suited algorithm to narrow down future research in the field of test automation and provide issues on the design of new proposals. The authors focus on the performance evaluation of different major meta-heuristic algorithms, namely hill climbing algorithm, particle swarm optimization, firefly algorithm, cuckoo search algorithm, bat algorithm, and artificial bee colony algorithm. Each algorithm is implemented to automatically generate test suites based on the program under test. Then, we develop a performance evaluation of each algorithm for five programs written in Java. The algorithms are compared using several process metrics and product metrics.

The seventh chapter presents a comparative study using three types of controllers, FLC, PI and PID, applied to the speed control of a robot built using the ev3 Lego Mindstorms kit. MATLAB and Simulink are used to validate the performance of the speed control obtained with the proposed controller.

The eighth chapter maintains a specific location and behavior for a robot that uses type-2 fuzzy logic for controlling its behavior. The authors propose a combination of behaviors by following a trajectory without leaving or losing it and avoiding obstacles in an omnidirectional mobile platform. The results of the simulation show the advantages of the proposed approach.

The ninth chapter reviews the recent applications and research papers of humanoid robots related to fuzzy control. Studies about the humanoid robots and fuzzy logic-based control are grouped under four major topics; the first one is stability and reliability control, the second one is walking pattern detection, the third one is navigation, and the final one is obstacle avoidance.

The tenth chapter explains how the concept of single-valued trapezoidal neutrosophic (SVTN) numbers can be applied in the field of humanoid robotics. To explain the concept of SVTN numbers, a multi-robot scenario is considered consisting of a central server and a group of mobile robots patrolling a given area for surveillance application. Using the correlation coefficient of SVTN numbers, the sensor readings are properly interpreted for proper identification of the problem faced by the robot.

The eleventh chapter aims at highlighting the recent developments in the field of neutrosophic graph (NG) theory and their generalizations including neutrosophic hypergraph, interval NGs, bipolar NGs, etc. Almost all the work based on the development of NGs and their applications are discussed thoroughly.

We would like to thank the authors for their efforts in writing their studies and the anonymous reviewers for their hard works in selecting high-quality chapters in this book. We would like to express our sincere thanks to Prof. Janusz Kacprzyk for his continuous supports and helps.

Maçka, Istanbul, Turkey
January 2021

Cengiz Kahraman
Eda Bolturk

Contents

Introduction to Humanoid Robots and Fuzzy Sets

Humanoid Robots and Fuzzy Sets

Eda Bolturk and Cengiz Kahraman

Abstract Humanoid robot is a research area in robotics. Humanoid robots are shaped in order to resemble human shape. These robots are used for different purposes in different industries such as health, space, education and manufacturing. In this chapter, we try to summarize the present status of humanoid robots in the world.

Keywords Fuzzy sets · Humanoid robots · Types of humanoid robots

1 Introduction

Artificial intelligence (AI) used in robots is the theory and development of computer systems able to perform tasks normally requiring human intelligence, such as visual perception, speech recognition, decision-making, and translation between languages. People think of AI as a robot gliding around and giving mechanical replies. There are many forms of AI but humanoid robots are one of the most popular forms. One of the earliest forms of humanoids was created in 1495 by Leonardo Da Vinci. It was an armor suit and it could perform a lot of human functions such as sitting, standing and walking. Initially, the major aim of AI for humanoids was for research purposes. Now, humanoids are being created for several purposes that are not limited to research. Classical logic (0-1 logic) has been used in most of these robot technologies, causing rigid and sharp mimics of human behaviors. Discrete logic based technologies cannot imitate the continuous structure of human intelligence.

Present humanoids are developed to carry out different human tasks and occupy different roles in the employment sector. Some of the roles they could occupy may be the roles of a personal assistant, receptionist, front desk officer and so on. The process of inventing a humanoid is quite complex since it is absolutely hard to resemble human behaviors. Most times, inventors and engineers face some challenges. Humanoids move, talk and carry out actions through certain features such as

E. Bolturk (✉) · C. Kahraman
Department of Industrial Engineering, Istanbul Technical University, 34367 Macka, Istanbul, Turkey
e-mail: bolturk@itu.edu.tr

C. Kahraman and E. Bolturk (eds.), *Toward Humanoid Robots: The Role of Fuzzy Sets*, Studies in Systems, Decision and Control 344,
https://doi.org/10.1007/978-3-030-67163-1_1

sensors and actuators. Humans behave based on a continuous decision making mechanism. Humans can rapidly select the appropriate behavior when they face any case among infinite number of possible cases. For instance, they can decide the degree of laughing when they face a comic case and the other body parts behave suitable to this degree.

Humanoid robots are designed to look like humans for intuitive collaboration, and the latest locomotion and AI technology is helping to speed up their development. Robots come in many shapes and sizes. But, perhaps, the most intriguing, endearing, and acceptable are the ones that resemble us, humans. Humanoid robots are used for research and space exploration, personal assistance and caregiving, education and entertainment, search and rescue, manufacturing and maintenance, public relations, and healthcare.

In the area of healthcare, it is quite clear that we need humanoid robots more and more today. As viruses like COVID 19 and chickenpox spread to the rest of the world, robots are being deployed in many countries. Some robots can help relieve tired nurses in the hospitals, do basic cleaning and deliveries.

Humanoids are usually either Androids or Gynoids [1]. An Android resembles a male human while gynoids resemble female humans. Humanoids have sensors that aid them in sensing their environments. These sensors can use crisp or fuzzy logic based Technologies. Fuzzy logic based control technologies are superior to the other discrete logic based technologies since every behavior of humans is a matter of degree which can be handled by a membership function. Some sensors have cameras that enable them to see clearly. Motors placed at strategic points are what guide them in moving and making gestures. These motors are usually referred to as actuators.

After the purpose of the humanoid robot is determined, robot researchers have to start the coding process which is one of the most vital stages in creating a humanoid. Coding is the stage whereby these inventors program the instructions and codes that would enable the humanoid to carry out its functions.

The robotics industry is improving day by day based on human requirements. Humanoid robotics is an emerging and challenging research field receiving significant attention during the past years. Professor Ichiro Kato from Waseda University and his colleagues developed the first humanoid robot in 1973. The name of the humanoid robot was WABOT-l [2]. Later, the WABOT-1 was differentiated to its different types: WABOT-2 developed in 1984 was a humanoid robot that could read a musical score and play an electronic keyboard and Hadaly-2 was intended to realize information interaction with humans by integrating environmental recognition with vision, conversation capability (voice recognition, voice synthesis) and the bipedal walking robot WABIAN was developed in 1997. Humanoid Robotics Institute was established for the connection of machines and humans in April 2000 [2] .

Honda started the humanoid robot research and development program in 1986. Honda's keys to the development of the robot were *intelligence* and *mobility*. The basic concept was that the robot should coexist and cooperate with human beings, by doing what a person cannot do and by cultivating a new dimension in mobility to ultimately benefit society. This provided a guideline for developing a new type of robot that would be used in daily life, rather than a robot purpose-built for special operations.

The first prototype of a man-like model with upper limbs and the body was called as P1. The robot could turn external electrical and computer switches on and off, grab doorknobs, and pick up and carry things. Coordination between arm and leg movements was also another research area. In December 1996, Honda introduced the P2, the most advanced humanoid robot ever built up to that time. The weight and height of P2 were 175 kg and 191.5 cm, respectively. P2 is considered the world's first self-regulating two-legged humanoid walking robot. P2's torso contained a computer, motor drives, battery, wireless radio and other necessary technology [2]. Independent walking, walking up and down stars, cart pushing and other operations were achieved without wires, allowing independent operations. Independent walking at 2 km/h, walking up and down stairs, cart pushing, and other operations were achieved without wires.

The weight and height of P3 were 210 kg and 182 cm, respectively. P3, which was the first completely independent, two-legged humanoid walking robot was completed in September 1997. The weight and height of P3 were 130 kg and 160 cm, respectively. The P3 is considered the world's first completely independent two-legged humanoid walking robot (2 km/h). The P3 was loaned to Japan's National Institute of Advanced Industrial Science and Technology (AIST) in 1998 to help kick start their Humanoid Robot Project. Further efforts were made to reduce size and weight, and improve dynamic performance and operability. ASIMO (Advanced Step in Innovative Mobility) is a humanoid robot created by Honda in 2000. It is currently Marikinain Tokyo, Japan. The name was chosen in honor of Isaac Asimov.

The aim of this chapter is to summarize the present status of the research on humanoid robots in the world. In addition, a literature review is presented in order to show the academic studies on humanoid robots.

The rest of the chapter is organized as follows. In Sect. 2, the types of robots are presented. In Sect. 3, humanoid robots and fuzzy logic relations are explained. In Sect. 4, today's humanoid robots are introduced. In Sect. 5, a literature review on humanoid robot publications is presented. In Sect. 6, books on humanoid robots are given. In Sect. 7, journals on humanoid robots are introduced. In the last section, the conclusions are presented.

2 Types of Robots

Robots are used for different reasons with different features. There are about 15 types of robots according to IEEE. They are presented as follows [3]:

- Aerospace
- Consumer
- Disaster Response
- Drones

- Education
- Entertainment
- Exoskeletons
- Humanoids
- Industrial
- Medical
- Military & Security
- Research
- Self-Driving Cars
- Telepresence
- Underwater

These types of robots are tried to explain shortly in the following.

Aerospace: They perform in space and flying robots are kind of these robots.

Consumer: These kind of robots are used in order to help in any chores and tasks.

Disaster Response: These robots perform in emergency situations for dangerous jobs and generally used for dangerous Works such as: earthquakes and tsunamis.

Drones: They can be called as remote-controlled aerial vehicles. Drones are shaped at various sizes.

Education: The education robots are generally used for the use in classrooms and at home. They are like teacher robots.

Entertainment: They are used in order to robots are designed to arouse emotional responses.

Exoskeletons: The robots are used for empowering paralyzed patient and physical rehabilitation.

Humanoids: These robots' appearance is like humans. Humanoid robots have androids and mechanical visual aspect.

Industrial: These robots are industrial and are used for execute iterative duties especially in warehouses and factories.

Medical: System of these robots is used in surgery and healthcare.

Military & Security: These robots include ground and autonomous mobile systems in order to transporting heavy gear.

Research: Research robots are used for doing functional things especially for helping researchers.

Self-Driving Cars: These robots are like cars as autonomous vehicles and they drive themselves.

Telepresence: These robots allow anyone to be a place without going to there via an avatar with the internet and anyone can communicate with others.

Underwater: These robots are featured for bio-inspired systems, diving humanoids, and, deep-sea submersiblest.

3 Humanoid Robots and Fuzzy Logic

Fuzzy logic presents opportunities for flexible modeling human behaviors. Otherwise, classical logic would cause more rigid and sharp movements on humanoid robots. Because of the usage of classical logic, inventors face a few challenges in creating fully functional and realistic behaviors. Some of these challenges include:

Actuators These are the motors that help in motion and making gestures. The human body is dynamic. To make a humanoid robot, you need strong, efficient fuzzy controlled actuators that can imitate these actions flexibly and within the same time frame or even less. The actuators should be efficient enough to carry a wide range of actions [1]. Fuzzy set theory provides a soft modeling tool similar to human beings.

Sensors Fuzzy logic is especially useful in the design of these components. These help humanoids to sense their environment. Humanoids need all the human senses [1]: touch, smell, sight, hearing and balance to function properly. The hearing sensor is important for the humanoid to hear instructions, decipher them and carry them out. The touch sensor prevents it from bumping into things and causing self-damage. The humanoid needs a sensor to balance movement and equally needs heat and pain sensors to know when it faces harm or is being damaged. Facial sensors also need to be intact for the humanoid to make facial expressions, and these sensors should be able to carry a wide range of expressions. In fuzzy sets, everything is a degree of matter. Their membership functions can model these expressions and movements through fuzzy control techniques in a continuous manner.

Humanoid robot technologies are secret technologies that what type of actuators and sensors, which artificial intelligence tools are used are not open. Fuzzy logic technologies are possibly used in the control of many humanoid robots of today. For instance, humanoid robot Pepper is so close to human thinking and perception style which is an indicator of the usage of fuzzy logic.

4 Today's Humanoid Robots

Today's humanoid robots are human-like and efficient to some degree. Here are a few of them.

The Kodomoroid TV Presenter This humanoid robot was invented in Japan. Her name is derived from the Japanese word for child—Kodomo—and the word 'Android' [4]. She speaks a number of languages and is capable of reading the news and giving weather forecasts. She has been placed at the Museum of Emerging Science and Innovation in Tokyo where she currently works.

Jia Jia This humanoid robot was worked on for three years by a team at the University of Science and Technology of China before its release. She is capable of making conversations but has limited motion and stilted speech. She does not have a full range of expressions but the team of inventors plans to make further developments and infuse learning abilities in her. Although her speech and vocabulary need further work, she is still fairly realistic [1].

ASIMO Advanced Step in Innovative Mobility, ASIMO for short, was created by HONDA in 2000, is a humanoid robot is designed in order to help to people. It is an entertainment and research humanoid robot and is from Japan. ASIMO which is Bipedal walking based on Zero Moment Point control approach, can dance, run, kick, jump, know faces, recognize speech and navigate human environments [5].

The first robot citizen Sophia Sophia is a gynoid humanoid robot and one of the popular humanoid robot in the world. She is the world's first robot citizen and granted Saudi Arabian citizenship. She was introduced to the United Nations on October 11, 2017. Hanson robotics created Sophia. She can do fifty facial expressions and her looking is very similar to human being. Sophia's AI aggregates cutting-edge operate in symbolic AI, adaptive motor control, neural networks, expert systems, cognitive architecture conversational natural language processing, and machine perception [6]. Sophia can identify faces of humans, see emotional expressions, and recognize various hand movements [6].

Ocean One This robot is invented by Stanford Robotics Lab and used in underwater. It is called bimanual underwater humanoid robot. Ocean one can go deeper in underwater and can explore places which humans cannot [7, 8]. It is stated that this robot dived 100 m below the Mediterranean sea in 2016. Ocean One investigated King Louis XIV's flagship which sank off the southern coast of France in 1664 [7]. It is stated that, Ocean One's system is the combination of artificial intelligence and haptic feedback systems and Ocean One's electronics are submersible in oil. This mechanisms aids to work under deep water [7].

Pepper This humanoid robot is created by SoftBank Robotics and a kind of educational robot. Pepper known as first social humanoid robot [9] Pepper can read human feeling for example sadness, anger and joy. It is stated that Pepper could communicate in intuitively and naturally and anyones' facial look and aspects. Different sensors are equipped in Pepper and this humanoid robot can discover and acquire about people [7]. Pepper offers an open and a platforms which is programmable. It has 20 degrees of independence in order to expressive and natural motion. It can speak in 15 languages such as English, German and Dutch and has the ability of speaking identification. There are perceptions modules in order to interaction with people. It has LEDs, touch sensors and microphones in order to interact multimodal and sonars, infrared sensors, one inertial unit, bumpers, 2D and 3D cameras, and sonars for navigation [9].

Nao It is a humanoid robot which is invented by Aldebaran Robotics and SoftBank obtained this humanoid robot [7]. NAO is used in research and education and assist visitors in healthcare centers in order to amuse, communicate and greet. NAO is worked with more than 600 best universities, labs, and secondary schools in the world [10]. NAO's 6 features are as follows. It offers an open and programmable platform. It has 25 degrees of independence in order to adapt and move. It can speak 20 languages such as English, German and Dutch and the ability of speaking identification. There are 4 directional microphones in order to interaction with people. There are 7 touch sensors which are placed in NAO's hands, head and feet. In order to understand NAO's environment, it uses 2 2D cameras, and sonars for navigation [11].

ATLAS It is created by Boston Dynamics and it is known as "the world's most dynamic humanoid" that is supervised by United States Defense Advanced Research Projects Agency (DARPA) [7]. The features of ATLAS are as follows [12]: 1.5 m height, 80 kg weight, 1.5 m/s speed, compact mobile hydraulic systems, 3D printed parts in order to add itself the strength-to-weight ratio necessary for tumbles and leaping, custom motors and valves, and advanced control system.

PETMAN PETMAN (Protection Ensemble Test Mannequin), was funded by US Department of Defense's Chemical and Biological Defense program, is created by Boston Dynamics and enhanced in order to examine chemical and biological teams for the US military [7]. The type of this humanoid robot's type is Military & Security humanoid robot [13]. This bipedal humanoid robot can squat and walk and do push-ups [13]. PETMAN's basic usage is for serving like a crash dummy of sorts for testing biological and chemical agents' effects and aid in any condition in biological or chemical war. PETMAN's features are as follows [13]: Anthropomorphic design, equipped with custom hydraulic actuators, 177.8 cm height, 79.4 kg weight, 6.44 km/h speed. It can alter its body temperature and humidity. Atlas is known as PETMAN's big brother. The other related humanoid robots with PETMAN are BEAR, TORO, HRP-4 and Robo Thespian.

ROBEAR Robear is a humanoid robot that was introduced by scientists from RIKEN and Sumitomo Riko Company Limited. This humanoid robot is bear shaped and is an an experimental nursing-care robot that is designed to lift patients out of beds and into wheelchairs, as well as helping those who need assistance to stand up [14]. The features of Robear are as follows [7]: 146 kg weigh, lighter than its earlier models RIBA (released in 2009) and RIBA II (released in 2011), integrated 3 kinds of sensors (includes smart rubber capacitance-type tactile sensors and torque sensors). In addition, Robear was seen as a potential solution to the problem of increasing lack of health care givers which Japan is set to observe.

Surena Surena is a research humanoid robot from Iran that is designed to in order to study artificial intelligence and bipedal locomotion. It's used to attract students to careers in engineering and provide entertainment at special events. The creator is University of Tehran. Surena can walk, pick up objects, dance, stand on one foot. climb, detect, recognize speech and do action imitation. The features of Surena are as follows [15]: 190 cm height, 46 cm length, 64 cm width, 98 kg weight, 0.7 km/h speed. It is said that Surena has 4 versions which are SURENA I (2008), Surena II (2010), Surena III (2015) and Surena IV [16]. Surena I had 8 degrees of freedom, Surena II had 22 degrees of freedom. SURENA III has 31 degrees of freedom [15].

Robotic Avatar Initially introduced by Toyota in 2017 [17], the T-HR3 is a humanoid robot that mimics the movements of its human operator, like a real-world avatar. Updated for Tokyo Olympics, the T-HR3 has improved controls and could walk more naturally. Envisioned as a mobility service, in the future these humanoids will be able to perform surgeries while their operators, human doctors, will be controlling them from another part of the world. It can also help caregivers to do their work remotely, or those in need of assistance to live a more independent life.

Delivery Robot Digit Ford became the first customer to incorporate Agility Robotics' Digit into a factory setting. The headless humanoid Digit has nimble limbs and is packed with sensors. It can navigate the stairs, various obstacles, and all kinds of terrains. It can balance on one foot, but usually it walks upright and is strong enough to pick up and stack boxes weighting up to 40 lb [18]. It can also fold itself for compact storage. Ford envisions that Digit will ride in a driverless car and deliver packages to customers, automating the whole delivery process [17].

Digital Humanoids Digital human beings look and act like humans but are entirely virtual. One example is Samsung Technology and Advanced Research (STAR) Labs' Neons [19], AI-powered beings with unique personalities and looks. These artificial humans are not designed to answer any questions like Alexa or Siri, but are supposed to show emotions, learn from experiences, and have real conversations. Each Neon is computer-generated and not necessisarly based on real people, and each can be customized for a different role, like a virtual doctor or a yoga instructor [17].

Robotic Bartender Kime is a food and beverage serving robot, developed by Macco Robotics in Spain. It has a human-like head and torso with two arms inside a kiosk. Tested at gas stations in Europe and in a Spanish brewery, Kime is known to be quite good at pouring beer and can serve up to 300 glasses per hour. The humanoid features 14 to 20 degrees of freedom, has smart sensors and uses machine learning to improve on its skills [17].

Robotic Actor Founded by director Will Jackson in 2004, Engineered Arts is a U.K.-based company that produces different entertainment humanoids through collaboration between artists, mechanical and computer engineers, and animators. For instance, their first humanoid—the well-known RoboThespian—is a robotic actor that comes with a library of impressions, greetings, songs, and gestures. The company is working on adding RoboThespian the ability to walk on its own, but for now the movement can be staged through a hidden system of tracks and dollies [17].

Robonauts Several countries have been working on humanoids for space exploration. In India, Vyommitra, a female humanoid robot, is set to launch on an uncrewed spaceflight in December 2020 [17]. The robot is scheduled to conduct microgravity experiments to help prepare for future crewed missions. NASA's Johnson Space

Center has worked on several humanoids, including Robonaut 2 (that spent seven years aboard the ISS) and Valkyrie. It's possible that future spacefaring humanoids will be designed to withstand harsh environments of the Moon or the Mars [17].

Fedor (Final Experimental Demonstration Object Research) It was a Russian remote-controlled humanoid that flew to the International Space Station (ISS) in 2019, where it simulated repairs during a spacewalk, and later returned back to Earth [17].

Collaborative Humanoids Most humanoids are intrinsically human collaborators. For instance, Nextage from Kawada Robotics is a humanoid research platform for industrial robots for Industry 4.0. Armar from Germany's Karlsruhe Institute of Technology was developed to perform maintenance tasks alongside human workers in industrial settings [18]. Walker by UBtech Robotics, on the other hand, is designed to collaborate with humans in their homes [17]. With seven degrees-of-freedom manipulators, the humanoid was developed to perform household tasks and smart home control.

HRP-2 It was created by Kawada Industries and AIST in 2002 in Japan. It is a research humanoid. Yutaka Izubuchi was created HRP-2's exterior. HRP-2 can crawl, stand up, sit, cooperate with humans in lifting heavy objects, walk on uneven surfaces, and even get up by itself if it falls over. HRP-2 doesn't need a "power backpack" because of having electrical system and spherical compact battery. The features of HRP-2 are as follows [20]: 154 cm height, 33.7 cm length, 65.4 cm width, 58 kg weight, 2 km/h speed. It includes 30 motors in order to shift its waist, legs, head, and arms.

Albert Hubo It was created by KAIST and Hanson Robotics in 2005 in South Korea, is a research humanoid robot. Sadness, sadness and anger can be displayed from Albert Hubo's face and it is an origination of Hubo series. Albert Hubo's heads made by Hanson Robotics and the project of this humanoid robot was in the leadership of Professor Jun-Ho Oh [21]. The features of Albert Hubo are as follows [21]: 125 cm height, 57 kg weight, and 1.37 km/h speed

Hubo 2 It was created by KAIST in 2009 in South Korea, is a research humanoid robot. This full size humanoid robot can do run, walk, grasp and dance. The features of Hubo 2 are as follows [22]: 125 cm height, 45 kg weight, 1.5 km/h walking speed, 3.6 km/h running speed, lightweight and modular design, and high performance actuation system.

Mahru It was created by Korea Institute of Science and Technology in 2005 in South Korea, is a research humanoid robot. This humanoid robot can walk, dance and do household tasks and Taekwondo. The features of Mahru are as follows [23]: 40 cm height, 65 kg weight, 1.2 km/h walking speed, recognizing objects. It faces and places, has autonomous and teleoperation modes, navigation rooms, avoidation obstacles.

AR-600 It was created by Android Technologies in 2012 in Russia, is a research humanoid robot. This full size humanoid robot can walk, motion tools and objects and talk. The features of AR-600 are as follows [24]: 145 cm height, 65 kg weight, 2 km/h speed, recognition of objects and faces. The former versions of AR600 are AR-400 in 2006, AR-400 M in 2007 and AR-600 in 2008, respectively. The owner of these robots is belonged to Android Technologies [24].

Erica Erica (Erato Intelligent Conversational Android), was created by Osaka University, Kyoto University, and ATR in 2015 in Japan, is a research humanoid robot. Erica is an android humanoid robot and can find out natural language. The features of Erica are as follows [25]: 166 cm height, find the natural language, recognition of objects and faces, has compositive human-like voice, expressing facial aspect.

Walker It was created by UBtech Robotics, is a collaborator humanoid robot. It is laid out for collaboration in humans' homes. It is introduced for performing household tasks and intelligent home control along with seven degrees-of-freedom manipulators [17]. Walker can walk up and down stairs. Walker can move on all types of surfaces. Walker can detect and recognize corresponding faces, objects and environments in complex and background heavy situations [26].

TALOS It was a humanoid robot created by PAL Robotics. This humanoid robot is fully electrical and the fields of application are IoT, rescue, exploration of space. The usage areas are as follows [27]: Human-Robot interaction, Manipulation, Perception, Navigation, Torque Control, Artificial Intelligence, Machine learning, Trajectory optimization, Multi-contact motion planning, and Dynamic walking. The features of TALOS are as follows [27]: 175 cm height, 95 kg weight, 1.5 h walking/3 h stand-by battery autonomy, 6 kg arm stretched – arm/gripper payload, full torque sensor feedback in all joints, bipedal humanoid robot, head and Gripper fully customizable.

REEM-C REEM-C is a humanoid biped robot created by PAL Robotics. This humanoid robot can make a step when a force from adventitious force for not to fall. The usage areas are as follows [28]: Artificial Intelligence/Machine Learning, Dynamic Walking, Human-Robot Interaction, Manipulation, Multi-contact motion planning, Navigation, Perception, and Trajectory Optimization. The features of REEM-C are as follows [28]: 165 cm height, 80 kg weight, 1 kg hand payload, 3 h walking/6 h stand by battery autonomy, operating System based on Real Time OS, ubuntu LTS, speak 30+ languages, get up a chair and sit on.

5 Literature Review on Humanoid Robots

We search the "Humanoid Robot" term in Scopus and find 13.991 results. In Figs. 1, 2, 3, 4, 5 and 6, the literature review results of Humanoid Robots are given.

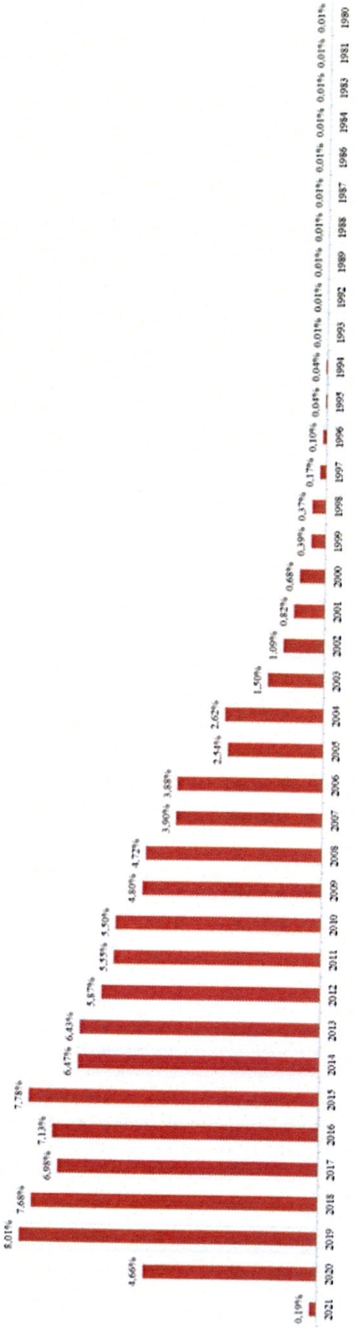

Fig. 1 Distribution based on humanoid robot papers with respect to years

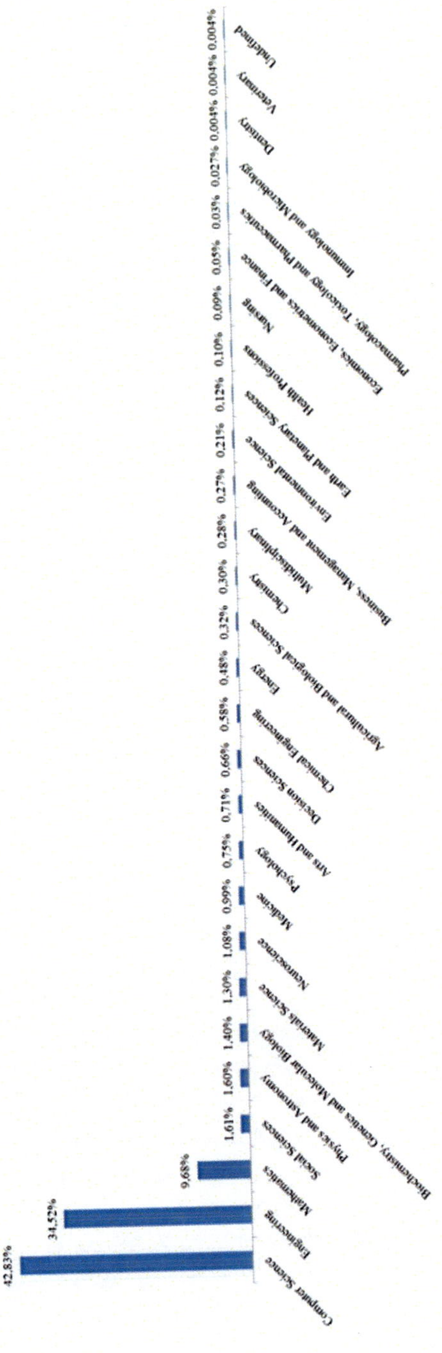

Fig. 2 Distribution of humanoid robot papers with respect to subject areas

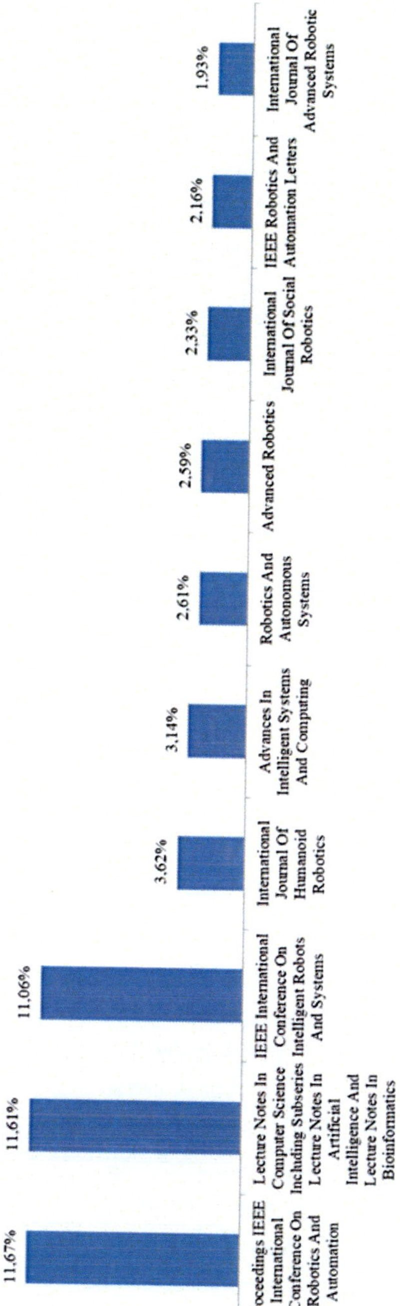

Fig. 3 Distribution of humanoid robot papers by their sources

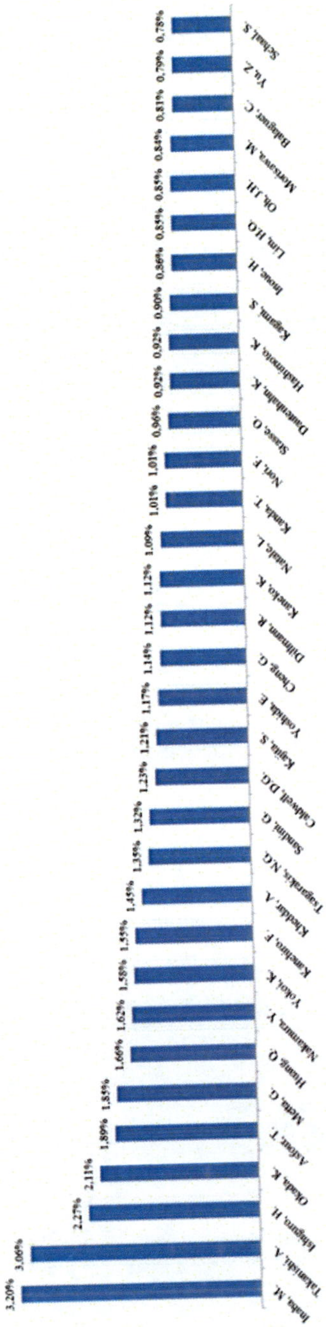

Fig. 4 Distribution of publication percentages of authors on humanoid robots

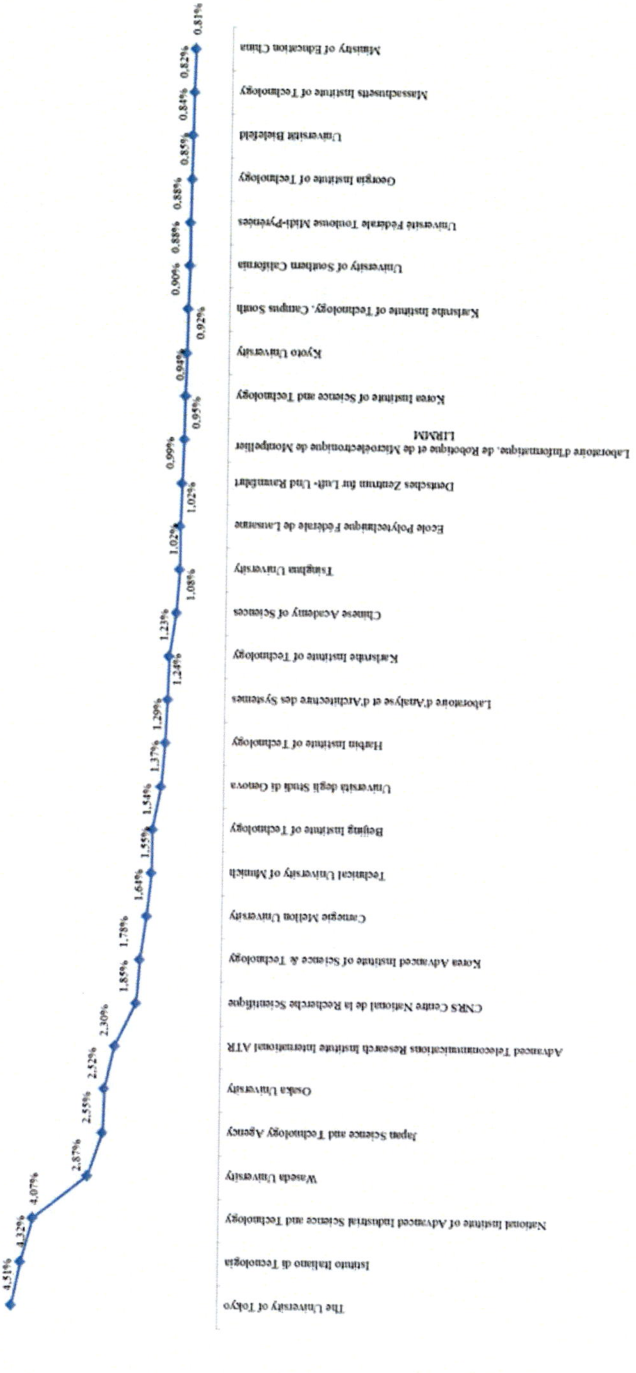

Fig. 5 Distribution of affiliations on humanoid robot papers

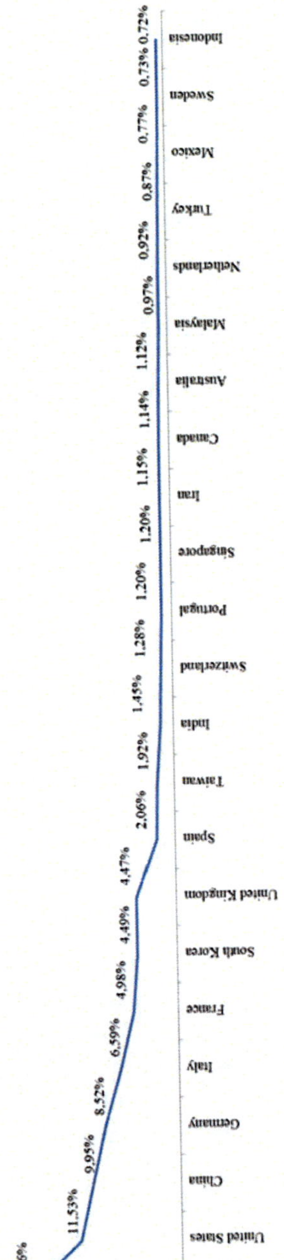

Fig. 6 Distribution of humanoid robot papers by their countries

In Fig. 1, the distribution of papers by years is given. First paper is published in 1980. The most papers were published in 2019 with a rate of 8.01%. The percentage of published papers based on humanoid robots is increased substantially since 1998.

In Fig. 2, the distribution of humanoid robot papers is illustrated by their subjects areas. The most published subjects are computer science, engineering and mathematics, respectively.

In Fig. 3, the percentages of humanoid robot papers based on their sources are given. Most of the publications have been published in Proceedings IEEE International Conference on Robotics and Automation and Lecture Notes in Computer Science including Subseries Lecture Notes in Artificial Intelligence and Lecture Notes in Bioinformatics.

In Fig. 4, the publication percentages and the corresponding numbers of authors on humanoid robot are presented. Inaba and Takanishi are the leaders among these authors.

In Fig. 5, the distribution of affiliations is presented based on humanoid robots papers. In addition to presented affiliations on Fig. 5, there are nearly 130 affiliations on these papers. The university of Tokyo and Istituto Italiano di Tecnologia are leaders among other affiliations.

In Fig. 6, the distribution of publications on humanoid robots with respect to their source countries is illustrated. Japan is the leading country on humanoid robots. United States and China are the next two countries after Japan, respectively.

The percentages and the corresponding numbers of humanoid robots are illustrated in Fig. 7. The document types on humanoid robots are: conference papers with a percentage of 64.16%, article with a percentage of 28.83%, conference review with a percentage of 3.17, book chapters with a percentage of 2.07, review with a percentage of 1, editorial with a percentage of 0.25, book a percentage of 0.16, note with a percentage of 0.11 and the others are erratum, short survey, letter and undefined ones.

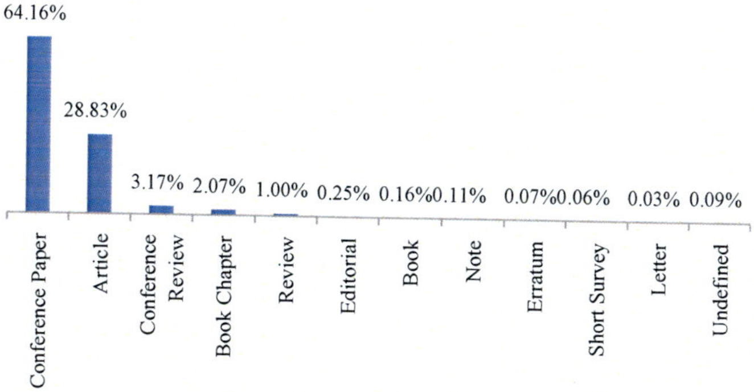

Fig. 7 Distribution of humanoid robot papers by their types

6 Books on Humanoid Robots

The Books on Humanoid Robots are listed as follows:

- Introduction to Humanoid Robotics (Authors: Dragomir Nenchev Atsushi Konno Teppei Tsujita)
- Humanoid Robotics: A Reference (Editors: Ambarish Goswami, Prahlad Vadakkepat)
- Motion Planning for Humanoid Robots (Editors: Kensuke Harada, Eiichi Yoshida, Kazuhito Yokoi)
- Consciousness in Humanoid Robots (Editors: Antonio Chella, Angelo Cangelosi, Giorgio Metta, Selmer Bringsjord)
- Humanoid Robots: modeling and Control (Authors: Dragomir N. Nenchev, Atsushi Konno, Teppei Tsujita
- Artificial Intelligence and Humanoid Robots: 4D an Augmented Reading Experience (Authors: Alicia Z. Klepeis)
- Humanoid Robots: Running Into the Future (Authors: Kathryn Clay)
- Bringing a Humanoid Robot Closer to Human Versatility (Authors: Berthold Bäuml)
- Software Architectures for Humanoid Robotics (Editors: Lorenzo Natale, Tamim Asfour, Fumio Kanehiro, Nikolaus Vahrenkamp)
- Biologically Inspired Control of Humanoid Robot Arms: Robust and Adaptive Approaches (Authors: Adam Spiers, Said Ghani Khan, Guido Herrmann)
- Door Opening by a Miniature Humanoid Robot (Author: Bharadwaj Ramesh)
- Introduction to Humanoid Robotics (Authors: Shuuji Kajita, Hirohisa Hirukawa, Kensuke Harada, Kazuhito Yokoi)
- A Roadmap for Cognitive Development in Humanoid Robots (Authors: David Vernon, Claes von Hofsten, Luciano Fadiga)
- Humanoid Robotics and Neuroscience: Science, Engineering and Society (Editor: Gordon Cheng)
- Visual Perception for Humanoid Robots: Environmental Recognition and Localization, from Sensor Signals to Reliable 6D Poses (Authors: David Israel González Aguirre)
- Humanoid Robots (Author: S L Hamilton)
- Humanoid Robot Control Policy and Interaction Design: A Study on Simulation to Machine Deployment (Author: Suman Deb)
- Experimental Robotics IX: Experimental Robotics IX: The 9th International Symposium on Experimental Robotics (Editors: Marcelo H. Ang, Oussama Khatib)
- Evolutionary Humanoid Robotics (Author: Malachy Eaton)
- Whole-Body Impedance Control of Wheeled Humanoid Robots (Author: Alexander Dietrich)
- Online-Learning in Humanoid Robots (Author: Jörg Conradt)
- Pursh Recovery for Humanoid Robots Using Linearized Double Inverted Pendulum (Author: Saurav Singh)

- Intelligent Robotics and Applications: First International Conference, ICIRA 2008 Wuhan, China, October 15-17, 2008 Proceedings, Part II [Lecture Notes in Computer Science (5315)] 2008th Edition (Editors: Caihua Xiong, Yongan Huang, Youlun Xiong)
- Design and Control of a Humanoid Robot (Author: Teck Chew Wee)
- The DARPA Robotics Challenge Finals: Humanoid Robots To The Rescue (Editors: Matthew Spenko, Stephen Buerger, Karl Iagnemma)
- Humanoid Robots: Human-like Machines (Author: Matthias Hackel)
- Research and Education in Robotics—EUROBOT 2008 (Editors: Achim Gottscheber, Stefan Enderle, David Obdrzalek)
- Advances in Robotics Research: Theory, Implementation, Application (Editors: Torsten Kröger, Friedrich Wahl)
- Simulation and Framework for the Humanoid Robot TigerBot (Author: Felisa Sze)
- Fundamentals of Robotics: Linking Perception to Action (Author: Min Xie)
- Development of Humanoid Robot (Authors: Che Noor Hajizul Che Ani, Mohd. Johan Elyas Mohd. Khairi)
- Build Your Own Humanoid Robot (Author: Karl P. Williams)
- Design and Realization of a Humanoid Robot for Fast and Autonomous Bipedal Locomotion (Author: Sebastian Lohmeier)
- R Code Reading Through a Humanoid Robot: Python Techniques for QR Code Reading (Author: Abigail Pop)
- Real-time SLAM for Humanoid Robot Navigation Using Augmented Reality (Author: Yixuan Zhang)
- Optimal Gait Generation in Biped Locomotion of Humanoid Robot to Improve Walking Speed (Author: Hanafiah Yussof)
- Implementation of a Wireless Haptic Controller for Humanoid Robot Walking (Author: Eun-Su Kim)
- Robotics Research: The Tenth International Symposium (Editors: Raymond Austin Jarvis, Alex Zelinsky)
- Service Robotics and Mechatronics (Editor: Keiichi Shirase, Seiji Aoyagi)
- Intelligent Robotics and Applications(Editor: Ming Xie, Youlun Xiong, Caihua Xiong, Zhencheng Hu)
- Build Your Own Humanoid Robot (Author: Karl P. Williams)
- Analysis And Energy Reduction Of Humanoid Robot Motions Stand Up And Sit Down (Author: Ercan Elibol)
- Tracking and Calibration for a Ball Catching Humanoid Robot (Author: Oliver Birbach)
- Design and Realization of a Humanoid Robot for Fast and Autonomous Bipedal (Author: Sebastian Lohmeier)
- Real-time SLAM for Humanoid Robot Navigation Using Augmented Reality (Author: Yixuan Zhang)
- Optimal Gait Generation in Biped Locomotion of Humanoid Robot to Improve Walking Speed (Author: Hanafiah Yussof)

- Robotics Research: The Tenth International Symposium (Editors: Raymond Austin Jarvis, Alex Zelinsky)
- Whole-Body Affordances for Humanoid Robots: A Computational Approach (Author: Kaiser, Peter)
- Humanoid Robots: Just Like Us?
- Human Centered Robot Systems: Cognition, Interaction, Technology (Editors: Helge Ritter, Gerhard Sagerer, Rüdiger Dillmann, Martin Buss)
- Humanoid Robot Motion in Unstructured Environment - Generation of Various Gait Patterns from a Single Nominal (Authors: Miomir Vukobratović, Dejan Andric, Branislav Borovac)
- Biologically Inspired Approaches for Locomotion, Anomaly Detection and Reconfiguration for Walking Robots (Author: Bojan Jakimovski)
- Semi-Passive Dynamic Walking Approach for Bipedal Humanoid Robot Based on Dynamic Simulation (Author: Aiman Omer)
- Human Centered Robot Systems: Cognition, Interaction, Technology (Editors: Helge Ritter, Gerhard Sagerer, Rüdiger Dillmann, Martin Buss)

7 Journals on Humanoid Robots

The journals on Humanoid Robots are given as follows:

- 2004 IEEE Conference on Robotics Automation and Mechatronics
- 2004 IEEE Rsj International Conference on Intelligent Robots and Systems Iros
- ACM International Conference Proceeding Series
- Adaptive Behavior
- Advanced Materials Research
- Advanced Robotics
- Advances in Intelligent Systems and Computing
- Advances in Mechanical Engineering
- Aip Conference Proceedings
- Applied Bionics and Biomechanics
- Applied Mechanics and Materials
- Applied Sciences Switzerland
- Artificial Life and Robotics
- At Automatisierungstechnik
- Autonomous Robots
- Bioinspiration and Biomimetics
- Biologically Inspired Cognitive Architectures
- Ceur Workshop Proceedings
- CISM International Centre for Mechanical Sciences Courses and Lectures
- Cognitive Systems Monographs
- Communications in Computer and Information Science
- Computers in Human Behavior
- Conference on Human Factors in Computing Systems Proceedings

- Connection Science
- Frontiers in Artificial Intelligence and Applications
- Frontiers in Neurorobotics
- Frontiers in Psychology
- Frontiers Robotics AI
- Harbin Gongye Daxue Xuebao Journal of Harbin Institute of Technology
- Huazhong Keji Daxue Xuebao Ziran Kexue Ban Journal of Huazhong University of Science and Technology Natural Science Edition
- IECON Proceedings Industrial Electronics Conference
- IEEE Access
- IEEE ASME International Conference on Advanced Intelligent Mechatronics AIM
- IEEE ASME Transactions on Mechatronics
- IEEE International Conference on Intelligent Robots and Systems
- IEEE Robotics and Automation Letters
- IEEE Robotics and Automation Magazine
- IEEE Sensors Journal
- IEEE Transactions on Autonomous Mental Development
- IEEE Transactions on Cognitive and Developmental Systems
- IEEE Transactions on Industrial Electronics
- IEEE Transactions on Robotics
- IEEE Transactions on Systems Man and Cybernetics Part B Cybernetics
- IFAC Papersonline
- IFAC Proceedings Volumes IFAC Papersonline
- Industrial Robot
- Intelligent Service Robotics
- Interaction Studies
- International Journal of Advanced Robotic Systems
- International Journal of Control Automation and Systems
- International Journal of Humanoid Robotics
- International Journal of Robotics Research
- International Journal of Social Robotics
- International Workshop on Advanced Motion Control AMC
- Jiqiren Robot
- Journal of Bionic Engineering
- Journal of Field Robotics
- Journal of Harbin Institute Of Technology New Series
- Journal of Institute of Control Robotics and Systems
- Journal of Intelligent and Robotic Systems Theory and Applications
- Journal of Mechanical Science and Technology
- Journal of Physics Conference Series
- Journal of Robotics and Mechatronics
- Knowledge Engineering Review
- Lecture Notes in Artificial Intelligence Subseries of Lecture Notes in Computer Science

- Lecture Notes in Computer Science Including Subseries Lecture Notes in Artificial Intelligence and Lecture Notes in Bioinformatics
- Lecture Notes in Electrical Engineering
- Matec Web of Conferences
- Mechanism and Machine Theory
- Mechanisms and Machine Science
- Mechatronics
- Neural Networks
- Neurocomputing
- Nihon Kikai Gakkai Ronbunshu C Hen Transactions of the Japan Society of Mechanical Engineers Part C
- Paladyn
- Plos One
- Procedia Computer Science
- Procedia Engineering
- Proceedings IEEE International Conference on Robotics and Automation
- Proceedings of SPIE The International Society for Optical Engineering
- Proceedings of the Annual International Conference of the IEEE Engineering in Medicine and Biology Society EMBS
- Proceedings of the ASME Design Engineering Technical Conference
- Proceedings of the IEEE International Conference on Systems Man and Cybernetics
- Proceedings of the International Joint Conference on Neural Networks
- Proceedings of the National Conference on Artificial Intelligence
- Proceedings of the SICE Annual Conference
- Proceedings of the World Congress on Intelligent Control and Automation WCICA
- Robotica
- Robotics
- Robotics and Autonomous Systems
- Robotics and Computer-Integrated Manufacturing
- Robotics and Systems
- Sensors Switzerland
- Smart Innovation Systems and Technologies
- Springer Tracts in Advanced Robotics
- Studies in Computational Intelligence
- VDI Berichte
- Xitong Fangzhen Xuebao Journal of System Simulation
- Zhongguo Jixie Gongcheng China Mechanical Engineering
- Zidonghua Xuebao Acta Automatica Sinica

8 Conclusion

The robots are today used in housework assistance, in space, in education, in entertainment, In the future, we guess and expect that humanoid robots will be used in daily life more commonly than other areas in order to simplify our life conditions. Flexible decisions and behaviors similar to humans seem to have successfully implemented in some humanoid robots. However, there is a long way for humanoid robots to exactly resemble humans. Fuzzy set theory is the most promising tool in modeling and controlling humanoid robots in order to achieve the real success.

References

1. Web-2, Dang, S.S.: PhD, 25 February 2019, https://www.forbes.com/sites/cognitiveworld/2019/02/25/artificial-intelligence-in-humanoid-robots/#719b664324c7
2. Web-1, http://www.humanoid.waseda.ac.jp/history.html
3. Web-3, https://robots.ieee.org/learn/types-of-robots/
4. Web-5, 26 February 2019, https://sciencefordemocracy.org/artificial-intelligence-in-humanoid-robots/
5. Web-4, https://robots.ieee.org/robots/asimo/?gallery=video2
6. Web-6, https://www.hansonrobotics.com/sophia/
7. Web-7, Biswas, J., 25 January 2018, https://analyticsindiamag.com/best-humanoid-robots-2018/
8. Web-8, https://cs.stanford.edu/group/manips/ocean-one.html
9. Web-9, https://www.softbankrobotics.com/emea/en/pepper
10. Web-11, https://robots.ieee.org/robots/nao/
11. Web-10, https://www.softbankrobotics.com/emea/en/nao
12. Web-12, https://www.bostondynamics.com/atlas
13. Web-13, https://robots.ieee.org/robots/petman/
14. Web-14, https://pedagogyeducation.com/Acute-Care-Campus/News-Blogs/Campus-News/News.aspx?news=531
15. Web-25, https://robots.ieee.org/robots/surena/?utm_source=spectrum
16. Web-27, Guizzo, E., 13 February 2020, https://spectrum.ieee.org/automaton/robotics/humanoids/iran-surena-iv-humanoid-robot
17. Web-21, Merkusheva, D., 25 March 2020, https://www.asme.org/topics-resources/content/10-humanoid-robots-of-2020
18. Web-29, https://asimo.honda.com/downloads/pdf/asimo-technical-information.pdf
19. Web-26, https://www.neon.life/meet-neon
20. Web-15, https://robots.ieee.org/robots/hrp2/?gallery=video2
21. Web-16, https://robots.ieee.org/robots/alberthubo/
22. Web-17, https://robots.ieee.org/robots/hubo/
23. Web-18, https://robots.ieee.org/robots/mahru/
24. Web-19, https://robots.ieee.org/robots/ar600/
25. Web-20, https://robots.ieee.org/robots/erica
26. Web-22, https://www.ubtrobot.com/collections/innovation-at-ubtech?ls=en
27. Web-23, https://pal-robotics.com/robots/talos/
28. Web-24, https://pal-robotics.com/robots/reem-c/

Fuzzy Sets and Extensions: A Literature Review

Eda Bolturk and Cengiz Kahraman

Abstract Humanoid robots generated by inspiring by human appearances and abilities have become essential in human society to improve the quality of their life. All over the world, there have been many researchers who have focused on humanoid robots to develop their capabilities. Generally, humanoid robot systems include mechanisms of decision making and information processing. Because of the uncertainty behind decision making and information processes, fuzzy sets can be used in humanoid systems efficiently. This study presents a comprehensive literature review on the recent developments and theories associated with fuzzy set models.

1 Introduction

Fuzzy logic has been generally employed for the automatic navigation of robots in the literature. This is since the capability of fuzzy logic to process large quantities of incomplete and vague input signals is very high for the automatic navigation of robots under uncertainty. Robots carry several sensors on them for sensing environmental information. The outputs of these sensors serve as inputs to the fuzzy controller. Fuzzification, fuzzy inference, and defuzzification generate decisions that control the robots' behaviors enabling robots to navigate automatically [1]. Fuzzy logic control systems provide an automatic navigation system very similar to the thinking style of humans.

Classical fuzzy logic is composed of only a membership degree where the non membership degree is the complement of this degree. Since human behaviour is the outcome of a complex thinking system, we need more parameters to define uncertainty, which the extensions of classical fuzzy logic aim at providing this capability. In the following, we will introduce these extensions since they are more suitable to

E. Bolturk (✉) · C. Kahraman
Industrial Engineering Department, Istanbul Technical University, Besiktas, Istanbul 34367, Turkey
e-mail: bolturk@itu.edu.tr

C. Kahraman
e-mail: kahramanc@itu.edu.tr

C. Kahraman and E. Bolturk (eds.), *Toward Humanoid Robots: The Role of Fuzzy Sets*, Studies in Systems, Decision and Control 344,
https://doi.org/10.1007/978-3-030-67163-1_2

model human behaviours than classical fuzzy logic. After the introduction of ordinary fuzzy sets (OFS) by Zadeh [2], fuzzy sets have been very popular in modeling the problems involving vagueness and impreciseness. Various researchers have proposed several extensions of ordinary fuzzy sets as given in Fig. 1 with a historical order. In this figure, large circles show that those extensions relatively have a larger impact than other extensions on the expansion of fuzzy set theory. In recent years, several researchers have used these extensions in the solution of various problems such as mathematical modeling and optimization, multicriteria decision making, data mining, and quality control. A classification of extensions of OFS is presented in the following.

1.1 Preliminaries: Extensions of Fuzzy Sets

In this section, the basic concepts and the mathematical operations of extensions of fuzzy sets have been briefly introduced.

1.1.1 Ordinary Fuzzy Sets

A way of describing a fuzzy set is to list ordered pairs: an object x and its membership degree $\mu_A(x)[0, 1]$ in a set $\tilde{A}$. To describe an ordinary fuzzy set, the following notation proposed by Zadeh [2] can be used:

$$\tilde{A} = \{ (x, \mu_{\tilde{A}}(x)) | x \in X \} \tag{1}$$

where X is the discrete universe. The non-membership degree of any x is calculated by the subtraction $1 - \mu_{\tilde{A}}(x)$.

1.1.2 Type-2 Fuzzy Sets (T2FS)

The concept of a type-2 fuzzy set was introduced by Zadeh [3] as an extension of the concept of an ordinary fuzzy set. Such sets are fuzzy sets whose membership grades themselves are fuzzy. They are veryseful in circumstances where it is difficult to determine an exact membership function for a fuzzy set.

A type 2 fuzzy set $\tilde{\tilde{A}}$ in the universe of discourse X can be represented by a type 2 membership function $\mu_{\tilde{\tilde{A}}}$, shown as follows [3]:

$$\tilde{\tilde{A}} = \left\{ ((x, u), \mu_{\tilde{\tilde{A}}}(x, u) | \forall x \in X, \forall u \in J_x \subseteq [0, 1], 0 \le \mu_{\tilde{\tilde{A}}}(x, u) \le 1 \right\}, \tag{2}$$

where J_x. denotes an interval [0,1].

Fig. 1 Fuzzy sets extensions

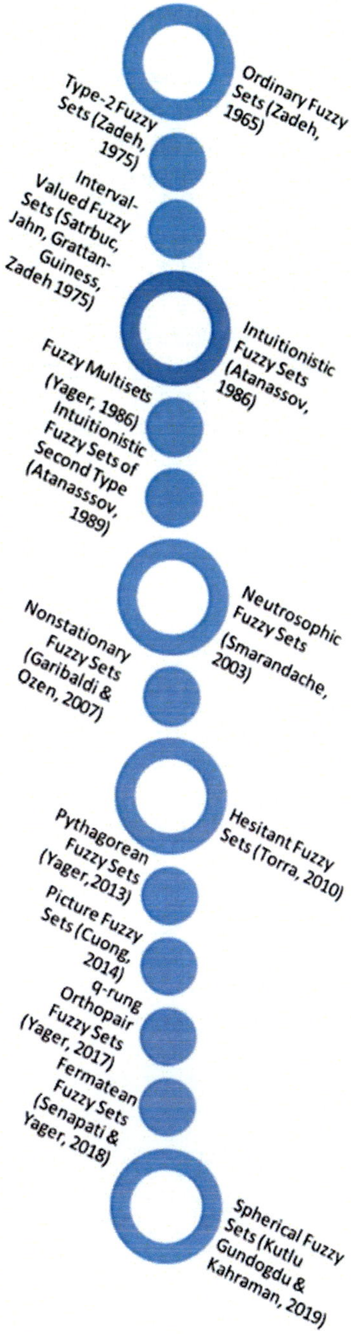

1.1.3 Interval-Valued Fuzzy Sets

Let us denote the set of all closed sub intervals in [0,1] by L([0,1]), that is, L([0,1]) = {x = $[x_L, x_U]$ > $(x_L, x_U) \in [0,1]^2$ and $x_L \leq x_U$} (1) An interval-valued fuzzy set (IVFS) $\tilde{A}$ on the universe U $\neq \emptyset$ is a mapping $\tilde{A}$: U $\rightarrow$ L([0,1]), such that the membership degree of u $\in$ U is given by A(u) = $[A_L(u), A_U(u)] \in$ L([0,1]), where A: U $\rightarrow$ [0,1] and A: U $\rightarrow$ [0,1] are mappings defining the lower and the upper bound of the membership interval A(u), respectively.

1.1.4 Intuitionistic Fuzzy Sets

Intuitionistic fuzzy sets introduced by Atanassov [4] enable defining both the membership and non-membership degrees of an element in a fuzzy set. Their sum can be equal to or less than 1. The difference from 1, if any, is called hesitancy. Let U be a universe of discourse. An IFS $\tilde{I}$ is defined as follows:

Definition 1 Let X be a non-empty set. An intuitionistic fuzzy set I in X is given by:

$$I = \{(x, \mu_I(x), \nu_I(x))|xX\} \tag{3}$$

where the function $\mu_I : X \rightarrow [0, 1]$ and $\upsilon_I : X \rightarrow [0, 1]$ defines the degree of membership and the degree of non-membership of element to the sets I, respectively, with the condition that

$$0 \leq \mu_I(x) + \nu_I(x) \leq 1, \quad for\ \forall xX \tag{4}$$

The degree of hesitancy is calculated as follows:

$$\pi_I(x) = 1 - \mu_I(x) - \nu_I(x) \tag{5}$$

Definition 2 Let $\tilde{\mathrm{A}} = \left(\mu_{\tilde{A}}, \nu_{\tilde{A}}\right)$ and $\tilde{\mathrm{B}} = \left(\mu_{\tilde{B}}, \nu_{\tilde{B}}\right)$ be two IFSs, then the addition and multiplication operations on these two IFSs are calculated as follows:

$$\tilde{\mathrm{A}} \oplus \tilde{B} = \left(\mu_{\tilde{\mathrm{A}}} + \mu_{\tilde{B}} - \mu_{\tilde{\mathrm{A}}}\mu_{\tilde{B}}\ , \nu_{\tilde{\mathrm{A}}}\nu_{\tilde{B}}\right) \tag{6}$$

$$\tilde{A} \otimes \tilde{B} = \left(\mu_{\tilde{A}}\mu_{\tilde{B},}\nu_{\tilde{A}} + \nu_{\tilde{B}} - \nu_{\tilde{A}}\nu_{\tilde{B}}\right) \tag{7}$$

Atanassov [4] introduced triangular intuitionistic fuzzy sets (IFS) by defining membership and non-membership functions independently with the constraint that

Fig. 2 A triangular intuitionistic fuzzy number

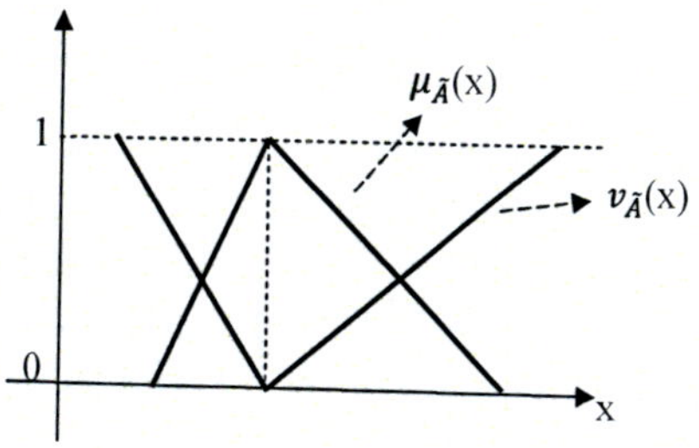

their sum must be at most one, letting a hesitancy degree be included. A triangular intuitionistic fuzzy number is shown in Fig. 2.

Defuzzification of triangular intuitionistic fuzzy numbers (IFNs) can be made by two defuzzification methods given in the following: one for triangular IFNs and one for trapezoidal IFNs [5]:

Let $I_i = (a_L, a_M, a_U; b_L, b_M, b_U)$ be a triangular intuitionistic fuzzy number. Then, defuzzification is realized by using the function defined in Eq. (8).

$$d_f = \frac{a_L + 2a_M + a_U}{4} + \frac{b_L + 2b_M + b_U}{\tau} \tag{8}$$

where τ is a very large number, such as 100, indicating the effect of non-membership function on the IFN.

Let $I_i = (a_L, a_{M1}, a_{M2}, a_U; b_L, b_{M1}, b_{M2}, b_U)$ be a trapezoidal intuitionistic fuzzy number. Then, the defuzzification is realized by using the function defined in Eq. (9).

$$d_f = \frac{a_L + 2(a_{M1} + a_{M2}) + a_U}{4} + \frac{b_L + 2(b_{M1} + b_{M2}) + b_U}{\tau} \tag{9}$$

Definition 3 Let X be a non-empty set. An interval-valued intuitionistic fuzzy (IVIF) set in X is an object $\tilde{A}$ given as in Eq. (10) [4]:

$$\tilde{A} = \left\{x, \left\langle \left[\mu_{\tilde{A}}^-, \mu_{\tilde{A}}^+\right], \left[v_{\tilde{A}}^-, v_{\tilde{A}}^+\right] \right\rangle; x \in X \right\} \tag{10}$$

where $0 \le \mu_{\tilde{A}}^+ + v_{\tilde{A}}^+ \le 1$ for every $x \in X$.

Definition 4 Let $\tilde{A} = \left(\left[\mu_{\tilde{A}}^-, \mu_{\tilde{A}}^+\right], \left[v_{\tilde{A}}^-, v_{\tilde{A}}^+\right]\right)$ and $\tilde{B} = \left(\left[\mu_{\tilde{B}}^-, \mu_{\tilde{B}}^+\right], \left[v_{\tilde{B}}^-, v_{\tilde{B}}^+\right]\right)$ be two IVIF numbers [6]. Then,

$$\tilde{A} \oplus \tilde{B} = \left(\left[\mu_{\tilde{A}}^- + \mu_{\tilde{B}}^- - \mu_{\tilde{A}}^-\mu_{\tilde{B}}^-, \mu_{\tilde{A}}^+ + \mu_{\tilde{B}}^+ - \mu_{\tilde{A}}^+\mu_{\tilde{B}}^+\right], \left[v_{\tilde{A}}^- v_{\tilde{B}}^-, v_{\tilde{A}}^+ v_{\tilde{B}}^+\right]\right) \tag{11}$$

$$\tilde{A} \otimes \tilde{B} = \left(\left[\mu_{\tilde{A}}^{-}\mu_{\tilde{B}}^{-}, \mu_{\tilde{A}}^{+}\mu_{\tilde{B}}^{+}\right], \left[v_{\tilde{A}}^{-} + v_{\tilde{B}}^{-} - v_{\tilde{A}}^{-}v_{\tilde{B}}^{-}, v_{\tilde{A}}^{+} + v_{\tilde{B}}^{+} - v_{\tilde{A}}^{+}v_{\tilde{B}}^{+}\right]\right) \quad (12)$$

Definition 5 Let $\tilde{r}_{ij}^{k} = \left(\left[\mu_{\tilde{r}}^{-}, \mu_{\tilde{r}}^{+}\right], \left[v_{\tilde{r}}^{-}, v_{\tilde{r}}^{+}\right]\right)$ be the IVIF numbers where $k = 1, 2, \ldots, n$. Then the aggregated IVIF number $(\tilde{r}_{ij}^{A})$ is obtained by using interval-valued intuitionistic fuzzy hybrid geometric (IIFHG) operator as in Eq. (13) [7]:

$$\tilde{r}_{ij}^{A} = \left\langle \left[\prod_{k=1}^{n}\left(\mu_{k}^{-}\right)^{\omega_k}, \prod_{k=1}^{n}\left(\mu_{k}^{+}\right)^{\omega_k}\right], \left[1 - \prod_{k=1}^{n}\left(1 - v_{k}^{-}\right)^{\omega_k}, 1 - \prod_{k=1}^{n}\left(1 - v_{k}^{+}\right)^{\omega_k}\right]\right\rangle \quad (13)$$

where ω_k is the weight vector of expert k where $\sum_{k=1}^{n}\omega_k = 1$.

Definition 6 Let $\tilde{r}_1 = \left(\left[\mu_{1}^{-}, \mu_{1}^{+}\right], \left[v_{1}^{-}, v_{1}^{+}\right]\right)$ and $\tilde{r}_2 = \left(\left[\mu_{2}^{-}, \mu_{2}^{+}\right], \left[v_{2}^{-}, v_{2}^{+}\right]\right)$ be two IVIF numbers. The distance between these two IVIF numbers is obtained by Hamming distance as in Eq. (14) [8]:

$$HD = \frac{1}{4}\sum(\left|\mu_{1}^{-} - \mu_{2}^{-}\right| + \left|\mu_{1}^{+} - \mu_{2}^{+}\right| + \left|v_{1}^{-} - v_{2}^{-}\right| + \left|v_{1}^{+} - v_{2}^{+}\right| \quad (14)$$

Defitiniton 7 Let $\tilde{r}_1 = \left(\left[\mu_{\tilde{x}}^{-}, \mu_{\tilde{x}}^{+}\right], \left[v_{\tilde{x}}^{-}, v_{\tilde{x}}^{+}\right]\right)$ is an IVIF number. Defuzzification formula $(\mathfrak{D}(\mathrm{x}))$ for $\tilde{r}_1$ is given in Eq. (15) [9].

$$\mathfrak{D}(\mathrm{x}) = \frac{\mu_{\tilde{x}}^{-} + \mu_{\tilde{x}}^{+} + \left(1 - v_{\tilde{x}}^{-}\right) + \left(1 - v_{\tilde{x}}^{+}\right) + \mu_{\tilde{x}}^{-} \times \mu_{\tilde{x}}^{+} - \sqrt{\left(1 - v_{\tilde{x}}^{-}\right) \times \left(1 - v_{\tilde{x}}^{+}\right)}}{4} \quad (15)$$

1.1.5 Fuzzy Multi Sets

Yager [10] first discussed fuzzy multi-sets, although he uses the term of fuzzy bag,an element of X may occur more than once with possibly the same or different membership values. Assume X is a set of elements. Then a fuzzy bag A drawn from X can be characterized by a function Count.Mem$_A$ such that

$$Count.Mem_A : X \rightarrow Q \quad (16)$$

where Q is the set of all crisp bags from the unit interval.

1.1.6 Neutrosophic Sets

Smarandache [11] developed neutrosophic logic and neutrosophic sets (NSs) as an extension of intuitionistic fuzzy sets. The neutrosophic set is defined as the set where each element of the universe has a degree of truthiness, indeterminacy and falsity. The sum of these degrees can be at most equal to 3 since each of them can be independently at most equal to 1. Let U be a universe of discourse.

In neutrosophic sets literature, a common specific symbol for a neutrosophic set has not been used up to now. Bolturk and Kahraman [12] propose the symbol $\tilde{\dddot{A}}$ for the neutrosophic set A, that the three dots represent the elements of a neutrosophic set,T, I, F and tilde represents that it is also a fuzzy set.

Definition 8 Smarandache [11]. Let E be a universe. A neutrosophic set $\tilde{\dddot{A}}$ in E is characterized by a truth-membership function T_A, a indeterminacy-membership function I_A, and a falsity-membership function F_A.

$T_A\ (x)$, $I_A\ (x)$ and $F_A\ (x)$ are real standart elements of [0,1]. A neutrosophic set $\tilde{\dddot{A}}$ can be given by Eq. (17):

$$\tilde{\dddot{A}} = \{\langle x, (T_A(x), I_A(x), F_A(x))\rangle : x \in E, (T_A(x), I_A(x), F_A(x) \in]^{-}0, 1[^{+} \quad (17)$$

There is no restriction on the sum of $T_A(x)$, $I_A\ (x)$ and $F_A\ (x)$, so that $0^- \leq T_A(x) + I_A(x) + F_A(x) \leq 3^+$.

Definition 9 Li et al. [13] X be a universe of discourse. An interval-valued neutrosophic set $\tilde{\dddot{N}}$ in X is independently defined by a truth-membership function $T_N(x)$, an indeterminacy-membership function $I_N(x)$, and a falsity-membership function $F_N(x)$ for each $x \in X$, where $T_N(x) = \left[T^L_{N(x)}, T^U_{N(x)}\right] \subseteq [0, 1]$, $I_N(x) = \left[I^L_{N(x)}, I^U_{N(x)}\right] \subseteq [0, 1]$, and $F_N(x) = \left[F^L_{N(x)}, F^U_{N(x)}\right] \subseteq [0, 1]$. They also meet the condition $0 \leq T^L_N(x) + I^L_N(x) + F^L_N(x) \leq 3$. So, the interval-valued neutrosophic set $\tilde{\dddot{\mathrm{N}}}$ can be given by Eq. (18):

$$\tilde{\dddot{\mathrm{N}}} = \{\langle \mathrm{x}, [\,\mathrm{T}^L_N(\mathrm{x}), \mathrm{T}^U_N(\mathrm{x})], [\,\mathrm{I}^L_N(\mathrm{x}), \mathrm{I}^U_N(\mathrm{x})], [\,\mathrm{F}^L_N(\mathrm{x}), \mathrm{F}^U_N(\mathrm{x})]\rangle | \mathrm{x} \in \mathrm{X}\} \quad (18)$$

Definition 10 Bolturk and Kahraman [12] propose a new deneutrosophication function of an interval-valued neutrosophic number which is given below:

$$\mathfrak{D}(\mathrm{x}) = \left(\frac{(\mathrm{T}^L_x + \mathrm{T}^U_x)}{2} + \left(1 - \frac{(\mathrm{I}^L_x + \mathrm{I}^U_x)}{2}\right) \times (\mathrm{I}^U_x) - \left(\frac{\mathrm{F}^L_x + \mathrm{F}^U_x}{2}\right) \times (1 - \mathrm{F}^U_x)\right) \quad (19)$$

where $\tilde{\dddot{\mathrm{x}}}_j = \langle [\mathrm{T}^L_x, \mathrm{T}^U_x], [\,\mathrm{I}^L_x, \mathrm{I}^U_x], [\mathrm{F}^L_x, \mathrm{F}^U_x]\rangle$.

Definition 11 Let $\tilde{\tilde{a}} = [T_a^L, T_a^U], [I_a^L, I_a^U], [F_a^L, F_a^U]$ and $\tilde{\tilde{b}} = \langle [T_b^L, T_b^U], [I_b^L, I_b^U], [F_b^L, F_b^U] \rangle$ be two interval-valued neutrosophic numbers. Their relations and arithmetic operations are given by Eqs. (20)–(24) [14]:

$$\tilde{\tilde{a}}^c = [F_a^L, F_a^U], [1 - I_a^U, 1 - I_a^L], [T_a^L, T_a^U] \tag{20}$$

$$\tilde{\tilde{a}} \subseteq \tilde{\tilde{b}} \text{ if and only if } T_a^L \le T_b^L;\ T_a^U \le T_b^U;\ I_a^L \ge I_b^L,\ I_a^U \ge I_b^U;\ F_a^L \ge F_b^L,\ F_a^U \ge F_b^U \tag{21}$$

$$\tilde{\tilde{a}} = \tilde{\tilde{b}} \text{ if and only if } \tilde{\tilde{a}} \subseteq \tilde{\tilde{b}} \text{ and } \tilde{\tilde{b}} \subseteq \tilde{\tilde{a}}. \tag{22}$$

$$\tilde{\tilde{a}} \oplus \tilde{\tilde{b}} = \langle [T_a^L + T_b^L - T_a^L T_b^L,\ T_a^U + T_b^U - T_a^U T_b^U], [I_a^L I_b^L, I_a^U I_b^U], [F_a^L F_b^L, F_a^U F_b^U] \rangle \tag{23}$$

$$\tilde{\tilde{a}} \otimes \tilde{\tilde{b}} = \langle [T_a^L T_b^L, T_a^U T_b^U][I_a^L + I_b^L - I_a^L I_b^L,\ I_a^U + I_b^U - I_a^U I_b^U], [F_a^L + F_b^L - F_a^L F_b^L,\ F_a^U + F_b^U - F_a^U F_b^U] \rangle \tag{24}$$

Definition 12 Subtraction operation of two interval-valued neutrosophic sets is given in Eq. (25) Karasan and Kahraman [15]:

$$\tilde{\tilde{x}} \ominus \tilde{\tilde{y}} = \left\langle [T_x^L - F_y^U,\ T_x^U - F_y^L], [\mathrm{Max}\left(I_x^L, I_y^L\right),\ \mathrm{Max}\left(I_x^U, I_y^U\right)], [F_x^L - T_y^U,\ F_x^U - T_y^L] \right\rangle \tag{25}$$

where $\tilde{\tilde{x}} = \langle [T_x^L, T_x^U], [I_x^L, I_x^U], [F_x^L, F_x^U] \rangle$ and $\tilde{\tilde{y}} = \langle \left[T_y^L, T_y^U\right], \left[I_y^L, I_y^U\right], \left[F_y^L, F_y^U\right] \rangle$.

Definition 13 Let $\tilde{\tilde{A}} = \left\{ \langle x, [T_N^L(x), T_x^U(x)], [I_x^L(x), I_x^U(x)], [F_x^L(x), F_x^U(x)] \rangle | x \in X \right\}$ be an interval-valued neutrosophic number. The following ranking formula is proposed for $\tilde{\tilde{A}}$ by Kahraman et al. [16]:

$$\mathcal{D}\left(\tilde{A}\right) = \frac{\left(T_N^L + T_N^U + \left(1 - F_N^L\right) + \left(1 - F_N^U\right) + T_N^L \times T_N^U - \sqrt{\left(1 - F_N^L\right) \times \left(1 - F_N^U\right)}\right)}{4} \times \left(1 - \frac{I_N^L + I_N^U}{8}\right) \tag{26}$$

Definition 14 Let $\tilde{A}=\langle[T_1^L, T_1^U], [I_1^L, I_1^U], [F_1^L, F_1^U]\rangle$; $\tilde{B}=\langle[T_2^L, T_2^U], [I_2^L, I_2^U], [F_2^L, F_2^U]\rangle$ be IVNNs where $T_2^L > 0; T_2^U > 0; I_2^L > 0; I_2^U > 0; F_2^L > 0; F_2^U > 0$. The division operation is proposed as in Eq. (27) [17]:

$$A \oslash B = \frac{\begin{pmatrix} \left[\min\left(\frac{T_1^L}{T_2^L}, \frac{T_1^L}{T_2^L}, \frac{T_1^U}{T_2^U}, \frac{T_1^U}{T_2^U}\right), \max\left(\frac{T_1^L}{T_2^L}, \frac{T_1^L}{T_2^L}, \frac{T_1^U}{T_2^U}, \frac{T_1^U}{T_2^U}\right)\right], \\ \left[\min\left(\frac{I_1^L}{I_2^L}, \frac{I_1^L}{I_2^L}, \frac{I_1^U}{I_2^U}, \frac{I_1^U}{I_2^U}\right), \max\left(\frac{I_1^L}{I_2^L}, \frac{I_1^L}{I_2^L}, \frac{I_1^U}{I_2^U}, \frac{I_1^U}{I_2^U}\right)\right], \\ \left[\min\left(\frac{F_1^L}{F_2^L}, \frac{F_1^L}{F_2^L}, \frac{F_1^U}{F_2^U}, \frac{F_1^U}{F_2^U}\right), \max\left(\frac{F_1^L}{F_2^L}, \frac{F_1^L}{F_2^L}, \frac{F_1^U}{F_2^U}, \frac{F_1^U}{F_2^U}\right)\right] \end{pmatrix}}{\max\begin{pmatrix} \left[\min\left(\frac{T_1^L}{T_2^L}, \frac{T_1^L}{T_2^L}, \frac{T_1^U}{T_2^U}, \frac{T_1^U}{T_2^U}\right), \max\left(\frac{T_1^L}{T_2^L}, \frac{T_1^L}{T_2^L}, \frac{T_1^U}{T_2^U}, \frac{T_1^U}{T_2^U}\right)\right], \\ \left[\min\left(\frac{I_1^L}{I_2^L}, \frac{I_1^L}{I_2^L}, \frac{I_1^U}{I_2^U}, \frac{I_1^U}{I_2^U}\right), \max\left(\frac{I_1^L}{I_2^L}, \frac{I_1^L}{I_2^L}, \frac{I_1^U}{I_2^U}, \frac{I_1^U}{I_2^U}\right)\right], \\ \left[\min\left(\frac{F_1^L}{F_2^L}, \frac{F_1^L}{F_2^L}, \frac{F_1^U}{F_2^U}, \frac{F_1^U}{F_2^U}\right), \max\left(\frac{F_1^L}{F_2^L}, \frac{F_1^L}{F_2^L}, \frac{F_1^U}{F_2^U}, \frac{F_1^U}{F_2^U}\right)\right] \end{pmatrix}} \tag{27}$$

Definition 15 Let $a_j = \left\langle\left[T_{a_j}^L, T_{a_j}^U\right], \left[I_{a_j}^L, I_{a_j}^U\right], \left[F_{a_j}^L, F_{a_j}^U\right]\right\rangle$, $j = 1, 2, \ldots, n$ be a collection of INNs. Based on the weighted aggregation operators of INNs, the interval neutrosophic number weighted average operator is given as below [14]

$$\begin{aligned} &\text{INNWA}\,(a_1, a_2, \ldots, a_n) \\ &= \sum_{j=1}^{n} w_j a_i = \left\langle\left[1-\prod_{j=1}^{n}\left(1-T_{aj}^L\right)^{w_j}, 1-\prod_{j=1}^{n}\left(1-T_{aj}^U\right)^{w_j}\right],\right. \\ &\left.\left[\prod_{j=1}^{n}\left(I_{aj}^L\right)^{w_j}, \prod_{j=1}^{n}\left(I_{aj}^U\right)^{w_j}\right], \left[\prod_{j=1}^{n}\left(F_{aj}^L\right)^{w_j}, \prod_{j=1}^{n}\left(F_{aj}^U\right)^{w_j}\right]\right\rangle \end{aligned} \tag{28}$$

where $w_j (j = 1, 2, \ldots, n)$ is the weight of $a_j (j = 1, 2, \ldots, n)$ with $w_j \in [0, 1]\, and \sum_{j=1}^{n} w_j = 1$.

Definition 16 The weighted aggregation operation (INNWA) for interval-valued neutrosophic numbers is given in Eq. (29) [14]

$$\begin{aligned} &INNWA(A_1, A_2, \ldots, A_n) \\ &= \left\langle\left[1-\prod_{i=1}^{n}(1-infT_{Ai})^{w_i}, 1-\prod_{i=1}^{n}\left(1-supT_{A_i}\right)^{w_i}\right],\right. \end{aligned}$$

$$\left[\prod_{i=1}^{n}(\inf I_{A_i})^{w_i}, \prod_{i=1}^{n}(\sup I_{A_i})^{w_i}\right], \left[\prod_{i=1}^{n}(\inf F_{A_i})^{w_i}, \prod_{i=1}^{n}(\sup F_{A_i})^{w_i}\right]\Bigg\rangle \quad (29)$$

where $W = (w_1, w_2, \ldots, w_n)$ is the weight vector of $A_j (j = 1, 2, \ldots, n)$, with $w_j \in [0, 1]$ and $\sum_{j=1}^{n} w_j = 1$.

There are few deneutrosophication methods to compare neutrosophic numbers [18, 19]. Bolturk et al. [17] proposed a new deneutrosophication method in order to compare the interval-valued neutrosophic numbers. It is given in Definition 17.

Definition 17 Let $A = \langle (T^L, T^U), (I^L, I^U), (F^L, F^U) \rangle$ be an interval-valued neutrosophic number. The deneutrosophicated A value ($\mathfrak{D}(A)$) is proposed in Eq. (30):

$$\mathfrak{D}(A) = \frac{\left(T^L + T^U + \left(1 - F^L\right) + \left(1 - F^U\right) + T^L \times T^U - \sqrt{(1 - F^L) \times (1 - F^U)}\right)}{4} \times \left(\left(1 - \frac{\left[\left(I^L\right) + \left(I^U\right)\right]}{2}\right) - \left(\sqrt{(I^L) \times (I^U)}\right)\right) \quad (30)$$

1.1.7 Nonstationary Fuzzy Sets

Nonstationary fuzzy sets are introduced by Garibaldi and Ozen [20]. Let A denote a fuzzy set of a universe of discourse X characterized by a membership function $\mu_{\dot{A}}$. Let T be a set of time points t_i (possibly infinite) and f: T→ < denote the perturbation function. Associates with each element (t, x) of $T \times X$ a time specific variation of $\mu_A(x)$. The nonstationary fuzzy set $\dot{A}$ is denoted by

$$\dot{A} = \int_{t \in T} \int_{x \in X} \mu_{\dot{A}}(t, x)/x/t. \quad (31)$$

1.1.8 Hesitant Fuzzy Sets

Hesitant fuzzy sets introduced by Torra [21] allow many potential degrees of membership of an element to be assigned to a set. These fuzzy sets force the membership degree of an element to be possible values between zero and one. A hesitant fuzzy set on X can be defined as in Eq. (49):

$$H = \{\langle X, h_H(x) \rangle | x \in X\} \quad (32)$$

where $h_H(x)$ is a set of hesitant fuzzy elements whose membership values are in [0,1].

1.1.9 Pythagorean Fuzzy Sets

Pythagorean fuzzy sets (PFSs) are an extension of intuitionistic fuzzy sets and it allows researchers to assign membership and nonmembership degrees in a wider area. Atanassov's intuitionistic fuzzy sets of second type (IFS2) or Yager's Pythagorean fuzzy sets [22] are characterized by a membership degree and a non-membership degree satisfying that their squared sum is equal to or less than one, which is a generalization of intuitionistic fuzzy sets. This provides a larger area than IFS in order to assign membership and non-membership degrees [23]. Let U be a universe of discourse. A PFS $\tilde{P}$ is an object having the form,

$$\tilde{P} = \{x, P(\mu_P(x), v_P(x))|x \in X\} \tag{33}$$

where $\mu_P : X \to [0, 1]$ is the membership degree and $v_P : X \to [0, 1]$ is the nonmembership degree. Then, Eq. (34) is valid:

$$(\mu_P(x))^2 + (v_P(x))^2 \leq 1 \tag{34}$$

The degree of indeterminancy is defined as follows:

$$\pi_P(x) = \sqrt{1 - (\mu_P(x))^2 - (v_P(x))^2} \tag{35}$$

A Single Valued Pythagorean Fuzzy Set (SVPFS) is defined as follows:

For two PFSs, $\tilde{P}_1 = \{x, P_1(\mu_{P1}(x), v_{P1}(x))|x \in X\}$ and $\tilde{P}_2 = \{x, P_2(\mu_{P2}(x), v_{P2}(x))|x \in X\}$, the following arithmetic operations are valid:

$$\tilde{P}_1 \oplus \tilde{P}_2 = P\left(\sqrt{\mu_{P_1}^2 + \mu_{P_2}^2 - \mu_{P_1}^2\mu_{P_2,}^2}, v_{P_1}v_{P_2}\right) \tag{36}$$

$$\tilde{P}_1 \otimes \tilde{P}_2 = P\left(\mu_{P_1}\mu_{P_2}, \sqrt{v_{P_1}^2 + v_{P_2}^2 - v_{P_1}^2 v_{P_2,}^2}\right) \tag{37}$$

$$\lambda\tilde{P} = P\left(\sqrt{1 - (1 - \mu_P^2)^\lambda, (v_p)^\lambda}\right), \ \lambda \geq 0 \text{ and } \lambda \in R \tag{38}$$

$$\tilde{P}^\lambda = P\left((\mu_P)^\lambda, \sqrt{1 - \left(1 - v_P^2\right)^\lambda}, \right), \lambda > 0 \tag{39}$$

Zhang and Xu [24] defined the Euclidean distance between two PFSs as in Eq. (40):

$$d\left(\tilde{P}_1, \tilde{P}_2\right) = \frac{1}{2}\left(\left|\mu_{P_1}^2 - \mu_{P_2}^2\right| + \left|v_{P_1}^2 - v_{P_2}^2\right| + \left|\pi_{P_1}^2 - \pi_{P_2}^2\right|\right) \tag{40}$$

The Taxican distance between two PFSs is defined by Eq. (41):

$$T\left(\tilde{P}_1, \tilde{P}_2\right) = \left|\mu_{P_1} - \mu_{P_2}\right| + \left|v_{P_1} - v_{P_2}\right| + \left|\pi_{P_1} - \pi_{P_2}\right| \tag{41}$$

Let $p_1 = (\mu_1, v_1)$ and $p_2 = (\mu_2, v_2)$ be two PFNs and $\rho > 0$. The following operations are presented for PFNs [24, 25].

$$\tilde{P}_1 \ominus \tilde{P}_2 = \left(\sqrt{\frac{\mu_1^2 - \mu_2^2}{1 - \mu_2^2}}, \frac{v_1}{v_2}\right), \quad \text{if} \mu_1 \geq \mu_2, v_1 \leq min\left\{v_2, \frac{v_2.\pi_1}{\pi_2}\right\} \tag{42}$$

$$\frac{\tilde{P}_1}{\tilde{P}_2} = \left(\frac{\mu_1}{\mu_2}, \sqrt{\frac{v_1^2 - v_2^2}{1 - v_2^2}}\right), \text{ if } \mu_1 \leq \min\left\{\mu_2, \frac{\mu_2.\pi_1}{\pi_2}\right\}, v_1 \geq v_2 \tag{43}$$

Definition 18 Let Int([0,1]) denote the set of all closed subintervals of [0,1], and X be a universe of discourse. An interval-valued PFS (IVPFS) $\widetilde{P}$ in X is given by Eq. (44) [26].

$$\widetilde{P} = \left\{ \langle x, \mu_p(x), v_p(x)\rangle \middle| x \in X\right\} \tag{44}$$

where the functions $\mu_p: X \to \text{Int}([0, 1])(x \in X \to \mu_p(x) \subseteq [0, 1])$ and $v_p: X \to \text{Int}([0, 1])$ $(x \in X \to v_p(x) \subseteq [0, 1])$ denote the membership degree and non-membership degree of the element $x \in X$ to the set $\tilde{P}$, respectively, and for every $x \in X$, $0 \leq \sup\left\{(\mu_p(x))^2\right\} + \sup\left\{(v_p(x))^2\right\} \leq 1$. Also, for each $x \in X$, $\mu_p(x)$ and $v_p(x)$ are closed intervals and their lower and upper bounds are denoted by $\mu_p^L(x)$, $\mu_p^U(x)$, $v_p^L(x)$, $v_p^U(x)$, respectively. Therefore, $\tilde{P}$ can also be expressed as follows:

$$\tilde{P} = \left\{\left\langle x, \left[\mu_p^L(x), \mu_p^U(x)\right], \left[v_p^L(x), v_p^U(x)\right]\right\rangle \middle| x \in X\right\} \tag{45}$$

The degree of indeterminacy is given by Eq. (46).

$$\pi_P(x) = \left[\begin{array}{c} \sqrt{1 - \left(\mu_p^U(x)\right)^2 - \left(v_p^U(x)\right)^2}, \\ \sqrt{1 - \left(\mu_p^L(x)\right)^2 - \left(v_p^L(x)\right)^2} \end{array}\right] \tag{46}$$

Definition 19 Let $\tilde{A} = \left\langle\left[\mu_A^L, \mu_A^U\right], \left[v_A^L, v_A^U\right]\right\rangle$, $\tilde{B} = \left\langle\left[\mu_B^L, \mu_B^U\right], \left[v_B^L, v_B^U\right]\right\rangle$ be two IVPF numbers, and $\lambda > 0$, then some operations of IVPF numbers are defined as follows [26].

$$\tilde{A} \oplus \tilde{B} = \left(\begin{array}{c} \left[\begin{array}{c} \sqrt{(\mu_A^L)^2 + (\mu_B^L)^2 - (\mu_A^L)^2 (\mu_B^L)^2}, \\ \sqrt{(\mu_A^U)^2 + (\mu_B^U)^2 - (\mu_A^U)^2 (\mu_B^U)^2} \end{array} \right], \\ \left[v_A^L v_B^L, v_A^U v_B^U \right] \end{array} \right) \tag{47}$$

$$\tilde{A} \otimes \tilde{B} = \left(\begin{array}{c} \left[\mu_A^L \mu_B^L, \mu_A^U \mu_B^U \right], \\ \left[\begin{array}{c} \sqrt{(v_A^L)^2 + (v_B^L)^2 - (v_A^L)^2 (v_B^L)^2}, \\ \sqrt{(v_A^U)^2 + (v_B^U)^2 - (v_A^U)^2 (v_B^U)^2} \end{array} \right] \end{array} \right) \tag{48}$$

$$\lambda \tilde{A} = \left(\left[\begin{array}{c} \sqrt{1 - \left(1 - (\mu_A^L)^2\right)^\lambda}, \\ \sqrt{1 - \left(1 - (\mu_A^U)^2\right)^\lambda} \end{array} \right], \left[(v_A^L)^\lambda, (v_A^U)^\lambda \right] \right) \tag{49}$$

$$\left(\tilde{A}\right)^\lambda = \left(\left[(\mu_A^L)^\lambda, (\mu_A^U)^\lambda \right], \left[\begin{array}{c} \sqrt{1 - \left(1 - (v_A^L)^2\right)^\lambda}, \\ \sqrt{1 - \left(1 - (v_A^U)^2\right)^\lambda} \end{array} \right] \right) \tag{50}$$

Definition 20 An IVPF number $\tilde{A} = ([a, b], [c, d])$ can be defuzzified by using Eq. (51).

$$Def\left(\tilde{A}\right) = \frac{a + b + (1 - c) + (1 - d) + a \times b - \sqrt{(1 - c) \times (1 - d)}}{4} \tag{51}$$

In Eq. (51), the terms $(1 - c)$ and $(1 - d)$ convert non-membership degrees to membership degrees while the term $\sqrt{(1 - c) \times (1 - d)}$ decreases the defuzzified value.

Definition 21 Score function for an IVPF number $\tilde{A} = ([a, b], [c, d])$ is given by Eq. (52) [26]:

$$S\left(\tilde{A}\right) = \frac{1}{2}(a^2 + b^2 - c^2 - d^2) \tag{52}$$

where $S\left(\tilde{A}\right) \in [-1, +1]$.

PFSs introduced by Yager [22] can be defined by both membership and non-membership degrees here the square sum of membership degree and non-membership degrees should be equal or less than 1. PFSs can be considered as a generalization of IFSs [4]. In order to provide some basic knowledge of IFSs and PFSs, Table 1 and Fig. 3 are given. Figure 3 shows the relation between IFSs and PFSs based on Table

Table 1 Membership and nonmembership data

IFS		PFSs		IFS		PFSs	
μ	v	μ	v	μ	v	μ	v
0	1	**0**	1	**0.55**	0.45	**0.55**	0.835
0.05	0.95	**0.05**	0.999	**0.6**	0.4	**0.6**	0.8
0.1	0.9	**0.1**	0.999	**0.65**	0.35	**0.65**	0.76
0.15	0.85	**0.15**	0.989	**0.7**	0.3	**0.7**	0.714
0.2	0.8	**0.2**	0.98	**0.75**	0.25	**0.75**	0.661
0.25	0.75	**0.25**	0.968	**0.8**	0.2	**0.8**	0.6
0.3	0.7	**0.3**	0.954	**0.85**	0.15	**0.85**	0.527
0.35	0.65	**0.35**	0.937	**0.9**	0.1	**0.9**	0.436
0.4	0.6	**0.4**	0.917	**0.95**	0.05	**0.95**	0.312
0.45	0.55	**0.45**	0.893	**1**	0	**1**	0
0.5	0.5	**0.5**	0.866				

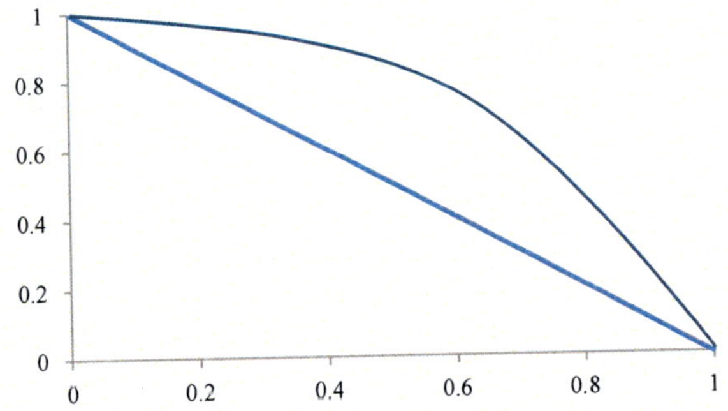

Fig. 3 Limits of membership and nonmembership degrees in IFS (blue) and PFSs (grey)

1. PFS is a kind of dilation operations over IFSs. Alternatively, we can say that IFS is a kind of concentration operation over PFSs (Kahraman et al. [28]).

Let $\tilde{A} = \langle[\mu_L, \mu_U], [V_L, V_U]\rangle$ be an interval-valued PFN, π_L and π_U are the hesitancy degree of the lower and upper points of $\tilde{A}$, respectively, can be calculated as in Eq. (53) and (54):

$$\pi_v^2 = 1 - (\mu_L^2 + v_L^2) \tag{53}$$

$$\pi_L^2 = 1 - (\mu_U^2 + v_U^2) \tag{54}$$

We know that the score functions or defuzzifying procedures are efficient when we compare PFNs in MADM problems. However, the score functions in the literature are insufficient to indicate which PFN is larger than the other since they don't associate the hesitancy properly. Motivated by the definition of hesitancy degree function of

PFNs and defuzzification function for IFNs [29], a defuzzification function as in Definition 22.

Definition 22 [30] Let $\tilde{A} = \langle[\mu_L, \mu_U], [V_L, V_U]\rangle$ be an interval-valued PFN and π_L, π_U are the hesitancy degree of the lower and upper points of $\tilde{A}$, then the defuzzifying procedure of this number is calculated by Eq. (55):

$$\mathfrak{H}\left(\tilde{A}\right) = \frac{\mu_L^2 + \mu_U^2 + \left(1 - \pi_L^4 - \upsilon_L^2\right) + \left(1 - \pi_U^4 - \upsilon_U^2\right) + \mu_L\mu_U + \sqrt[4]{\left(1 - \pi_L^4 - \upsilon_L^2\right)\left(1 - \pi_U^4 - \upsilon_U^2\right)}}{6} \quad (55)$$

A larger value of $\mathfrak{H}$ indicates a larger $\tilde{A}$. Since $0 \le \mu_\upsilon^2 + \upsilon_u^2 \le 1$, $\mathfrak{H}\left(\tilde{A}\right) \epsilon [0, 1]$.

1.1.10 Picture Fuzzy Sets

Intuitionistic fuzzy set (IFS) theory has been introduced by Atanassov [4, 23]. Although it has been used in many different fields, it has not provided to be a realistic approach for some situations. Hence, IFS has been extended to picture fuzzy sets (PiFS).

PiFS based approaches are more effective methods to meet different human views such as yes, abstain, no, and refusal. PiFS based models are successful in symbolizing uncertain information in different processes such as cluster analysis and pattern recognition [31]. Cuong [31]introduced picture fuzzy sets (PiFS) which are direct extensions of intuitonistic fuzzy sets. A picture fuzzy set

A picture fuzzy set $\tilde{A}$ on the universe X is an object of the form

$$\tilde{A} = \{\langle x; \mu_A(x), \eta_A(x), \vartheta_A(x)\rangle | x \in X\} \quad (56)$$

where $\mu_{\tilde{A}}(x) \in [0, 1]$ is called the "degree of positive membership of $\tilde{A}$", $\eta_{\tilde{A}}(x) \in [0, 1]$ is called the "degree of neutral membership of $\tilde{A}$" and $\nu_{\tilde{A}}(x) \in [0, 1]$ is called the "degree of negative membership of $\tilde{A}$", and $\mu_{\tilde{A}}(x)$, $\eta_{\tilde{A}}(x)$, and $\nu_{\tilde{A}}(x)$ satisfy the following condition: $0 \le \mu_{\tilde{A}}(x) + \eta_{\tilde{A}}(x) + \nu_{\tilde{A}}(x) \le 1, \forall x \in X$. Then for $x \in X$, $\pi_{\tilde{A}}(x) = 1 - \mu_{\tilde{A}}(x) - \eta_{\tilde{A}}(x) - \nu_{\tilde{A}}(x))$ could be called the degree of refusal membership of x in $\tilde{A}$. Thereafter, $\langle x; \mu_{\tilde{A}}(x), \eta_{\tilde{A}}(x), \vartheta_{\tilde{A}}(x)\rangle$ will be given as $\left(x; \mu_{\tilde{A}}, \eta_{\tilde{A}}, \vartheta_{\tilde{A}}\right)$.

Voting can be a good illustration of such a situation as the human voters may be divided into four groups of those who: vote for, hesitant, and vote against, refusal of the voting [32]. Some definitions and theorems of PiFS are given in the following [33].

Definition 23 A PFSs on a $\tilde{A}_p$ of the universe of discourse U is given by;

$$\tilde{A}_p = \left\{\left\langle u, (\mu_{\tilde{A}_p}(u), \nu_{\tilde{A}_p}(u), \pi_{\tilde{A}_p}(u)) \right| u \in U\right\} \quad (57)$$

where

$$\mu_{\tilde{A}_p}(u) : U \to [0, 1], \quad \nu_{\tilde{A}_p}(u) : U \to [0, 1], \quad \pi_{\tilde{A}_p}(u) : U \to [0, 1] \quad (58)$$

and

$$0 \leq \mu_{\tilde{A}_p}(u) + \nu_{\tilde{A}_p}(u) + \pi_{\tilde{A}_p}(u) \leq 1 \qquad \forall u \in U \quad (59)$$

Then, for each u, the numbers $\mu_{\tilde{A}_S}(u)$, $\nu_{\tilde{A}_S}(u)$ and $\pi_{\tilde{A}_S}(u)$ are the degree of membership, non-membership and hesitancy of u to $\tilde{A}_S$, respectively. $\rho = 1 - \left(\mu_{\tilde{A}_P}(u) + v_{\tilde{A}_P}(u) + \pi_{\tilde{A}_P}(u)\right)$ is called as a refusal degree [34].

Definition 24 Let $\tilde{\alpha} = (\mu_a, \eta_\alpha, \vartheta_\alpha)$ be a picture fuzzy number (PiFN). Then, the score function S of a picture fuzzy number can be given as follows [35]:

$$S\left(\tilde{\alpha}\right) = \mu_a - \vartheta_\alpha, \; S\left(\tilde{\alpha}\right) \in [-1, 1] \quad (60)$$

Definition 25 Basic operators of Single-valued PFSs;

$$\tilde{A}_p \oplus \tilde{B}_p = \left\{\mu_{\tilde{A}_p} + \mu_{\tilde{B}_p} - \mu_{\tilde{A}_p}\mu_{\tilde{B}_p}, \; \pi_{\tilde{A}_p}\pi_{\tilde{B}_p}, \; \upsilon_{\tilde{A}_p}\upsilon_{\tilde{B}_p}\right\} \quad (61)$$

$$\tilde{A}_p \otimes \tilde{B}_p = \left\{\mu_{\tilde{A}_p}\mu_{\tilde{B}_p}, \; \pi_{\tilde{A}_p} + \pi_{\tilde{B}_p} - \pi_{\tilde{A}_p}\pi_{\tilde{B}_p}, \; \upsilon_{\tilde{A}_p} + \upsilon_{\tilde{B}_p} - \upsilon_{\tilde{A}_p}\upsilon_{\tilde{B}_p}\right\} \quad (62)$$

$$\lambda \cdot \tilde{A}_p = \left\{\left(1 - \left(1 - \mu_{\tilde{A}_p}\right)^\lambda\right), \; \pi^\lambda_{\tilde{A}_p}, \; \upsilon^\lambda_{\tilde{A}_p}\right\} \quad for\,\lambda > 0 \quad (63)$$

$$\tilde{A}^\lambda_p = \left\{\mu^\lambda_{\tilde{A}_p}, \; \left(1 - \left(1 - \upsilon_{\tilde{A}_p}\right)^\lambda\right), \; \left(1 - \left(1 - \pi_{\tilde{A}_p}\right)^\lambda\right)\right\} \quad for\ \lambda > 0 \quad (64)$$

Picture fuzzy arithmetic aggregation operators are used for aggregating the different evaluations of multiexperts. Picture fuzzy weighted averaging (PiFWA) operator and picture fuzzy weighted geometric (PiFWG) operator are aggregation operators of PiFS as arithmetic and geometric aggregation operators which are developed by Guiwu [33].

Definition 26 Let α_j ($j = 1, 2, \ldots, n$) be a collection of PFNs. The picture fuzzy weighted averaging (PFWA) operator is a mapping $P^n \to P$ such that

$$PFWA_{\omega}(\alpha_1, \alpha_2, \ldots \alpha_n) = \oplus_{j=1}^{n}(\omega_j \alpha_j) \quad (65)$$

where $\omega = (\omega_1,\omega_2,\ldots,\omega_n)^T$ be the weight vector of $\alpha_j (j = 1, 2,\ldots,n)$, and $\omega_j > 0$, $\sum_{j=1}^{n} \omega_j = 1$

Theorem 1 *The aggregated value by using PFWA operator is also a PiFN, where*

$$PFWA_{\omega}(\alpha_1, \alpha_2, \ldots, \alpha_n) = \left(1 - \prod_{j=1}^{n}(1 - \mu_{\alpha_j})^{\omega_j}, \prod_{j=1}^{n}(\eta_{\alpha_j})^{\omega_j}, \prod_{j=1}^{n}(\vartheta_{\alpha_j})^{\omega_j}\right) \quad (66)$$

where $\omega = (\omega_1,\omega_2,\ldots,\omega_n)^T$ be the weight vector of $\alpha_j (j = 1, 2,\ldots,n)$, and $\omega_j > 0$, $\sum_{j=1}^{n} \omega_j = 1$

Definition 27 Let α_j (j = 1, 2, …, n) be a collection of PiFNs. The picture fuzzy weighted geometric (PiFWG) operator is a mapping $P^n \rightarrow P$ such that

$$PFWG_{\omega}(\alpha_1, \alpha_2, \ldots, \alpha_n) = \otimes_{j=1}^{n}\left(\alpha_j^{\omega_j}\right) \quad (67)$$

where $\omega = (\omega_1,\omega_2, \ldots,\omega_n)^T$ be the weight vector of $\alpha_j (j = 1, 2,\ldots,n)$, and $\omega_j > 0, \sum_{j=1}^{n} \omega_j = 1$

Theorem 2 *The aggregated value by using PFWG operator is also a PFN, where*

$$PFWG_{\omega}(\alpha_1, \alpha_2, \ldots, \alpha_n) = \left(\prod_{j=1}^{n}\left(\mu_{\alpha_j}\right)^{\omega_j}, 1 - \prod_{j=1}^{n}\left(1 - \eta_{\alpha_j}\right)^{\omega_j}, 1 - \prod_{j=1}^{n}\left(1 - \vartheta_{\alpha_j}\right)^{\omega_j}\right) \quad (68)$$

where $\omega = (\omega_1,\omega_2, \ldots,\omega_n)^T$ be the weight vector of $\alpha_j (j = 1, 2, \ldots,n)$, and $\omega_j > 0$, $\sum_{j=1}^{n} \omega_j = 1$

Definition 28 Single-valued Picture Fuzzy Weighted Averaging operator (PFWA) with respect to,$w = (w_1, w_2 \ldots, w_n)$; $w_i \in [0, 1]$; $\sum_{i=1}^{n} w_i = 1$, is defined as;

$$\begin{aligned} PFWA_w(\tilde{A}_1, \ldots \tilde{A}_n) &= w_1\tilde{A}_1 + w_2\tilde{A}_2 + \cdots + w_n\tilde{A}_n \\ &= \left\{1 - \prod_{i=1}^{n}(1 - \mu_{\tilde{A}_i})^{w_i}, \prod_{i=1}^{n} v_{\tilde{A}_i}^{w_i}, \prod_{i=1}^{n} \pi_{\tilde{A}_i}^{w_i}\right\} \end{aligned} \quad (69)$$

Definition 29 Score functions and Accuracy functions of sorting picture fuzzy numbers are defined by;

$$Score\left(\tilde{A}_p\right) = \frac{1}{2}\left(1 + 2\mu_{\tilde{A}_p} - \upsilon_{\tilde{A}_p} - \pi_{\tilde{A}_p}/2\right) \tag{70}$$

$$Accuracy\left(\tilde{A}_p\right) = \mu_{\tilde{A}_p} + \upsilon_{\tilde{A}_p} + \pi_{\tilde{A}_p} \tag{71}$$

Note that: $\tilde{A}_p < \tilde{B}_p$ if and only if
or

(a) $Score(\tilde{A}_p) < Score(\tilde{B}_p)$or and
(b) $Score(\tilde{A}_p) = Score(\tilde{B}_p)$ and $Accuracy(\tilde{A}_p) < Accuracy(\tilde{B}_p)$

Definition 30 Distance formulas for picture fuzzy numbers $\left(\tilde{A}_p, \tilde{B}_p\right)$ are defined as follows [36],

The normalized Hamming Distance;

$$D_H(\tilde{A}_p, \tilde{B}_p) = \frac{1}{2n}\sum_{i=1}^{n}\begin{bmatrix} \left|\mu_{\tilde{A}_p}(x_i) - \mu_{\tilde{B}_p}(x_i)\right| + \left|\upsilon_{\tilde{A}_p}(x_i) - \upsilon_{\tilde{B}_p}(x_i)\right| \\ +\left|\pi_{\tilde{A}_p}(x_i) - \pi_{\tilde{B}_p}(x_i)\right| + \left|\rho_{\tilde{A}_p}(x_i) - \rho_{\tilde{B}_p}(x_i)\right| \end{bmatrix} \tag{72}$$

The normalized Euclidean Distance;

$$D_E(\tilde{A}_p, \tilde{B}_p) = \sqrt{\frac{1}{2n}\sum_{i=1}^{n}\begin{bmatrix} \left(\mu_{\tilde{A}_p}(x_i) - \mu_{\tilde{B}_p}(x_i)\right)^2 + \left(\upsilon_{\tilde{A}_p}(x_i) - \upsilon_{\tilde{B}_p}(x_i)\right)^2 \\ +\left(\pi_{\tilde{A}_p}(x_i) - \pi_{\tilde{B}_p}(x_i)\right)^2 + \left(\rho_{\tilde{A}_p}(x_i) - \rho_{\tilde{B}_p}(x_i)\right)^2 \end{bmatrix}} \tag{73}$$

1.1.11 Q-Rung Orthopair Fuzzy Sets (Q-RPFSs)

Q-ROFSs introduced by Yager [37] are represented with the degree of membership and non-membership. In q-ROFSs, the sum of the qth power of the membership and non-membership degrees must be at most equal to one [37]. In Fig. 1, it is easily observed that q-ROFSs have an acceptable membership grade space larger than of IFSs and PFSs. Q-ROFSs are described as in Definition 31 and geometric space for IFNs, PFNs and q-ROFs are shown in Fig. 4.

Definition 31 A q-ROFS Q in a finite universe of discourse X is defined as follows by Yager [37].

Fig. 4 Geometric space range of IFNs, PFNs, and q-ROFNs

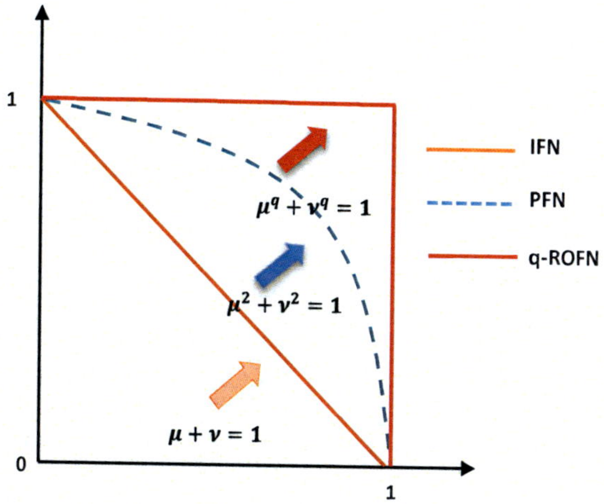

$$Q = \{(x, \mu_Q(x), \nu_Q(x))|x \epsilon X\} \tag{74}$$

where the function $\mu_Q : X \to [0, 1]$ denotes the degree of membership and $\upsilon_Q : X \to [0, 1]]$ denotes the degree of non-membership of the element $x \in X$ to the set Q, respectively, with the condition that $0 \le \mu_Q(x) + \nu_Q(x) \le 1$, $(q \ge 1)$ for every $x \epsilon X$. The degree of indeterminacy is given as $\pi_P(x) = \sqrt[q]{1 - \mu_P(x)^q - \nu_P(x)^q}$ [38].

Definition 32 Let $Q = (\mu_Q, \nu_Q)$, $Q_1 = (\mu_{Q_1}, \nu_{Q_1})$ and $Q_2 = (\mu_{Q_2}, \nu_{Q_2})$ be three q-rung orthopair fuzzy numbers (q-ROFNs), then their operations can be defined as follows [37].

$$Q_1 \cap Q_2 = (min\{\mu_{Q_1}, \mu_{Q_2}\}, max\{\nu_{Q_1}, \nu_{Q_2}\}) \tag{75}$$

$$Q_1 \cup Q_2 = (max\{\mu_{Q_1}, \mu_{Q_2}\}, min\{\nu_{Q_1}, \nu_{Q_2}\}) \tag{76}$$

$$Q_1 \oplus Q_2 = \left(\left(\mu_{Q_1^q} + \mu_{Q_2^q} - \mu_{Q_1^q}\mu_{Q_2^q}\right)^{1/q}, \nu_{Q_1}\nu_{Q_2}\right) \tag{77}$$

$$Q_1 \otimes Q_2 = \left(\mu_{Q_1}\mu_{Q_2}, \left(\nu_{Q_1^q} + \nu_{Q_2^q} - \nu_{Q_1^q}\nu_{Q_2^q}\right)^{1/q}\right) \tag{78}$$

$$\lambda Q = \left(\left(1 - (1 - \mu_Q^q)^\lambda\right)^{1/q}, \nu_Q^\lambda\right), \lambda > 0 \tag{79}$$

$$Q^{\lambda} = \left(\mu_Q^{\lambda}, \left(1 - \left(1 - \nu_Q^q\right)^{\lambda}\right)^{1/q}\right), \lambda > 0 \tag{80}$$

Definition 33 Let $Q = (\mu_Q, \nu_Q)$ be a q-ROFN, then the score function $s(\alpha)$ and accuracy function $H(\alpha)$ of a can be defined as follows [37]:

$$s(\alpha) = {\mu_Q}^q - {\nu_Q}^q \tag{81}$$

$$H(\alpha) = {\mu_Q}^q + {\nu_Q}^q \tag{82}$$

Definition 34 Let $Q_i = (\mu_{Q_i}, \nu_{Q_i})$ $(i = 1, 2, \ldots, n)$ be a set of q-ROFNs and $w = (w_1, w_2, \ldots, w_n)^T$ be weight vector of Q_i with$\sum_{i=1}^{n} w_i = 1$, then a q-rung orthopair fuzzy weighted average (q − ROFWA) operator is [37]:

$$q - ROFWA(Q_1, Q_2, \ldots, Q_n) = \left(\left(1 - \prod_{i=1}^{n}\left(1 - \mu_{Q_i}^q\right)^{w_i}\right)^{1/q}, \prod_{i=1}^{n} \upsilon_{\underset{\sim}{\mathcal{F}}_i}^{w_i}\right) \tag{83}$$

Definition 35 Let $Q_i = (\mu_{Q_i}, \nu_{Q_i})$ $(i = 1, 2, \ldots, n)$ be a set of q-ROFNs and $w = (w_1, w_2, \ldots, w_n)^T$ be weight vector of Q_i with$\sum_{i=1}^{n} w_i = 1$, then a q-rung orthopair fuzzy weighted geometric $(q - ROFWG)$ operator is [37]:

$$q - ROFWG(Q_1, Q_2, \ldots, Q_n) = \left(\prod_{i=1}^{n} \mu_{\underset{\sim}{\mathcal{F}}_i}^{w_i}, \left(1 - \prod_{i=1}^{n}\left(1 - \upsilon_{Q_i}^q\right)^{w_i}\right)^{1/q}\right) \tag{84}$$

1.1.12 Fermatean Fuzzy Sets

[37] have introduced q-rung orthopair fuzzy sets being a general class of IFSs and PFSs. The sum of the qth power of membership degree and nonmembersip degree q-rung orthopair fuzzy sets is bounded by one. When q = 3, Senapati and Yager [39] have called q-rung orthopair fuzzy sets as fermatean fuzzy sets (FFSs). In Fig. 5, the relation between IFSs, PFSs and FFSs is given

Definition 36 Let X be a universe of discourse. A Fermatean fuzzy sets $\mathcal{F}$ in X is an object having the form Senapati and Yager [39]:

Fig. 5 Comparision of IFSs, PFSs and FFSs

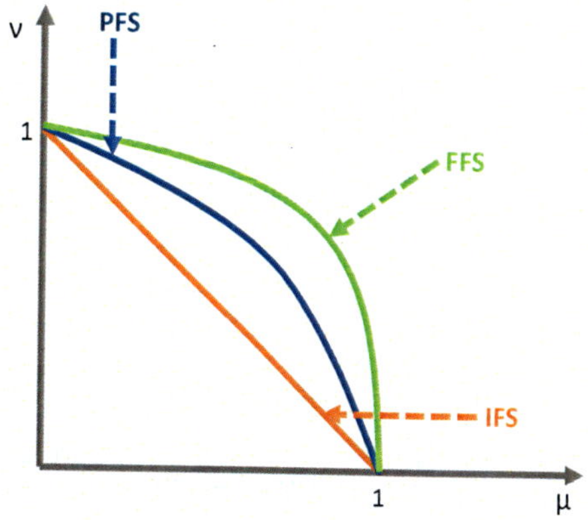

$$\mathcal{F} = \{\langle x, \mu_F(x), \nu_F(x)\rangle | x \epsilon X\} \tag{85}$$

where $\mu_F : X \to [0, 1]$ and $\nu_F : X \to [0, 1]$

$$0 \le \left(\mu_{\tilde{\mathcal{F}}}(x)\right)^3 + \left(\upsilon_{\tilde{F}}(x)\right)^3 \le 1 \tag{86}$$

for all $x \epsilon X$. The numbers $\mu_F(x)$ and $\nu_F(x)$ indicate, respectively, the degree of membership and the degree of non-membership of the element x in the set $\mathcal{F}$.

For any FFS $\mathcal{F}$ and $x \epsilon X$, the degree of hesitancy is calculated as follows:

$$\pi_F(X) = \sqrt[3]{1 - \mu_F(x)^3 - \nu_F(x)^3} \tag{87}$$

Definition 37 Let $\mathcal{F} = (\mu_F, \nu_F)$, $\mathcal{F}_1 = \left(\mu_{F_1}, \nu_{F_1}\right)$ and $\mathcal{F}_2 = \left(\mu_{F_2}, \nu_{F_2}\right)$be three FFSs, then their operations are defined as follows [39]:

$$\mathcal{F}_1 \cap \mathcal{F}_2 = \left(min\left\{\mu_{F_1}, \mu_{F_2}\right\}, max\left\{\nu_{F_1}, \nu_{F_2}\right\}\right) \tag{88}$$

$$\mathcal{F}_1 \cup \mathcal{F}_2 = \left(max\left\{\mu_{F_1}, \mu_{F_2}\right\}, min\left\{\nu_{F_1}, \nu_{F_2}\right\}\right) \tag{89}$$

$$\mathcal{F}^c = (\nu_F, \mu_F)$$

Definition 38 Let $\mathcal{F} = (\mu_F, \nu_F)$, $\mathcal{F}_1 = \left(\mu_{F_1}, \nu_{F_1}\right)$ and $\mathcal{F}_2 = \left(\mu_{F_2}, \nu_{F_2}\right)$be three FFSs and$\lambda > 0$, then the operations of these three FFNs are interpreted in these ways 39:

$$\mathcal{F}_1 \oplus \mathcal{F}_2 = \left(\sqrt[3]{{\mu_{F1}}^3 + {\mu_{F2}}^3 - {\mu_{F1}}^3{\mu_{F2}}^3}, \nu_{F1}\nu_{F2}\right) \tag{90}$$

$$\mathcal{F}_1 \otimes \mathcal{F}_2 = \left(\mu_{F1}\mu_{F2}, \sqrt[3]{\nu_{F1}{}^3 + \nu_{F2}{}^3 - \nu_{F1}{}^3\nu_{F2}{}^3}\right) \tag{91}$$

$$\lambda\mathcal{F} = \left(\sqrt[3]{1 - (1 - \mu_F{}^3)^\lambda}, \nu_F{}^\lambda\right), \lambda > 0 \tag{92}$$

$$\mathcal{F}^\lambda = \left(\mu_F{}^\lambda, \sqrt[3]{1 - (1 - \nu_F{}^3)^\lambda}\right), \lambda > 0 \tag{93}$$

Definition 39 Let $\mathcal{F}_i = (\mu_{F_i}, \nu_{F_i})$ $(i = 1, 2, \ldots, n)$ be a set of FFNs and $w = (w_1, w_2, \ldots, w_n)^T$ be weight vector of $\mathcal{F}_i$ with$\sum_{i=1}^n w_i = 1$, then a fermatean fuzzy weighted average (FFWA) operator is [40]:

$$FFWA(\mathcal{F}_1, \mathcal{F}_2, \ldots, \mathcal{F}_n) = \left(\sum_{I=1}^n w_i\mu_{\mathcal{F}_i}, \sum_{i=1}^n w_i\upsilon_{\mathcal{F}_i}\right) \tag{94}$$

Definition 40 Let $\mathcal{F}_i = (\mu_{F_i}, \nu_{F_i})$ $(i = 1, 2, \ldots, n)$ be a set of FFNs and $w = (w_1, w_2, \ldots, w_n)^T$ be weight vector of $\mathcal{F}_i$ with$\sum_{i=1}^n w_i = 1$, then a fermatean fuzzy weighted geometric (FFWG) operator is [40]:

$$FFWG(\mathcal{F}_1, \mathcal{F}_2, \ldots, \mathcal{F}_n) = \left(\prod_{i=1}^n \mu_{\mathcal{F}_i}{}^{w_i}, \prod_{i=1}^n \upsilon_{\mathcal{F}_i}{}^{w_i}\right) \tag{95}$$

Definition 41 Let X be an interval-valued fermatean fuzzy set (IVFFS). IVFFS $\tilde{\mathcal{F}}$ in X is an object having the form

$$\tilde{\mathcal{F}} = \left\{\langle x, \mu_{\tilde{\mathcal{F}}}(x), \nu_{\tilde{\mathcal{F}}}(x)\rangle | x \epsilon X\right\} \tag{96}$$

where $\mu_{\tilde{\mathcal{F}}}(x) \subseteq [0, 1]$ and $\upsilon_{\tilde{\mathcal{F}}}(x) \subseteq [0, 1]$ denote the membership degree and non-membership degree of the element $x \epsilon X$ to the set $\tilde{\mathcal{F}}$, respectively. Also, for each $x \epsilon X$, $\mu_{\tilde{\mathcal{F}}}(X)$ and $\upsilon_{\tilde{\mathcal{F}}}(X)$ are closed intervals and their lower and upper bounds are denoted by $\mu^L_{\tilde{\mathcal{F}}}(x)$, $\mu^U_{\tilde{\mathcal{F}}}(x)$, $\upsilon^L_{\tilde{\mathcal{F}}}(x)$,$\upsilon^U_{\tilde{\mathcal{F}}}(x)$, respectively. Therefore, $\tilde{\mathcal{F}}$ can also be expressed as follows:

$$\mu_{\tilde{\mathcal{F}}}(x) = \left[\mu^L_{\tilde{\mathcal{F}}}(x), \mu^U_{\tilde{\mathcal{F}}}(x)\right] \subset [0, 1] \tag{97}$$

$$\nu_{\tilde{\mathcal{F}}}(x) = \left[\upsilon^{L}_{\tilde{\mathcal{F}}}(x), \upsilon^{U}_{\tilde{\mathcal{F}}}(x)\right] \subset [0, 1] \tag{98}$$

where the expression is subject to the condition

$$0 \le \left(\mu^{U}_{\tilde{\mathcal{F}}}(x)\right)^3 + \left(\upsilon^{U}_{\tilde{F}}(x)\right)^3 \le 1 \tag{99}$$

For every $x \epsilon X$, $\pi_{\tilde{\mathcal{F}}}(x) = \left[\pi^{L}_{\tilde{\mathcal{F}}}(x), \pi^{U}_{\tilde{\mathcal{F}}}(x)\right]$ is called as the degree of hesitancy in IVFFSs, where $\pi^{L}_{\tilde{\mathcal{F}}}(x) = \sqrt[3]{1 - \left(\mu^{U}_{\tilde{\mathcal{F}}}(x)\right)^3 - \left(\upsilon^{U}_{\tilde{\mathcal{F}}}(x)\right)^3}$ and $\pi^{U}_{\tilde{\mathcal{F}}}(x) = \sqrt[3]{1 - \left(\mu^{L}_{\tilde{\mathcal{F}}}(x)\right)^3 - \left(\upsilon^{L}_{\tilde{\mathcal{F}}}(x)\right)^3}$.

Definition 42 Let $\tilde{\mathcal{F}} = \left(\left[\mu^{L}_{\tilde{\mathcal{F}}}, \mu^{U}_{\tilde{\mathcal{F}}}\right], \left[\upsilon^{L}_{\tilde{\mathcal{F}}}, \upsilon^{U}_{\tilde{\mathcal{F}}}\right]\right)$, $\tilde{\mathcal{F}}_1 = \left(\left[\mu^{L}_{\tilde{\mathcal{F}}_1}, \mu^{U}_{\tilde{\mathcal{F}}_1}\right], \left[\upsilon^{L}_{\tilde{\mathcal{F}}_1}, \upsilon^{U}_{\tilde{\mathcal{F}}_1}\right]\right)$ and $\tilde{\mathcal{F}}_2 = \left(\left[\mu^{L}_{\tilde{\mathcal{F}}_2}, \mu^{U}_{\tilde{\mathcal{F}}_2}\right], \left[\upsilon^{L}_{\tilde{\mathcal{F}}_2}, \upsilon^{U}_{\tilde{\mathcal{F}}_2}\right]\right)$ be three FFSs and $\lambda > 0$, then their operations are defined as follows:

$$\tilde{\mathcal{F}}_1 \oplus \tilde{\mathcal{F}}_2 = \left(\left[\begin{array}{c} \sqrt[3]{\left(\mu^{L}_{\tilde{\mathcal{F}}_1}\right)^3 + \left(\mu^{L}_{\tilde{\mathcal{F}}_2}\right)^3 - \left(\mu^{L}_{\tilde{\mathcal{F}}_1}\right)^3\left(\mu^{L}_{\tilde{\mathcal{F}}_2}\right)^3}, \\ \sqrt[3]{\left(\mu^{U}_{\tilde{\mathcal{F}}_1}\right)^3 + \left(\mu^{U}_{\tilde{\mathcal{F}}_2}\right)^3 - \left(\mu^{U}_{\tilde{\mathcal{F}}_1}\right)^3\left(\mu^{U}_{\tilde{\mathcal{F}}_2}\right)^3} \end{array}\right], \left[\upsilon^{L}_{\tilde{\mathcal{F}}_1}\upsilon^{L}_{\tilde{\mathcal{F}}_2}, \upsilon^{U}_{\tilde{\mathcal{F}}_1}\upsilon^{U}_{\tilde{\mathcal{F}}_2}\right]\right) \tag{100}$$

$$\tilde{\mathcal{F}}_1 \otimes \tilde{\mathcal{F}}_2 = \left(\left[\mu^{L}_{\tilde{\mathcal{F}}_1}\mu^{L}_{\tilde{\mathcal{F}}_2}, \mu^{U}_{\tilde{\mathcal{F}}_1}\mu^{U}_{\tilde{\mathcal{F}}_2}\right], \left[\begin{array}{c} \sqrt[3]{\left(\upsilon^{L}_{\tilde{\mathcal{F}}_1}\right)^3 + \left(\upsilon^{L}_{\tilde{\mathcal{F}}_2}\right)^3 - \left(\upsilon^{L}_{\tilde{\mathcal{F}}_1}\right)^3\left(\upsilon^{L}_{\tilde{\mathcal{F}}_2}\right)^3}, \\ \sqrt[3]{\left(\upsilon^{U}_{\tilde{\mathcal{F}}_1}\right)^3 + \left(\upsilon^{U}_{\tilde{\mathcal{F}}_2}\right)^3 - \left(\upsilon^{U}_{\tilde{\mathcal{F}}_1}\right)^3\left(\upsilon^{U}_{\tilde{\mathcal{F}}_2}\right)^3} \end{array}\right]\right) \tag{101}$$

$$\lambda\,\tilde{\mathcal{F}} = \left(\left[\sqrt[3]{1 - \left(1 - \left(\mu^{L}_{\tilde{\mathcal{F}}}\right)^3\right)^{\lambda}}, \sqrt[3]{1 - \left(1 - \left(\mu^{U}_{\tilde{\mathcal{F}}}\right)^3\right)^{\lambda}}\right], \left[\left(\upsilon^{L}_{\tilde{\mathcal{F}}}\right)^{\lambda}, \left(\upsilon^{U}_{\tilde{\mathcal{F}}}\right)^{\lambda}\right]\right) \tag{102}$$

$$\tilde{\mathcal{F}}^{\lambda} = \left(\left[\left(\mu^{L}_{\tilde{\mathcal{F}}}\right)^{\lambda}, \left(\mu^{U}_{\tilde{\mathcal{F}}}\right)^{\lambda}\right], \left[\sqrt[3]{1 - \left(1 - \left(\upsilon^{L}_{\tilde{\mathcal{F}}}\right)^3\right)^{\lambda}}, \sqrt[3]{1 - \left(1 - \left(\upsilon^{U}_{\tilde{\mathcal{F}}}\right)^3\right)^{\lambda}}\right]\right) \tag{103}$$

Definition 43 Let $\tilde{\mathcal{F}}_i = \left(\left[\mu^L_{\tilde{\mathcal{F}}_i}, \mu^U_{\tilde{\mathcal{F}}_i}\right], \left[\upsilon^L_{\tilde{\mathcal{F}}_i}, \upsilon^U_{\tilde{\mathcal{F}}_i}\right]\right)(i = 1, 2, \ldots, n)$ be a set of IVFFSs and $w = (w_1, w_2, \ldots, w_n)^T$ be weight vector of $\mathcal{F}_i$ with$\sum_{i=1}^n w_i = 1$, then an interval-valued Fermatean fuzzy weighted average (IVFFWA) operator is a mapping IVFFWA:$\tilde{\mathcal{F}}^n \to \tilde{\mathcal{F}}$, where

$$IVFFWA\left(\tilde{\mathcal{F}}_1, \tilde{\mathcal{F}}_2, \ldots, \tilde{\mathcal{F}}_n\right) = \left(\begin{array}{c}\left[\sqrt[3]{\left(1 - \prod_{i=1}^n \left(1 - \left(\mu^L_{\tilde{\mathcal{F}}_i}\right)^3\right)^{w_i}\right)}, \sqrt[3]{\left(1 - \prod_{i=1}^n \left(1 - \left(\mu^U_{\tilde{\mathcal{F}}_i}\right)^3\right)^{w_i}\right)}\right], \\ \left[\prod_{i=1}^n \left(\upsilon^L_{\tilde{\mathcal{F}}_i}\right)^{w_i}, \prod_{i=1}^n \left(\upsilon^U_{\tilde{\mathcal{F}}_i}\right)^{w_i}\right]\end{array}\right) \tag{104}$$

Definition 44 Let $\tilde{\mathcal{F}}_i = \left(\left[\mu^L_{\tilde{\mathcal{F}}_i}, \mu^U_{\tilde{\mathcal{F}}_i}\right], \left[\upsilon^L_{\tilde{\mathcal{F}}_i}, \upsilon^U_{\tilde{\mathcal{F}}_i}\right]\right)(i = 1, 2, \ldots, n)$ be a set of IVFFSs and $w = (w_1, w_2, \ldots, w_n)^T$ be weight vector of $\tilde{\mathcal{F}}_i$ with$\sum_{i=1}^n w_i = 1$, then an interval-valued Fermatean fuzzy weighted geometric (IVFFWG) operator is a mapping IVFFWG:$\tilde{\mathcal{F}}^n \to \tilde{\mathcal{F}}$, where

$$IVFFWG\left(\tilde{\mathcal{F}}_1, \tilde{\mathcal{F}}_2, \ldots, \tilde{\mathcal{F}}_n\right) = \left(\left[\prod_{i=1}^n \left(\mu^L_i\right)^{w_i}, \prod_{i=1}^n \left(\mu^U_i\right)^{w_i}\right], \left[\begin{array}{c}\sqrt[3]{\left(1 - \prod_{i=1}^n \left(1 - \left(\upsilon^L_{\tilde{\mathcal{F}}_i}\right)^3\right)^{w_i}\right)}, \\ \sqrt[3]{\left(1 - \prod_{i=1}^n \left(1 - \left(\upsilon^U_{\tilde{\mathcal{F}}_i}\right)^3\right)^{w_i}\right)}\end{array}\right]\right) \tag{105}$$

Definition 45 Deffuzzification of $\tilde{\mathcal{F}}_i = \left(\left[\mu^L_{\tilde{\mathcal{F}}_i}, \mu^U_{\tilde{\mathcal{F}}_i}\right], \left[\upsilon^L_{\tilde{\mathcal{F}}_i}, \upsilon^U_{\tilde{\mathcal{F}}_i}\right]\right)(i = 1, 2, \ldots, n)$ is given as in Eq. (106):

$$\text{Deff}(\tilde{\mathcal{F}}_i)$$

$$= \begin{cases} \frac{1+\left|(\mu_i^L)^3-(\nu_i^L)^3\right|+1+\left|(\mu_i^U)^3-(\nu_i^U)^3\right|-\left(\pi_{ij}^L\right)^3-\left(\pi_{ij}^U\right)^3}{4} \times 10, EI \leq IVFFN \leq CHI \\ \frac{1}{\left(\frac{1+\left|(\mu_{ij}^L)^3-(\nu_{ij}^L)^3\right|+1+\left|(\mu_{ij}^U)^3-(\nu_{ij}^U)^3\right|-(\pi_{ij}^L)^3-(\pi_{ij}^U)^3}{4}\times 10\right)}, SLI \leq IVFFN \leq CLI \end{cases} \quad (106)$$

This defuzzification operation is based on Saaty's classical 1–9 scale so that the defuzzification produces values between 1 and 9 for $EI \leq IVFFN \leq CHI$ and 1/9 – 1 for $SLI \leq IVFFN \leq CLI$.

1.1.13 Spherical Fuzzy Sets

Spherical fuzzy sets (SFS) have been recently introduced by Kutlu Gundogdu and Kahraman (2019). These ets are based on the fact that the hesitancy of a decision maker can be assigned inde pendently satisfying the condition that the squared sum of membership, non-membership and hesitancy degrees is at most equal to 1. Thus, SFS are a mixture of PFS and NS theories. In the following, definition of SFS is presented:

Definition 46 Single valued Spherical Fuzzy Sets (SFS) $\tilde{A}_S$ of the universe of discourse U is given by

$$\tilde{A}_S = \left\{\langle u, (\mu_{\tilde{A}_S}(u), \nu_{\tilde{A}_S}(u), \pi_{\tilde{A}_S}(u))\middle| u \in U\right\} \quad (107)$$

where

$$\mu_{\tilde{A}_S}(u) : U \to [0, 1], \quad \nu_{\tilde{A}_S}(u) : U \to [0, 1], \quad \pi_{\tilde{A}_S}(u) : U \to [0, 1]$$

and

$$0 \leq \mu^2_{\tilde{A}_S}(u) + \nu^2_{\tilde{A}_S}(u) + \pi^2_{\tilde{A}_S}(u) \leq 1 \qquad \forall u \in U \quad (108)$$

For each u, the numbers $\mu_{\tilde{A}_S}(u)$, $\nu_{\tilde{A}_S}(u)$ and $\pi_{\tilde{A}_S}(u)$ are the degree of membership, non-membership and hesitancy of u to $\tilde{A}_S$, respectively.

The novel concept of Spherical Fuzzy Sets provides a larger preference domain for decision makers to assign membership degrees since the squared sum of the spherical parameters is allowed to be at most 1.0. DMs can determine their hesitancy degree independently under spherical fuzzy environment. Spherical fuzzy Sets (SFS) are a generalization of Pythagorean Fuzzy Sets, picture fuzzy sets and neutrosophic sets [41–45]. Neutrosophic sets are defined with three parameters *truthiness, falsity,* and *indeterminacy.* The values of these parameters are between 0 and 1, and the

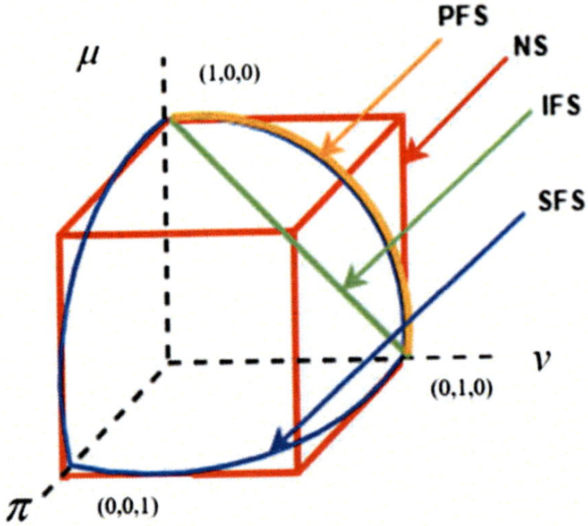

Fig. 6 Geometric representations of IFS, PFS, NS, and SFS [43]

sum of these parameters can be between 0 and 3. The parameters can be defined independently in neutrosophic sets.

In spherical fuzzy sets, the squared sum of membership, nonmembership, and hesitancy parameters can be between 0 and 1, and each of them can be defined between 0 and 1 independently. Figure 6 illustrates the differences between IFS, PFS, NS, and SFS [46–56].

Definition 47 Basic operators of Single-valued SFS;

$$\tilde{A}_S \oplus \tilde{B}_S = \left\{ \begin{array}{l} \left(\mu_{\tilde{A}_S}^2 + \mu_{\tilde{B}_S}^2 - \mu_{\tilde{A}_S}^2 \mu_{\tilde{B}_S}^2\right)^{1/2}, \ v_{\tilde{A}_S} v_{\tilde{B}_S}, \\ \left(\left(1 - \mu_{\tilde{B}_S}^2\right)\pi_{\tilde{A}_S}^2 + \left(1 - \mu_{\tilde{A}_S}^2\right)\pi_{\tilde{B}_S}^2 - \pi_{\tilde{A}_S}^2 \pi_{\tilde{B}_S}^2\right)^{1/2} \end{array} \right\} \tag{109}$$

$$\tilde{A}_S \otimes \tilde{B}_S = \left\{ \begin{array}{l} \mu_{\tilde{A}_S} \mu_{\tilde{B}_S}, \ \left(v_{\tilde{A}_S}^2 + v_{\tilde{B}_S}^2 - v_{\tilde{A}_S}^2 v_{\tilde{B}_S}^2\right)^{1/2}, \\ \left(\left(1 - v_{\tilde{B}_S}^2\right)\pi_{\tilde{A}_S}^2 + \left(1 - v_{\tilde{A}_S}^2\right)\pi_{\tilde{B}_S}^2 - \pi_{\tilde{A}_S}^2 \pi_{\tilde{B}_S}^2\right)^{1/2} \end{array} \right\} \tag{110}$$

$$\lambda \cdot \tilde{A}_S = \left\{ \begin{array}{l} \left(1 - \left(1 - \mu_{\tilde{A}_S}^2\right)^{\lambda}\right)^{1/2}, \ v_{\tilde{A}_S}^{\lambda}, \\ \left(\left(1 - \mu_{\tilde{A}_S}^2\right)^{\lambda} - \left(1 - \mu_{\tilde{A}_S}^2 - \pi_{\tilde{A}_S}^2\right)^{\lambda}\right)^{1/2} \end{array} \right\} \ for \, \lambda \geq 0 \tag{111}$$

$$\lambda \cdot \tilde{A}_S = \left\{ \begin{array}{l} \left(1 - \left(1 - \mu_{\tilde{A}_S}^2\right)^{\lambda}\right)^{1/2}, \quad v_{\tilde{A}_S}^{\lambda}, \\ \left(\left(1 - \mu_{\tilde{A}_S}^2\right)^{\lambda} - \left(1 - \mu_{\tilde{A}_S}^2 - \pi_{\tilde{A}_S}^2\right)^{\lambda}\right)^{1/2} \end{array} \right\} for\ \lambda \geq 0 \tag{112}$$

Definition 48 For these SFS $\tilde{A}_S = (\mu_{\tilde{A}_S}, v_{\tilde{A}_S}, \pi_{\tilde{A}_S})$ and $\tilde{B}_S = (\mu_{\tilde{B}_S}, v_{\tilde{B}_S}, \pi_{\tilde{B}_S})$, the followings are valid under the condition $\lambda, \lambda_1, \lambda_2 \geq 0$.

$$\tilde{A}_S \oplus \tilde{B}_S = \tilde{B}_S \oplus \tilde{A}_S \tag{113}$$

$$\tilde{A}_S \otimes \tilde{B}_S = \tilde{B}_S \otimes \tilde{A}_S \tag{114}$$

$$\lambda(\tilde{A}_S \oplus \tilde{B}_S) = \lambda\tilde{A}_S \oplus \lambda\tilde{B}_S \tag{115}$$

$$\lambda_1\tilde{A}_S \oplus \lambda_2\tilde{A}_S = (\lambda_1 + \lambda_2)\tilde{A}_S \tag{116}$$

$$(\tilde{A}_S \otimes \tilde{B}_S)^{\lambda} = \tilde{A}_S^{\lambda} \otimes \tilde{B}_S^{\lambda} \tag{117}$$

$$\tilde{A}_S^{\lambda_1} \otimes \tilde{A}_S^{\lambda_2} = \tilde{A}_S^{\lambda_1+\lambda_2} \tag{118}$$

Definition 49 Single-valued Spherical Weighted Arithmetic Mean (SWAM) with respect to,$w = (w_1, w_2 \ldots, w_n)$; $w_i \in [0, 1]$; $\sum_{i=1}^{n} w_i = 1$, SWAM is defined as;

$$\begin{aligned} & SWAM_w(\tilde{A}_{S1}, \ldots, \tilde{A}_{Sn}) \\ & = w_1\tilde{A}_{S1} + w_2\tilde{A}_{S2} + \cdots + w_n\tilde{A}_{Sn} \\ & = \left\{ \begin{array}{l} \left[1 - \prod_{i=1}^{n}(1 - \mu_{\tilde{A}_S}^2)^{w_i}\right]^{1/2} \\ \prod_{i=1}^{n} v_{\tilde{A}_S}^{w_i}, \left[\prod_{i=1}^{n}(1 - \mu_{\tilde{A}_S}^2)^{w_i} - \prod_{i=1}^{n}(1 - \mu_{\tilde{A}_S}^2 - \pi_{\tilde{A}_{Si}}^2)^{w_i}\right]^{1/2} \end{array} \right\} \end{aligned} \tag{119}$$

Definition 50 Single-valued Spherical Weighted Geometric Mean (SWGM) with respect to,$w = (w_1, w_2 \ldots, w_n)$; $w_i \in [0, 1]$; $\sum_{i=1}^{n} w_i = 1$, SWGM is defined as

$$SWGM_w(\tilde{A}_1, \ldots, \tilde{A}_n) = \tilde{A}_{S1}^{w_1} + \tilde{A}_{S2}^{w_2} + \cdots + \tilde{A}_{Sn}^{w_n}$$

$$= \left\{ \begin{array}{l} \prod_{i=1}^{n} \mu_{\tilde{A}_{Si}}^{w_i}, \left[1 - \prod_{i=1}^{n} (1 - v_{\tilde{A}_{Si}}^{2})^{w_i}\right]^{1/2}, \\ \left[\prod_{i=1}^{n} (1 - v_{\tilde{A}_{Si}}^{2})^{w_i} - \prod_{i=1}^{n} (1 - v_{\tilde{A}_{Si}}^{2} - \pi_{\tilde{A}_{Si}}^{2})^{w_i}\right]^{1/2} \end{array} \right\} \tag{120}$$

Definition 51 Score functions and Accuracy functions of sorting SFS are defined by;

$$Score\left(\tilde{A}_S\right) = (\mu_{\tilde{A}_S} - \pi_{\tilde{A}_S}/2)^2 - (v_{\tilde{A}_S} - \pi_{\tilde{A}_S}/2)^2 \tag{121}$$

$$Accuracy\left(\tilde{A}_S\right) = \mu_{\tilde{A}_S}^2 + v_{\tilde{A}_S}^2 + \pi_{\tilde{A}_S}^2 \tag{122}$$

Note that: $\tilde{A}_S < \tilde{B}_S$ if and only if

i. $Score(\tilde{A}_S) < Score(\tilde{B}_S)$ or
ii. $Score(\tilde{A}_S) = Score(\tilde{B}_S)$ and $Accuracy(\tilde{A}_S) < Accuracy(\tilde{B}_S)$

1.2 Literature Review

In order to reach the appropriate number of samples, we have emphasized the two main features of the fuzzy sets articles available in the literature; the number of citations and been published in the recent years. To identify the articles, the most comprehensive academic search engines, SCOPUS and ScienceDirect, are used. The used search pattern consists of the keywords "Neutrosophic Sets", "Pyhtagorean Fuzzy Sets", "Spherical Fuzzy Sets" and etc. After the analysis, we introduce brief information about the studies which is given in Fig. 7.

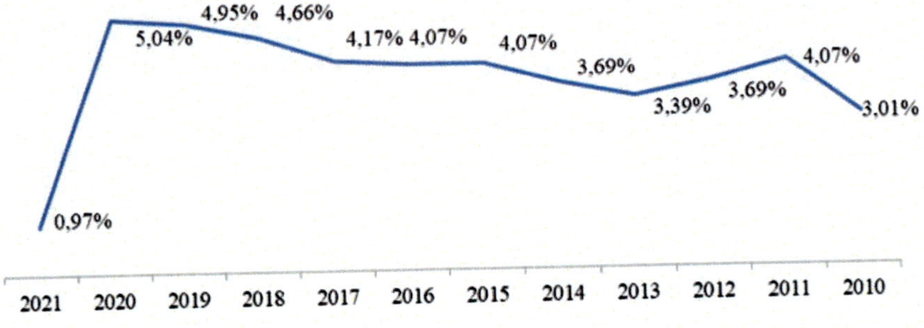

Fig. 7 Distribution of ordinary fuzzy set papers with respect to years

1.2.1 Literature Review on Ordinary Fuzzy Sets

We search the "*Ordinary* Fuzzy" Sets term in scopus and find 1031 results. The analysis of *Ordinary* fuzzy sets are given in Fig. 7. In this figure, the distribution of papers by years is given.

Figure 7 shows that the distribution of "Ordinary fuzzy" sets papers with respect to years. It is seen that most of the studies have been published in 2020 with a rate of 5.04%.

In Fig. 8, ordinaryfuzzy sets papers are summarized with their subject areas. The other subject areas with smaller percentages are Social Sciences, Earth and Planetary Sciences, Agricultural and Biological Sciences, Chemistry, Business, Management and Accounting, Economics, Econometrics and Finance, Energy, Biochemistry, Genetics and Molecular Biology and Medicine, etc.

Figure 9 illustrates the sources of the ordinary fuzzy set papers with their percentages.

Most of the publications on ordinary fuzzy sets have been published in Fuzzy Sets and Systems, Lecture Notes in Computer Science Including Subseries Lecture Notes in Artificial Intelligence and Lecture Notes in Bioinformatics.

In Fig. 10, the publication percentages of authors on ordinary fuzzy sets are presented. Kahraman and Belohlavek are the leader among these authors.

In Fig. 11, the distribution of publications on ordinary fuzzy sets with respect to their source countries is illustrated. China is the leading country on ordinary fuzzy publications. United States and Turkey are the next two countries after China.

In Fig. 12, the document types on ordinary fuzzy sets are of seven types: articles with a percentage of 64.7, book chapters with a percentage of 2.2, conference papers with a percentage of 31.4 and conference review with a percentage of 0.3, review with a percentage of 0.6 and editorial with the percentage of 0.1.

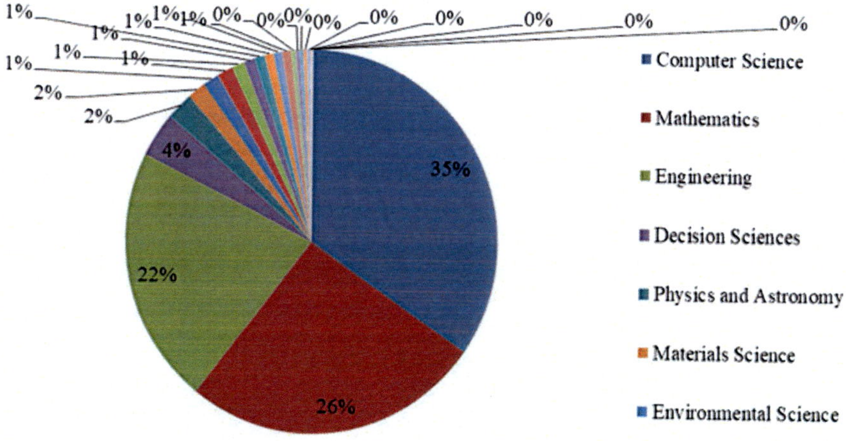

Fig. 8 Distribution of ordinary fuzzy set papers with respect to subject areas

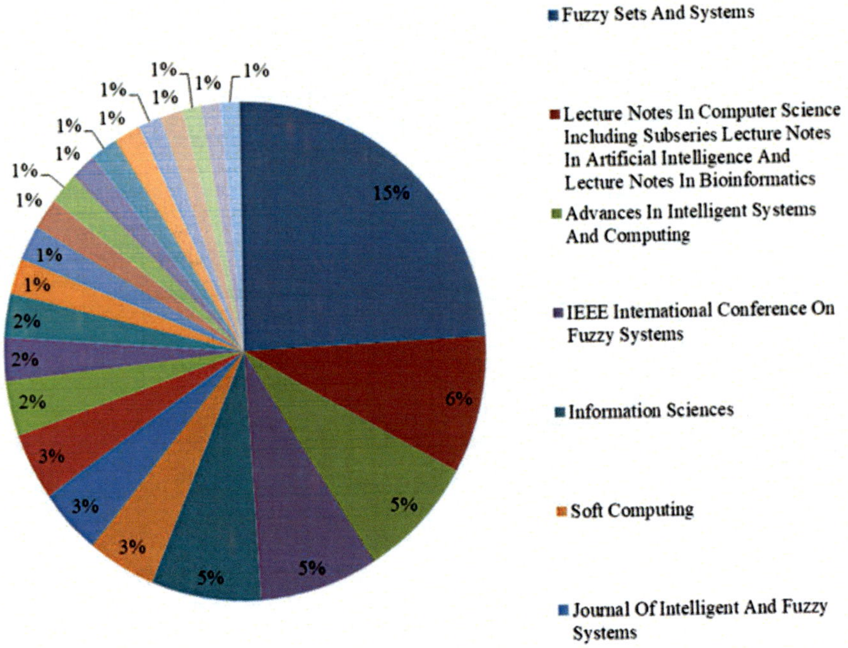

Fig. 9 Ordinary fuzzy set papers by their sources

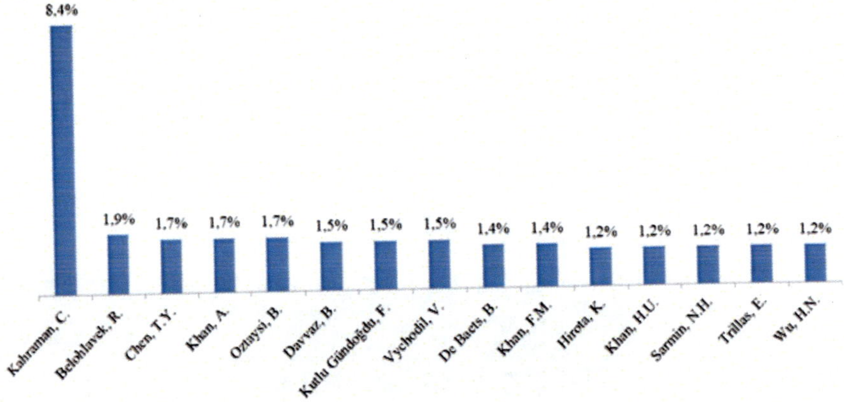

Fig. 10 The distribution of publication percentages of authors on ordinary fuzzy sets

1.2.2 Literature Review on Type 2 Fuzzy Sets

We search the "Type 2 Fuzzy" sets term in Scopus and find 4064 results. The analysis of type 2 fuzzy sets are given in Figs. 13, 14, 15, 16, 17 and 18. In Fig. 13, the distribution of type 2 papers by years is given.

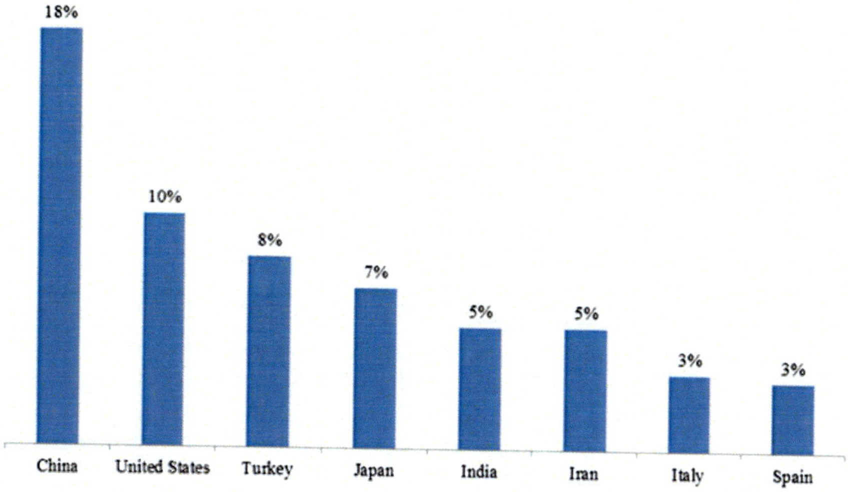

Fig. 11 Distribution of ordinary fuzzy sets papers by their countries

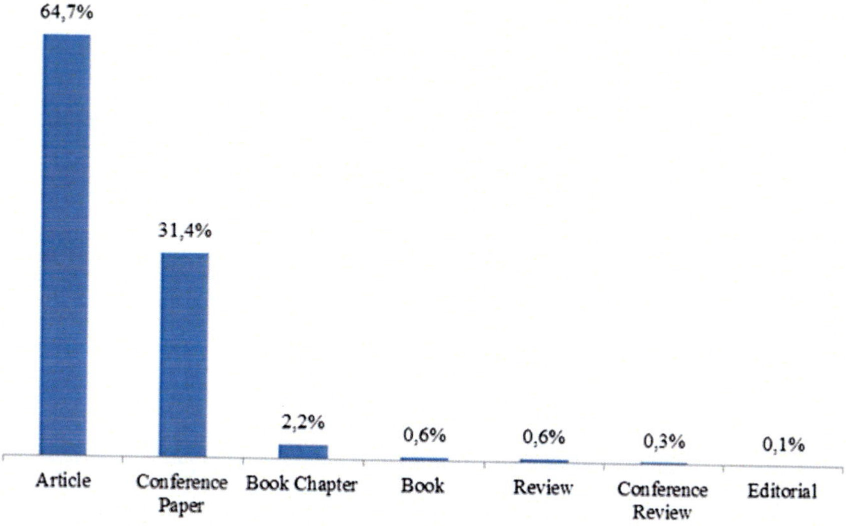

Fig. 12 Distribution of ordinary fuzzy sets papers by their types

Figure 13 shows that the distribution of "Type 2 fuzzy" set papers with respect to years. It is seen that most of the studies have been published in 2019 with a rate of 9.1%.

In Fig. 14, type 2 fuzzy set papers are summarized by their subject areas. In Fig. 14, the distribution of type 2 fuzzy sets papers by their subject areas is illustrated. The

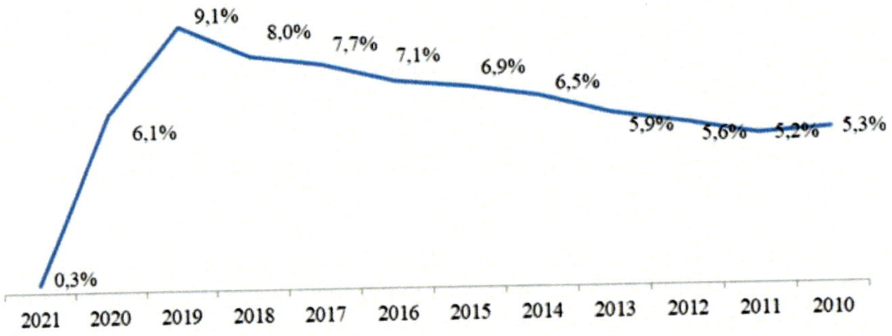

Fig. 13 Distribution of Type 2 fuzzy sets papers with respect to years

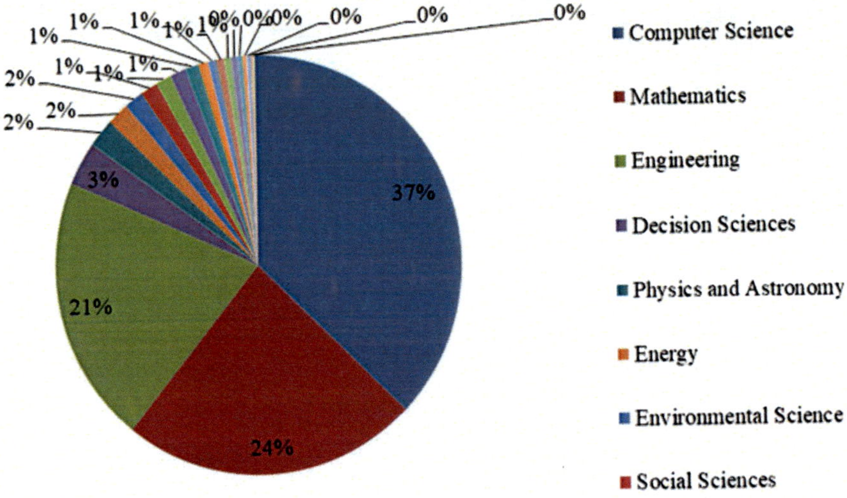

Fig. 14 Distribution of type 2 fuzzy set papers with respect to subject areas

other subject areas with lower percentages are Materials Science, Business, Management and Accounting, Earth and Planetary Sciences, Medicine, Chemical Engineering, Agricultural and Biological Sciences, Biochemistry, Genetics and Molecular Biology, Chemistry, Neuroscience, Economics, Econometrics and Finance, Multidisciplinary, Arts and Humanities, Health Professions, Psychology, Immunology and Microbiology, and Nursing.

Figure 15 presents type 2 fuzzy set papers by their sources. Most of the publications on type 2 fuzzy sets have been published in IEEE International Conference on Fuzzy Systems. The total number of journals is 130.

In Fig. 16, the publication percentages and the corresponding numbers of authors on type 2 fuzzy sets are presented. Mendel and Castillo are the leader among these authors.

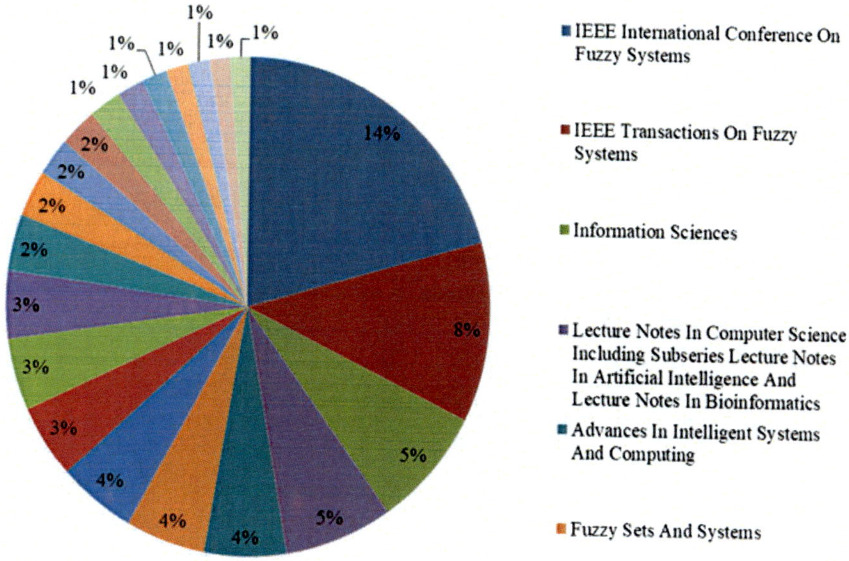

Fig. 15 Type 2 fuzzy set papers by their sources

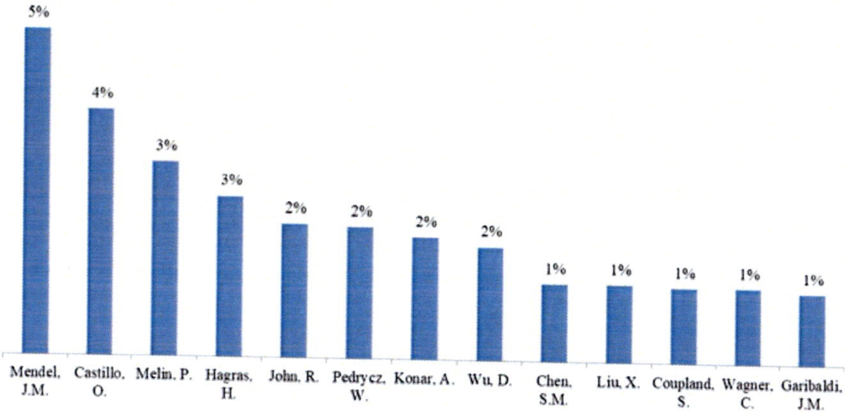

Fig. 16 Distribution of publication percentages of authors on Type 2 fuzzy sets

In Fig. 17, the distribution of publications on type 2 fuzzy sets with respect to their source countries is illustrated. China is the leading country on type 2 fuzzy publications. United States and India are the next two countries after China.

The percentages and the corresponding numbers of Type 2 fuzzy papers are illustrated in Fig. 18. The document types on type 2 fuzzy sets are article with a percentage of 54.55%, book chapter with a percentage of 1.99, conference paper with a percentage of 40.03% and conference review with a percentage of 1.8, editorial with a percentage of 0.15, book with a percentage of 0.27.

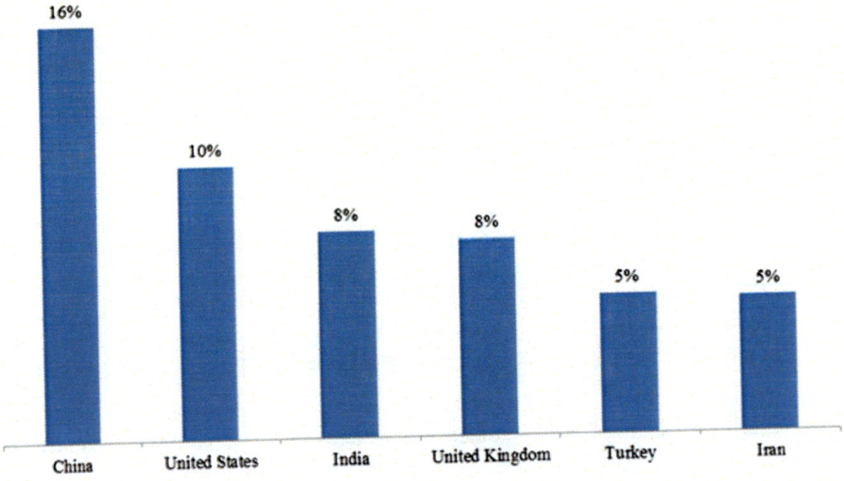

Fig. 17 Distribution of Type 2 fuzzy sets papers by their countries

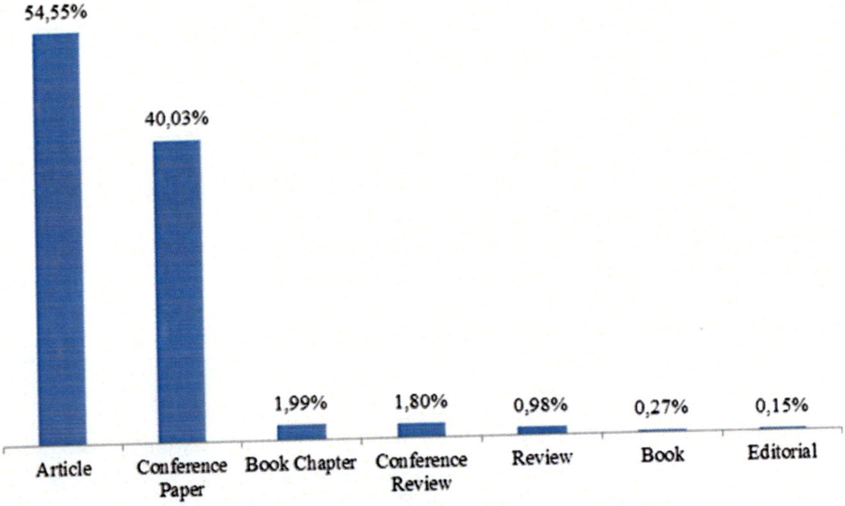

Fig. 18 Distribution of type 2 fuzzy sets papers by their types

1.2.3 Literature Review on Interval Valued Fuzzy Sets

We search the "Interval Valued Fuzzy Sets" term in Scopus and find 2989 results. The analysis of interval valued fuzzy sets is given in Figs. 19 and 20. In Fig. 19, the distribution of papers by years is given.

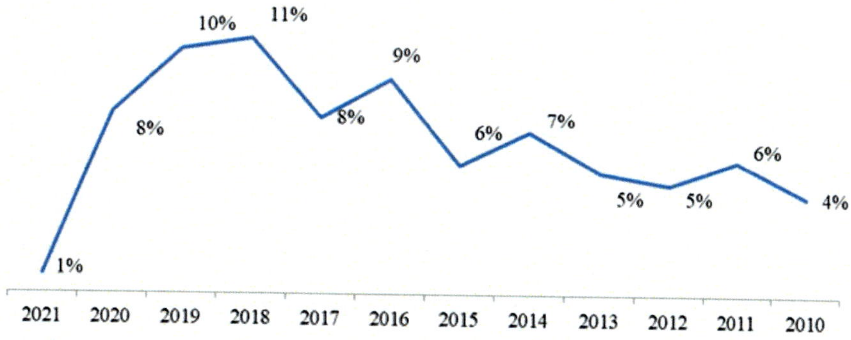

Fig. 19 Distribution of interval valued fuzzy set papers with respect to years

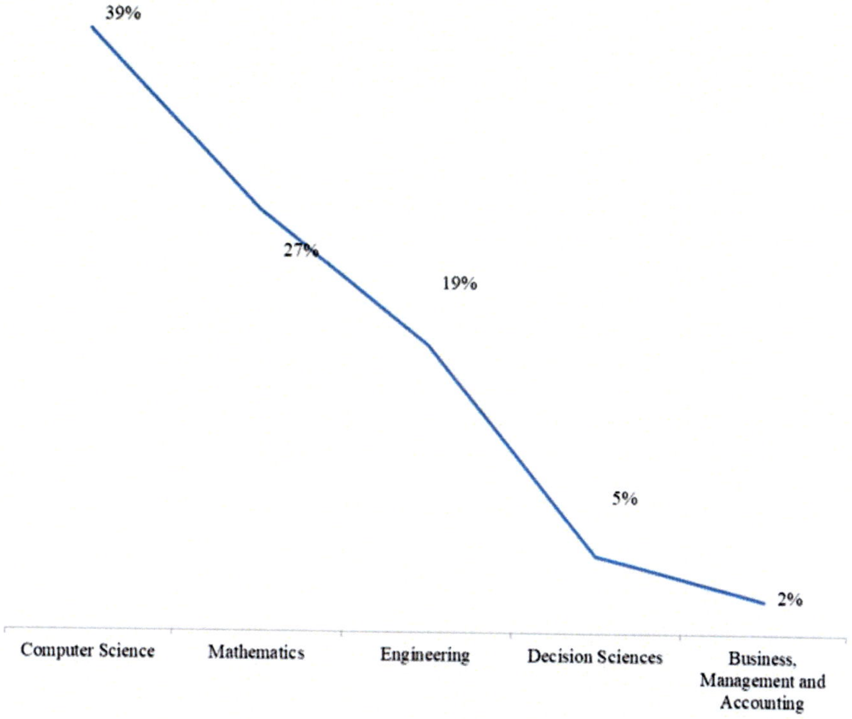

Fig. 20 Distribution of interval valued fuzzy sets papers with respect to subject areas

Figure 19 shows that the distribution of "Interval Valued fuzzy" sets papers with respect to years. It is seen that most of the studies have been published in 2018 with a rate of 11%.

In Fig. 20, interval valued fuzzy set papers are summarized with respect to their subject areas. The other subject areas with lower percentages are Physics and Astronomy, Environmental Science, Social Sciences, Materials Science, Multidisciplinary, Economics, Econometrics and Finance, Earth and Planetary Sciences, Energy, Chemistry, Biochemistry, Genetics and Molecular Biology, Arts and Humanities, Neuroscience, Medicine, Chemical Engineering, Agricultural and Biological Sciences, Pharmacology, Toxicology and Pharmaceutics, Immunology and Microbiology.

Figure 21 illustrates interval valued fuzzy set papers by their sources. Most of the publications on interval valued fuzzy sets have been published in Fuzzy Sets and Systems and Journal of Intelligent and Fuzzy Systems. The total of other journals is 135.

In Fig. 22, the publication percentages of authors on interval valued fuzzy sets are presented. Bustince and Xu are the leader among these authors.

In Fig. 23, the distribution of publications on interval valued fuzzy sets with respect to their source countries is illustrated. China is the leading country on interval valued fuzzy publications. India and United States are the next two countries after China.

The percentages and the corresponding numbers of interval valued fuzzy papers are illustrated in Fig. 24. The document types on interval valued fuzzy sets are articles with a percentage of 67.21%, book chapters with a percentage of 1.84, conference papers with a percentage of 27.53% and conference review with a percentage of 1.84.

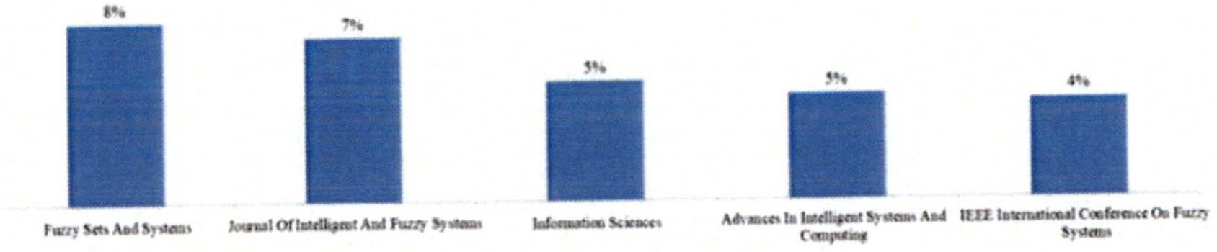

Fig. 21 Interval valued fuzzy set papers by their sources

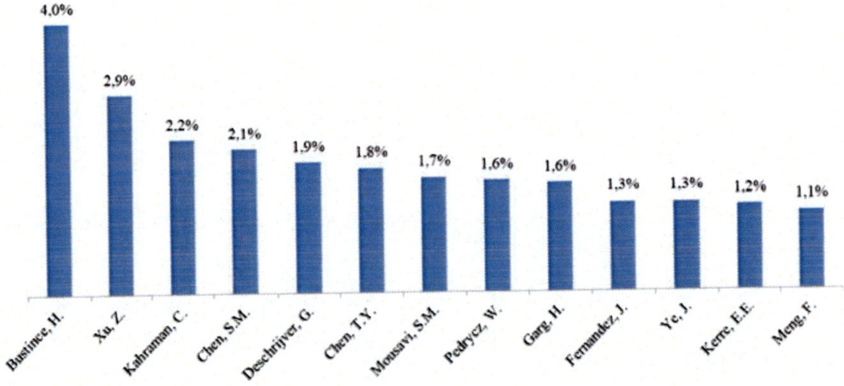

Fig. 22 Distribution of publication percentages of authors on interval valued fuzzy sets

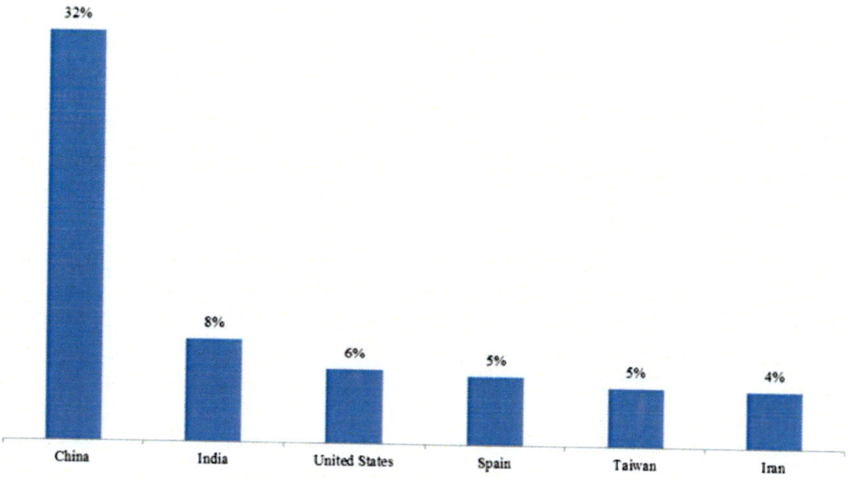

Fig. 23 Distribution of interval valued fuzzy sets papers by their countries

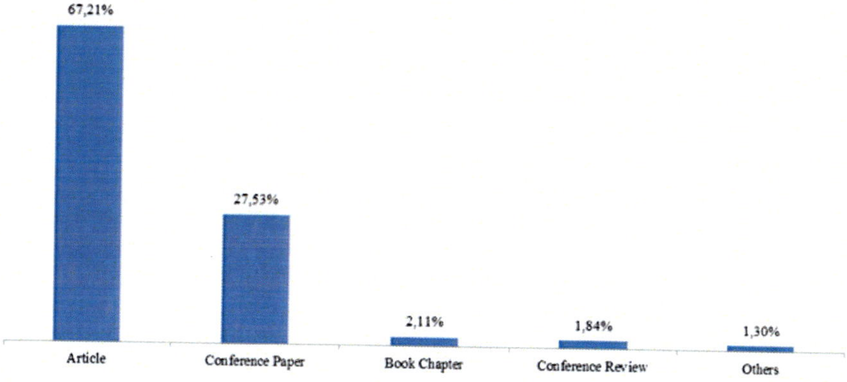

Fig. 24 Distribution of interval valued fuzzy sets papers by their types

1.2.4 Literature Review on Intuitionistic Fuzzy Sets

We search the "intuitionistic fuzzy sets" term in Scopus and find 5893 results. The analysis of intuitionistic fuzzy sets is given in Figs. 25, 26, 28, 29 and 30. In Fig. 25, the distribution of papers by years is given. It is seen that most of the studies have been published in 2019 with a rate of 12%.

In Fig. 26, intuitionistic fuzzy set papers are summarized with their contents, types of picture and applied methods. In Fig. 26, the distribution of intuitionistic fuzzy sets papers by their subject areas. The other subject areas with lower percentages

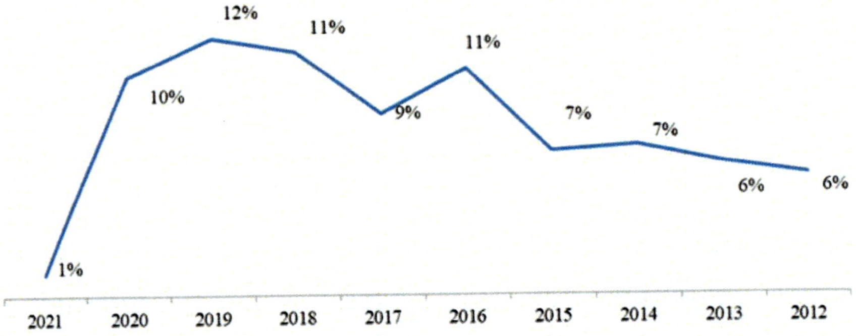

Fig. 25 Distribution of intuitionistic fuzzy set papers with respect to years

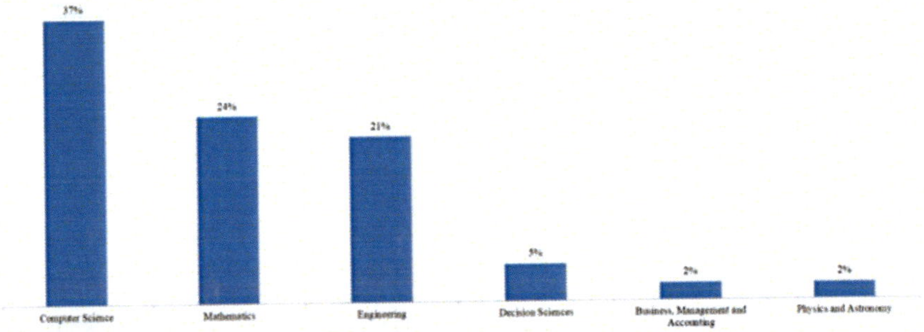

Fig. 26 Distribution of intuitionistic fuzzy set papers with respect to subject areas

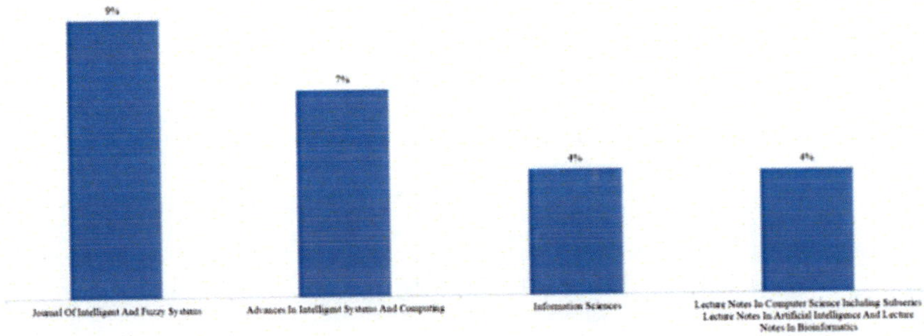

Fig. 27 Intuitionistic fuzzy set papers by their sources

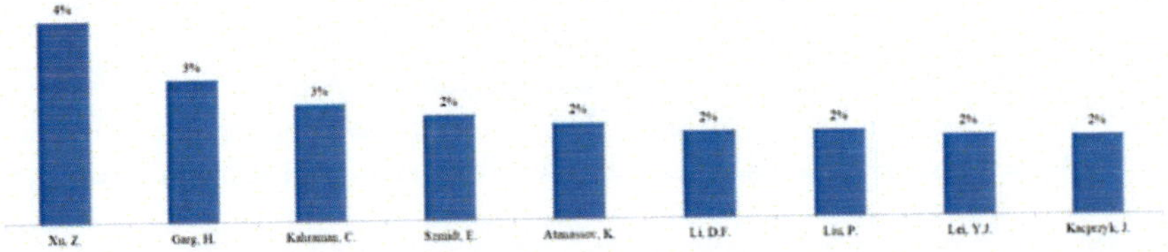

Fig. 28 Distribution of publication percentages of authors on intuitionistic fuzzy sets

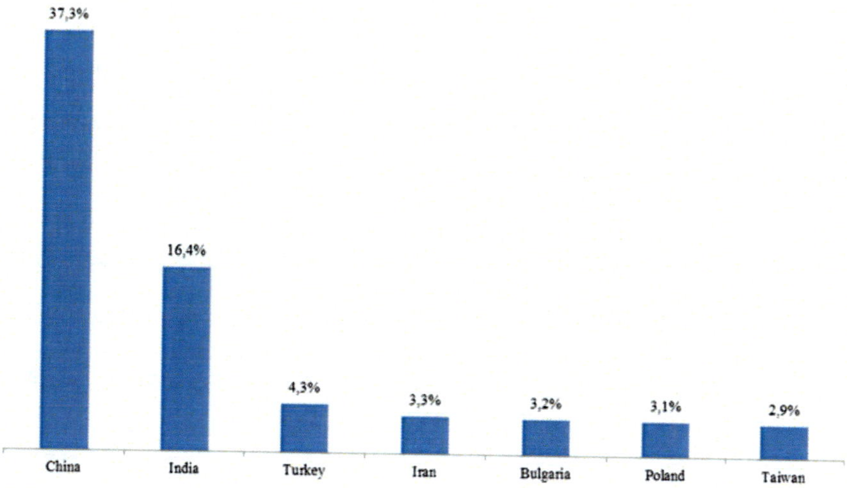

Fig. 29 Distribution of intuitionistic fuzzy sets papers by their countries

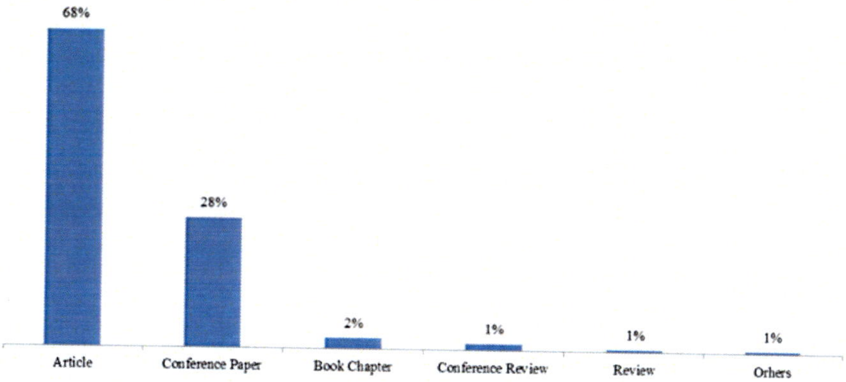

Fig. 30 Distribution of intuitionistic fuzzy sets papers by their types

are Materials Science, Social Sciences, Energy, Environmental Science, Multidisciplinary, Chemistry, Earth and Planetary Sciences, Biochemistry, Genetics and Molecular Biology Economics, Econometrics and Finance Medicine, Chemical Engineering, Agricultural and Biological Sciences, Arts and Humanities, Neuroscience, Pharmacology, Toxicology and Pharmaceutics, Psychology, Health Professions, and Immunology and Microbiology.

Figure 27 illustrates the sources of intuitionistic fuzzy papers. Most of the publications on intuitionistic fuzzy sets have been published in Journal of Intelligent and Fuzzy Systems and Advances in Intelligent Systems and Computing. The total of other journals is 143.

In Fig. 28, the publication percentages and the corresponding authors on İntuitionistic fuzzy sets are presented. Xu and Garg are the leader among these authors.

In Fig. 29, the distribution of publications on Intuitionistic fuzzy sets with respect to their source countries is illustrated. China is the leading country on Nonstationary fuzzy publications. India and Turkey are the next two countries after China.

The percentages and the corresponding numbers of intuitionistic fuzzy papers are illustrated in Fig. 30. The document types on intuitionistic fuzzy sets are articles with a percentage of 68%, book chapters with a percentage of 2, conference papers with a percentage of 28% and conference review with a percentage of 1.

1.2.5 Literature Review on Fuzzy Multisets

We search the "Fuzzy Multisets" term in Scopus and find 129 results. The analysis of Fuzzy Multisets is given in Figs. 31, 32, 33, 34, 35 and 36. In Fig. 31, the distribution of papers by years is given. Figure 31 shows that the distribution of Fuzzy Multisets papers with respect to years. It is seen that most of the studies have been published in 2014 with a rate of 10%.

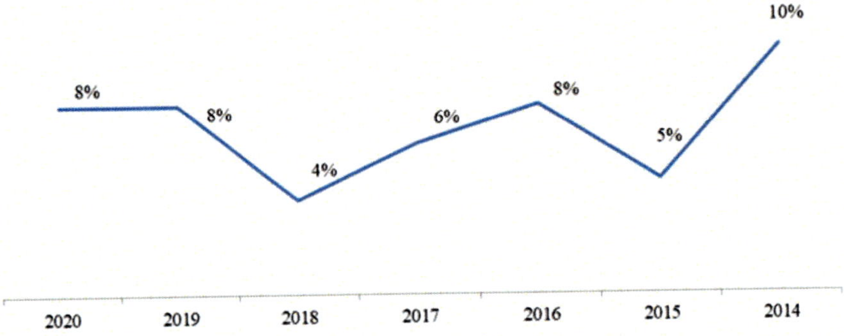

Fig. 31 Distribution of fuzzy multisets papers with respect to years

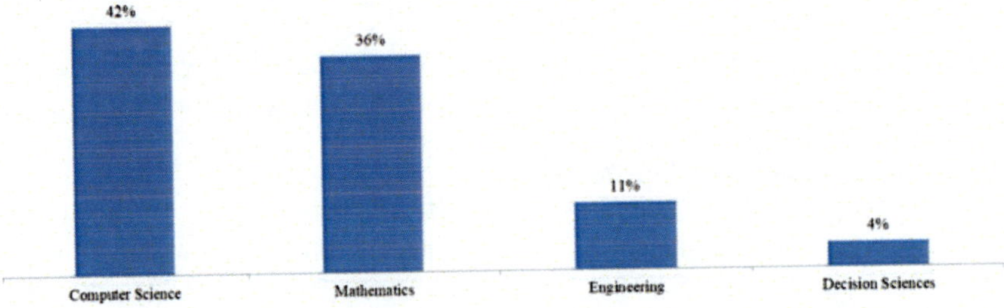

Fig. 32 Distribution of fuzzy multisets papers with respect to subject areas

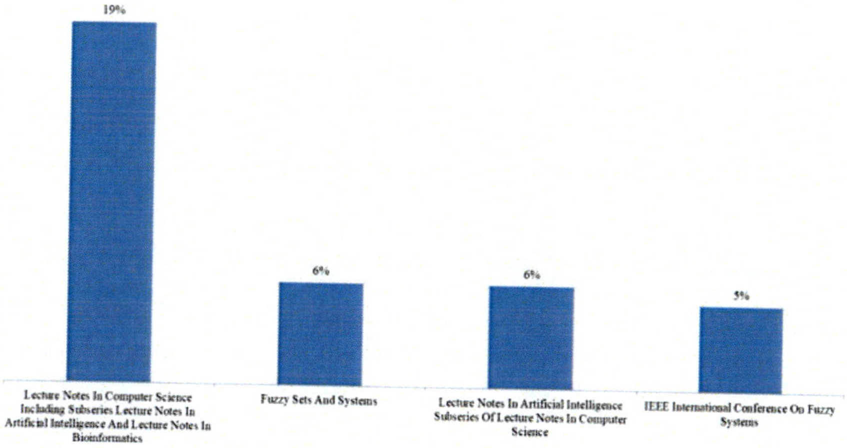

Fig. 33 Fuzzy multisets papers by their sources

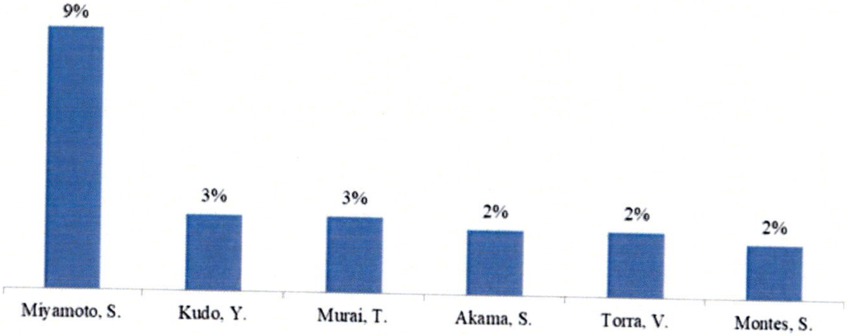

Fig. 34 The distribution of publication percentages of authors on Fuzzy Multisets

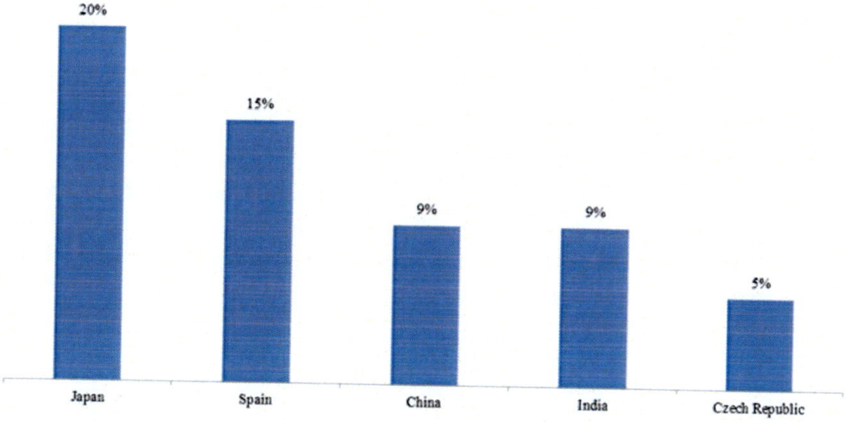

Fig. 35 Distribution of fuzzy multisets papers by their countries

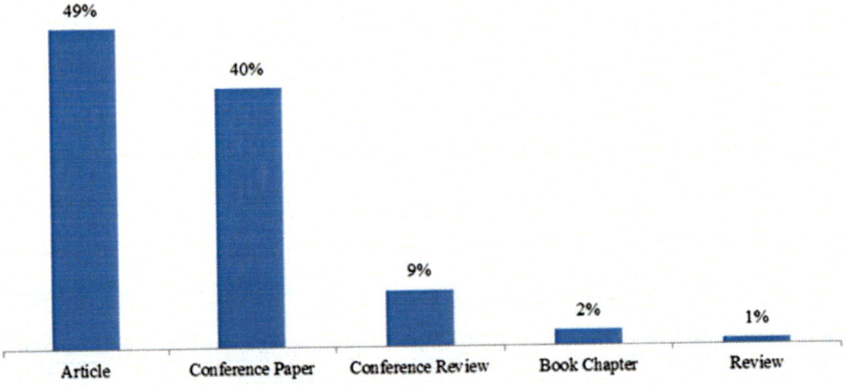

Fig. 36 Distribution of Fuzzy Multisets papers by their types

In Fig. 32, Fuzzy Multisets papers are summarized with respect to their subjects. The other subjects with lower percentages are Physics and Astronomy, Chemistry, Business, Management and Accounting, Agricultural and Biological Sciences, Chemical Engineering, Economics, Econometrics and Finance, Environmental Science, Materials Science, Social Sciences,Medicine, Chemical Engineering, Agricultural and Biological Sciences, Arts and Humanities, Neuroscience, Pharmacology, Toxicology and Pharmaceutics, Psychology, Health Professions, Immunology and Microbiology.

Figure 33 shows the sources of the fuzzy multisets publications. Most of the publications on Fuzzy Multisets have been published in Lecture Notes in Computer Science Including Subseries Lecture Notes in Artificial Intelligence and Lecture Notes in Bioinformatics and Fuzzy Sets and Systems. The total of other journals is 51.

In Fig. 34, the publication percentages and the corresponding numbers of authors on Fuzzy Multisets are presented. Miyamoto and Kudo are the leader among these authors.

In Fig. 35, the distribution of publications on Fuzzy Multisets with respect to their source countries is illustrated. Japan is the leading country on Fuzzy Multisets publications. Spain and China are the next two countries after Japan.

The percentages and the corresponding numbers of Fuzzy Multisets papers are illustrated in Fig. 36. The document types on Fuzzy Multisets are of five types: articles with a percentage of 49%, book chapters with a percentage of 2, conference papers with a percentage of 40%, conference review with a percentage of 9 and review with a percentage of 1.

1.2.6 Literature Review on Intuitionistic Fuzzy Sets of Second Type

We search the "Intuitionistic Fuzzy Sets of Second Type" sets term in Scopus and find 54 results. The analysis of intuitionistic Fuzzy Sets of Second Type is given in Figs. 37, 38, 39, 40, 41 and 42. In Fig. 37, the distribution of papers by years is given. It is seen that the largest percentage has been published in 2016 with a rate of 17%.

In Fig. 38, Intuitionistic Fuzzy Sets of Second Type papers are summarized with respect to their subject areas. The other subject areas with lower percentages are Arts

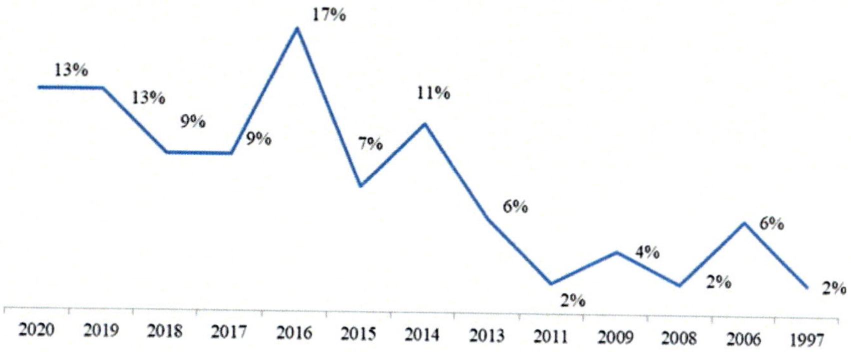

Fig. 37 Distribution of intuitionistic fuzzy sets of second type papers with respect to years

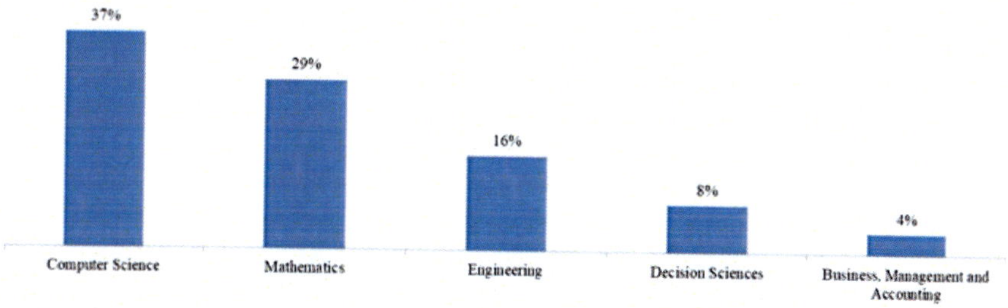

Fig. 38 Distribution of intuitionistic fuzzy sets of second type papers with respect to subject areas

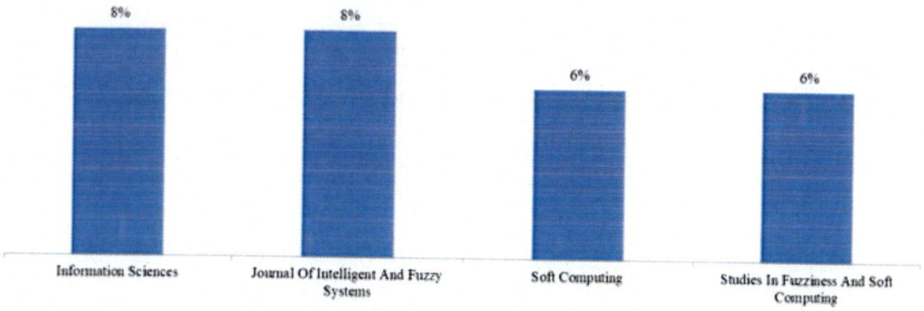

Fig. 39 Intuitionistic fuzzy sets of second type papers by their sources

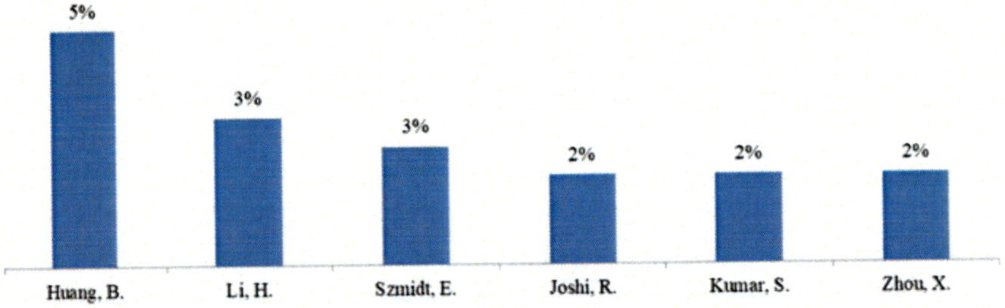

Fig. 40 Distribution of publication percentages of authors on Intuitionistic Fuzzy Sets of second type

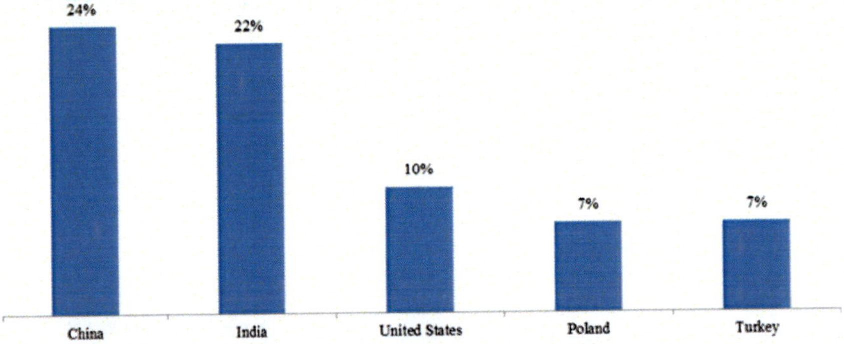

Fig. 41 Distribution of intuitionistic fuzzy sets of Second Type papers by their countries

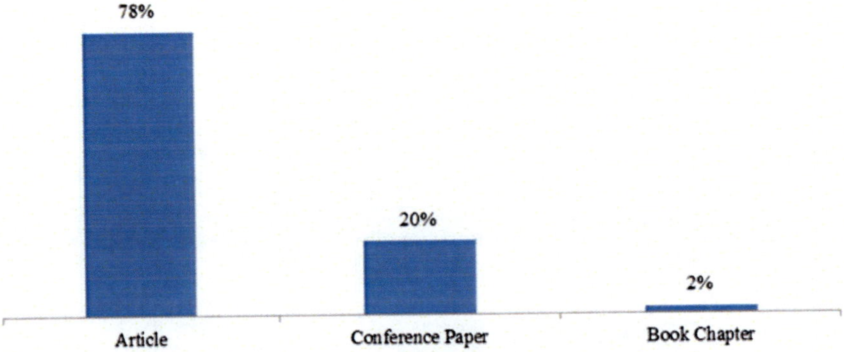

Fig. 42 Distribution of intuitionistic fuzzy set of second type by their types

and Humanities, Health Professions, Multidisciplinary, Neuroscience, Physics and Astronomy and Social Sciences.

Figure 39 shows the sources of the published papers. Most of the publications on Intuitionistic Fuzzy Sets of Second Type have been published in Information Sciences and Journal of Intelligent and Fuzzy Systems.

In Fig. 40, the publication percentages and the corresponding numbers of authors on Intuitionistic Fuzzy Sets of Second Type are presented. Huang and Li are the leaders among these authors.

In Fig. 41, the distribution of publications on Intuitionistic Fuzzy Sets of Second Type with respect to their source countries is illustrated. China is the leading country on intuitionistic fuzzy sets of second type publications. India and United States are the next two countries before Turkey.

The percentages and the corresponding numbers of intuitionistic fuzzy sets of second type papers are illustrated in Fig. 42. The document types on intuitionistic fuzzy sets of second type are of three types: articles with a percentage of 78%, book chapters with a percentage of 2, conference papers with a percentage of 20%.

1.2.7 Literature Review on Neutrosophic Fuzzy Sets

We search the "Neutrosophic Sets" term in Scopus and find 456 results. The analysis of neutrosophic sets are given in Figs. 43, 44, 45, 46 and 48. In Fig. 43, the distribution of papers by years is given.

In Fig. 44, neutrosophic sets papers are summarized with respect to their subject areas. Other subject areas with lower percentages are Multidisciplinary Business, Management and Accounting Energy Economics, Econometrics and Finance Medicine Earth and Planetary Sciences Arts and Humanities Neuroscience Agricultural and Biological Sciences.

Figure 45 illustrates that the most of the publications on neutrosophic sets have been published in International Journal of Intelligent Systems and Journal of Intelligent and Fuzzy Systems.

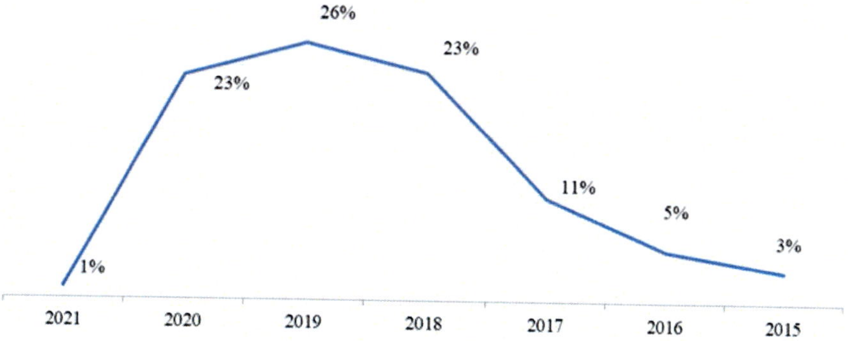

Fig. 43 Distribution of neutrosophic sets papers with respect to years

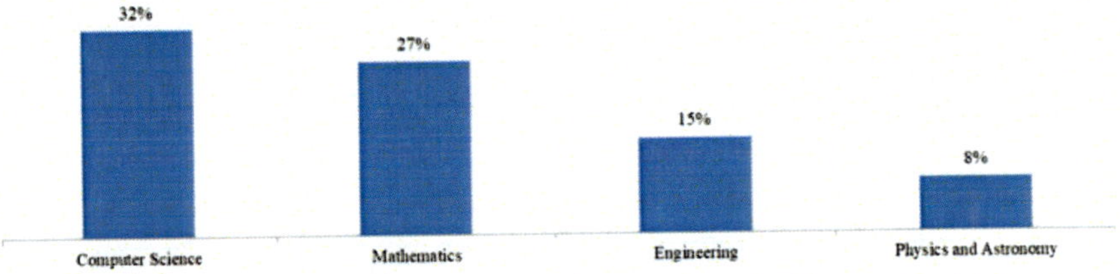

Fig. 44 Distribution of neutrosophic sets papers with respect to subject areas

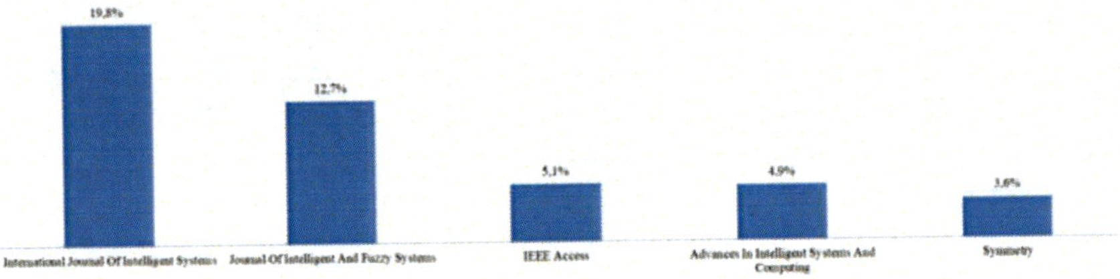

Fig. 45 Neutrosophic sets papers by their sources

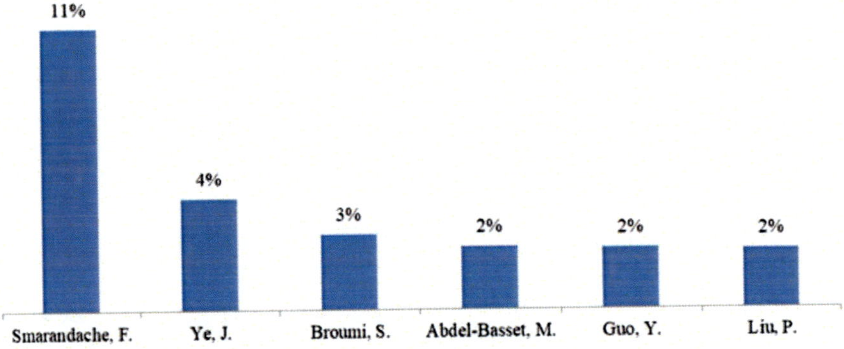

Fig. 46 Distribution of publication percentages of authors on neutrosophic sets

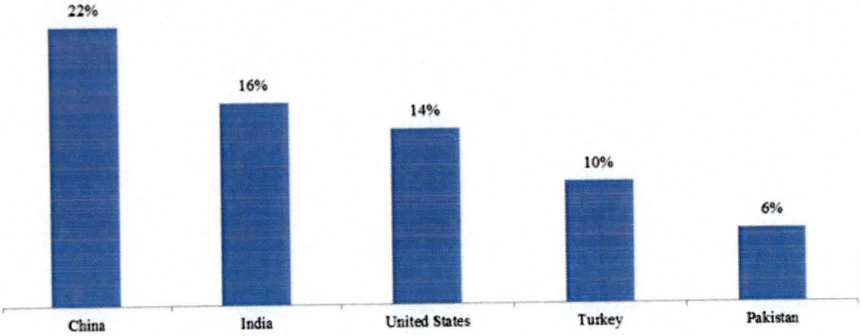

Fig. 47 Distribution of neutrosophic sets papers by their countries

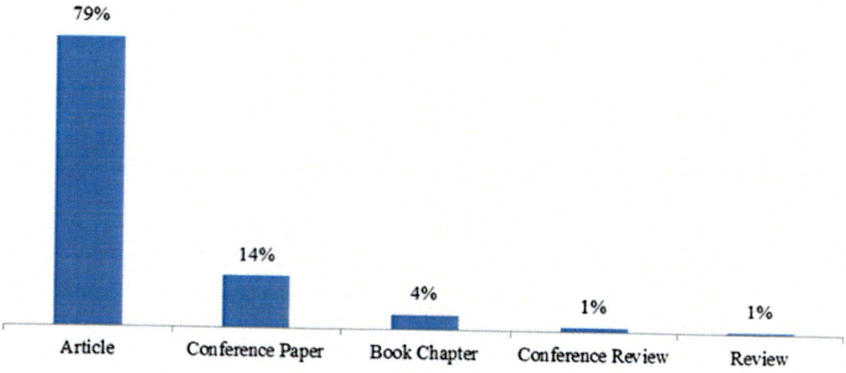

Fig. 48 Distribution of neutrosophic sets papers by their document types

In Fig. 46, the publication percentages and the corresponding numbers of authors on neutrosophic sets are presented. Smarandache and Ye are the leader among these authors.

In Fig. 47, the distribution of publications on neutrosophic fuzzy sets with respect to their source countries is illustrated. China and Turkey are the leading countries on neutrosophic fuzzy publications. India and United States are the next two countries after China.

The percentages and the corresponding numbers of neutrosophic papers are illustrated in Fig. 48. The document types on neutrosophic sets are of five types: articles with a percentage of 79 book chapters with a percentage of 4, conference papers with a percentage of 14, conference review with a percentage of 1 and review with a percentage of 1.

1.2.8 Literature Review on Hesitant Fuzzy Sets

We search the "Hesitant Fuzzy" Sets term in Scopus and find 1583 results. The analysis results of Hesitant fuzzy sets are given in Figs. 49, 50, 51, 52, 53 and 54. Figure 49 shows that the distribution of "Hesitant fuzzy" sets papers with respect to years.

In Fig. 50, Hesitant fuzzy sets papers are classified with respect to their subject areas. The other subject areas with lower percentages are Medicine, Earth and Planetary Sciences, Neuroscience, Biochemistry, Genetics and Molecular Biology, Arts and Humanities, Chemical Engineering, Agricultural and Biological Sciences, Pharmacology, Toxicology and Pharmaceutics, Health Professions, Immunology and Microbiology and Psychology.

Figure 51 shows the sources of hesitant fuzzy publications. Most of the publications on Hesitant fuzzy have been published in International Journal of Intelligent Systems and Journal of Intelligent and Fuzzy Systems.

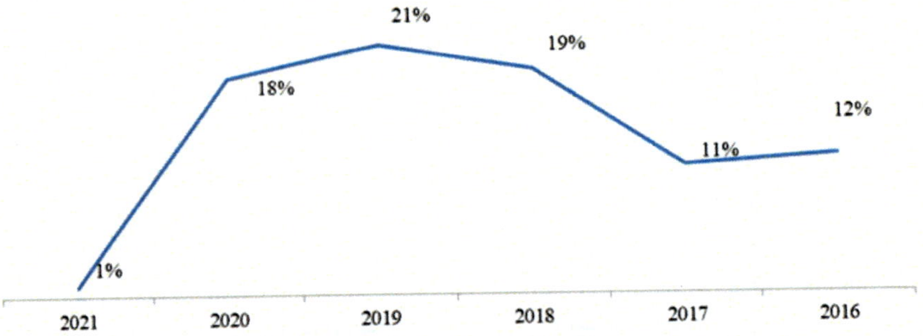

Fig. 49 Distribution of Hesitant fuzzy sets papers with respect to years

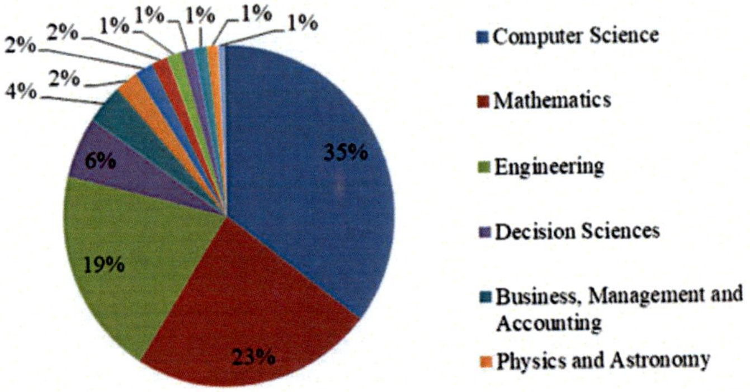

Fig. 50 Distribution of Hesitant fuzzy sets papers with respect to subject areas

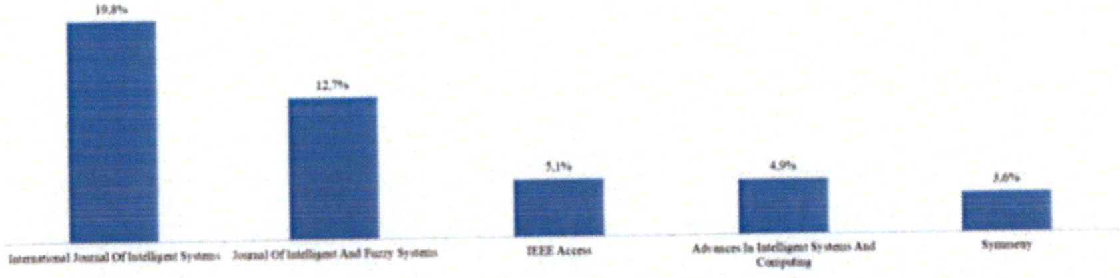

Fig. 51 Hesitant fuzzy sets papers by their sources

In Fig. 52, the publication percentages and the corresponding numbers of authors on Nonstationary fuzzy sets are presented. Xu and Liao are the leaders among these authors.

In Fig. 53, the distribution of publications on Hesitant fuzzy sets with respect to their source countries is illustrated. China and Spain are the leading country on Hesitant fuzzy publications. Turkey and India are the next two countries after Turkey.

The percentages and the corresponding numbers of Hesitant fuzzy papers are illustrated in Fig. 54. The document types on Hesitant fuzzy sets are of five types:

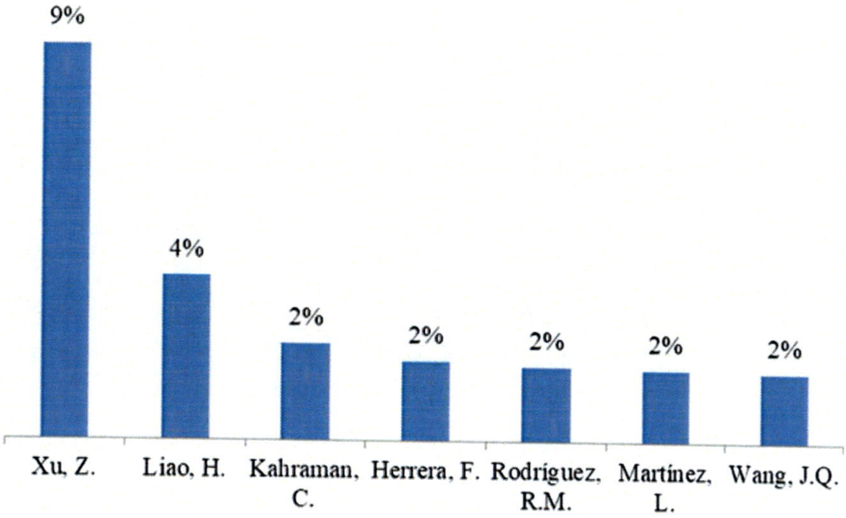

Fig. 52 Distribution of publication percentages of authors on Hesitant fuzzy sets

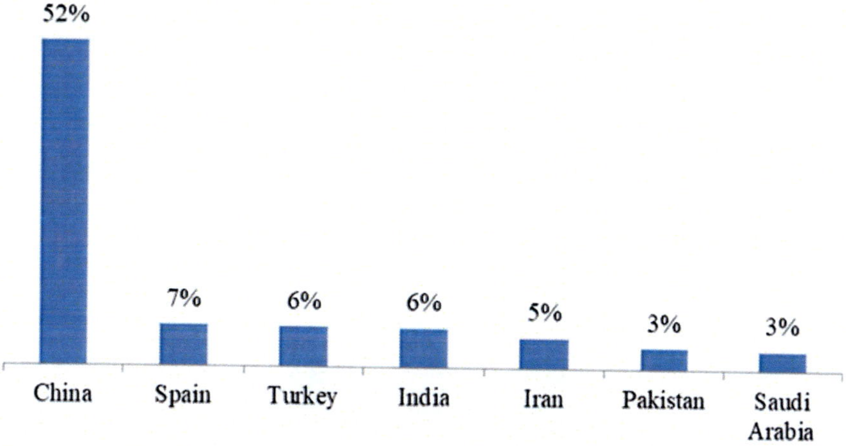

Fig. 53 Distribution of hesitant fuzzy sets papers by their countries

articles with a percentage of 81, book chapters with a percentage of 1, conference papers with a percentage of 15 and conference review with a percentage of 2.

1.2.9 Literature Review on Nonstationary Fuzzy Sets

We search the *Nonstationary Fuzzy Sets* term in Scopus and find 456 results. The analysis of Nonstationary fuzzy sets is given in Figs. 55, 56, 57, 58, 59 and 60.

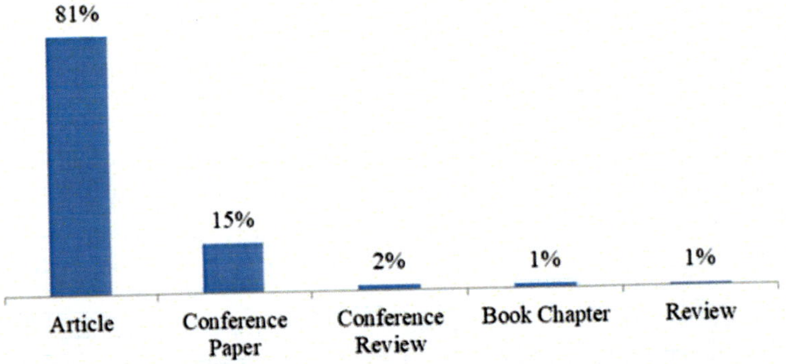

Fig. 54 Distribution of hesitant fuzzy sets papers by their document types

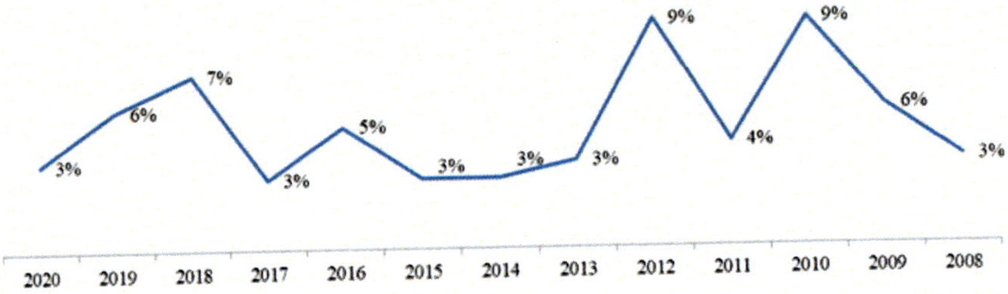

Fig. 55 Distribution of nonstationary fuzzy sets papers with respect to years

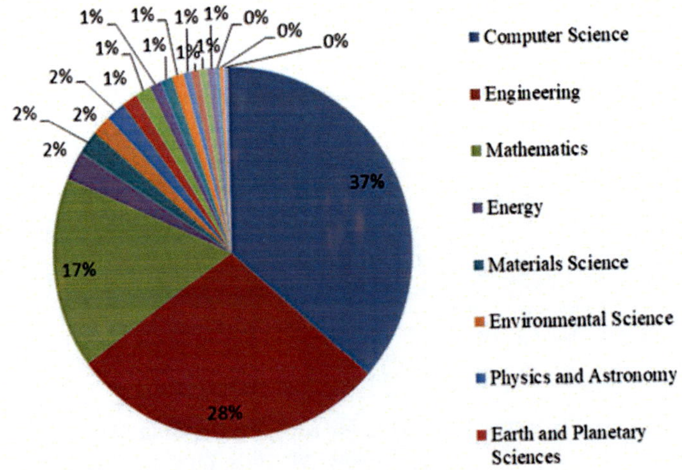

Fig. 56 Distribution of nonstationary fuzzy sets papers with respect to subject areas

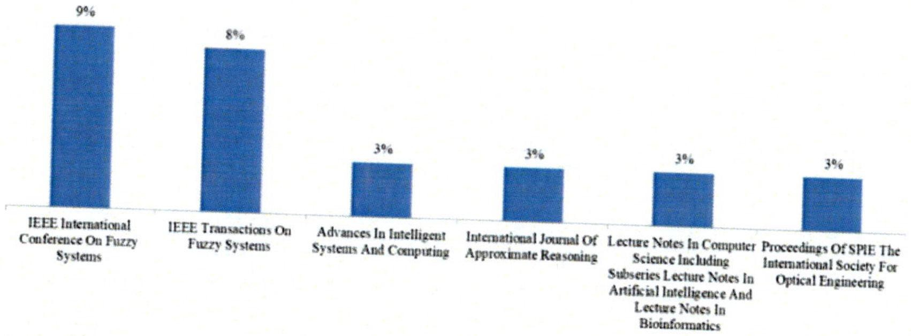

Fig. 57 Nonstationary fuzzy sets papers by their sources

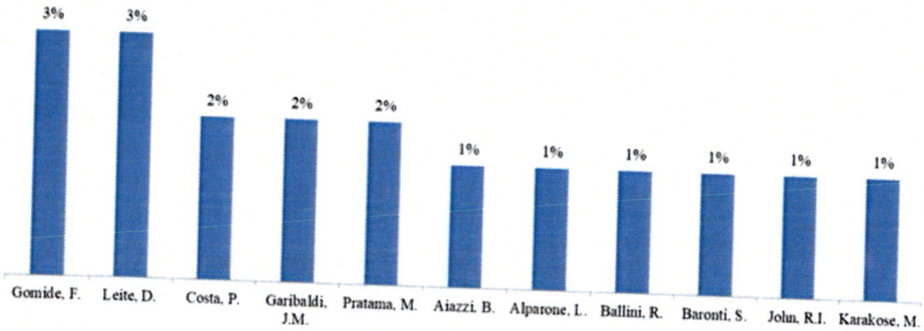

Fig. 58 Distribution of publication percentages of authors on Nonstationary fuzzy sets

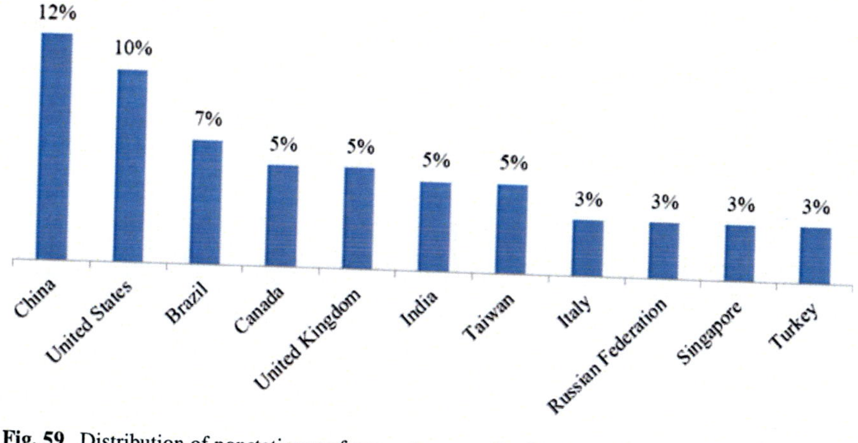

Fig. 59 Distribution of nonstationary fuzzy sets papers by their countries

Figure 55 shows that the distribution of "Nonstationary fuzzy" sets papers with respect to years.

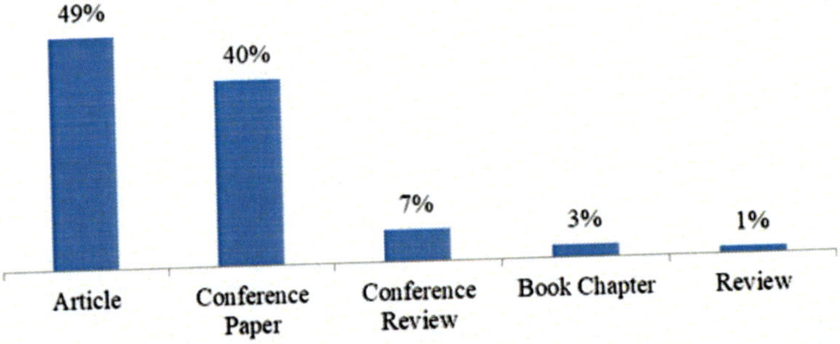

Fig. 60 Distribution of nonstationary fuzzy sets papers by their document types

In Fig. 56, the distribution of Nonstationary fuzzy sets papers is given by their subjects areas. The other subject areas with lower percentages are Medicine, Business, Management and Accounting, Decision Sciences, Social Sciences, Biochemistry, Genetics and Molecular Biology, Chemical Engineering, Health Professions, Immunology and Microbiology, Agricultural and Biological Sciences, Economics, Econometrics and Finance and Multidisciplinary.

Figure 57 shows the sources of Nonstationary fuzzy publications. Most of the publications on Nonstationary have been published in IEEE International Conference on Fuzzy Systems and IEEE Transactions on Fuzzy Systems.

In Fig. 58, the publication percentages and the corresponding numbers of authors on Nonstationary fuzzy sets are presented. Gomide and Leite are the leaders among these authors.

In Fig. 59, the distribution of publications on Nonstationary fuzzy sets with respect to their source countries is illustrated. China is the leading country on Nonstationary fuzzy publications. USA and Brazil are the next two countries after China.

The percentages and the corresponding numbers of Nonstationary fuzzy papers are illustrated in Fig. 60. The document types on Nonstationary fuzzy sets are of five types: articles with a percentage of 49%, book chapters with a percentage of 3, conference papers with a percentage of 40 and conference review with a percentage of 7.

1.2.10 Literature Review on Pythagorean Fuzzy Sets

We search the "Pythagorean Fuzzy Sets" term in Scopus and find 456 results. The analyses of Pythagorean fuzzy sets are given in Figs. 61, 62, 63, 64, 65 and 66.

In Fig. 61, the distribution of papers by years is given. It shows that the distribution of Pythagorean fuzzy set papers with respect to years. It is seen that most of the studies have been published in 2020 with a rate of 35%.

In Fig. 62, the distribution of Pythagorean fuzzy sets papers by their subjects areas is presented. The other subject areas with lower percentages are Business,

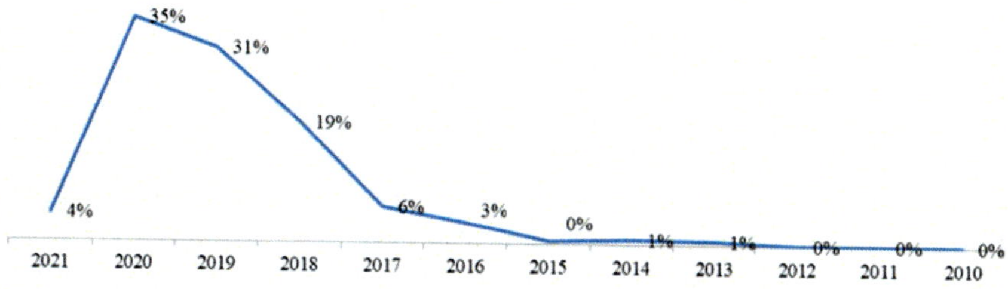

Fig. 61 Distribution of Pythagorean fuzzy sets papers with respect to years

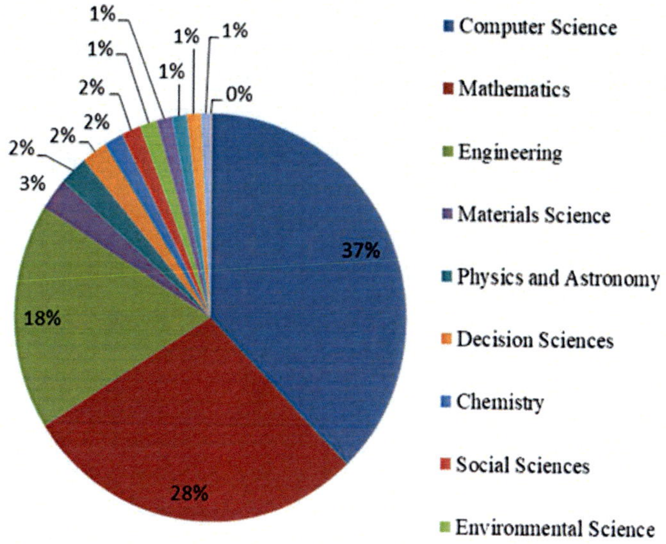

Fig. 62 Distribution of pythagorean fuzzy sets papers with respect to subject areas

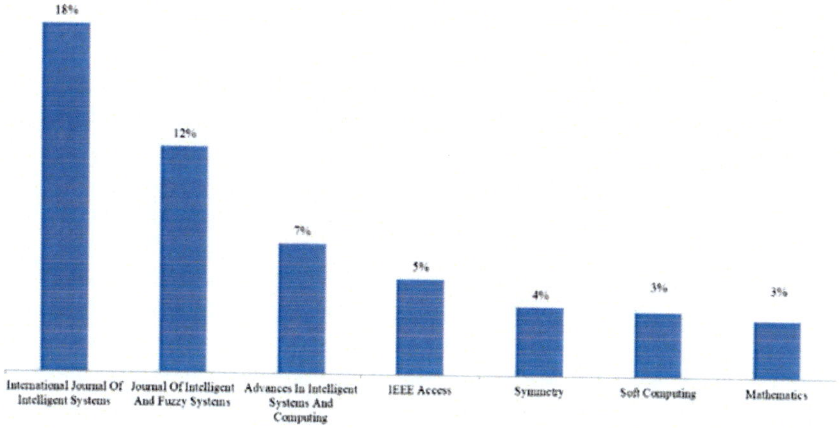

Fig. 63 Pythagorean fuzzy sets papers by their sources

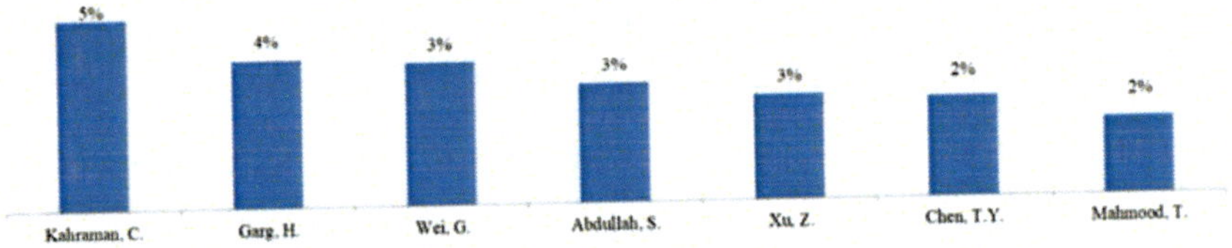

Fig. 64 Distribution of publication percentages of authors on Pythagorean fuzzy sets

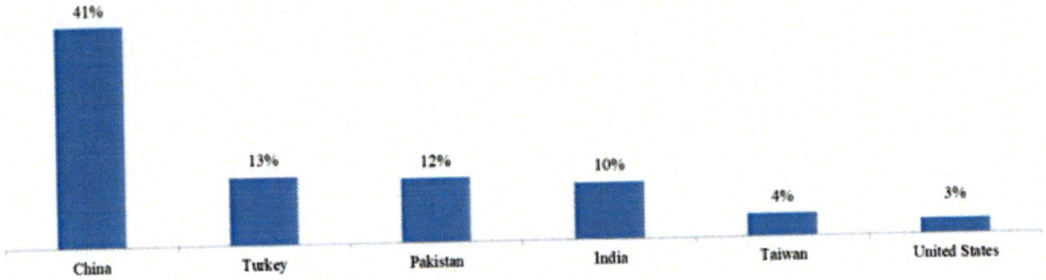

Fig. 65 Distribution of pythagorean fuzzy sets papers by their countries

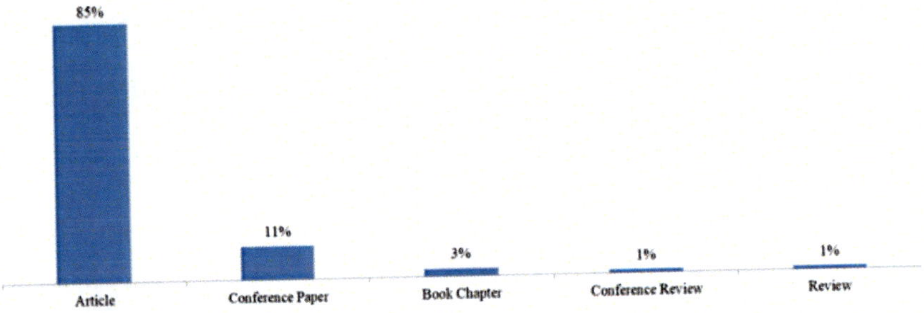

Fig. 66 Distribution of Pythagorean fuzzy sets papers by their types

Management and Accounting, Multidisciplinary, Energy, Economics, Econometrics and Finance, Medicine, Earth and Planetary Sciences, Neuroscience, Arts and Humanities, Agricultural and Biological Sciences.

Figure 63 shows the sources of Pythagorean fuzzy set publications. Most of the publications on Pythagorean fuzzy sets have been published in International Journal of Intelligent Systems and Journal of Intelligent and Fuzzy Systems.

In Fig. 64, the publication percentages and the corresponding numbers of authors on Pythagorean fuzzy sets are presented. Kahraman and Garg are the leaders among these authors.

In Fig. 65, the distribution of publications on Pythagorean fuzzy sets with respect to their source countries is illustrated. China is the leading country on Pythagorean fuzzy publications. Turkey and Pakistan are the next two countries after China.

The document types of Pythagorean fuzzy papers are illustrated in Fig. 66. The document types on Pythagorean fuzzy sets are of five types: articles with a percentage of 85, book chapters with a percentage of 3, conference papers with a percentage of 11 and conference review with a percentage of 1.

1.2.11 Literature Review on Picture Fuzzy Sets

We search the "Picture Fuzzy Sets" term in Scopus and find 298 results. The analyses of Picture fuzzy sets are given in Figs. 67, 68, 69, 70, 71 and 72.

Figure 67 shows that the distribution of "q-rung Orthopair fuzzy" set papers with respect to years. It is seen that most of the studies have been published in 2019 with a rate of 25.2%.

In Fig. 68, Picture fuzzy sets papers are summarized by their subject areas. The other subject areas with lower percentages are Decision Sciences Environmental

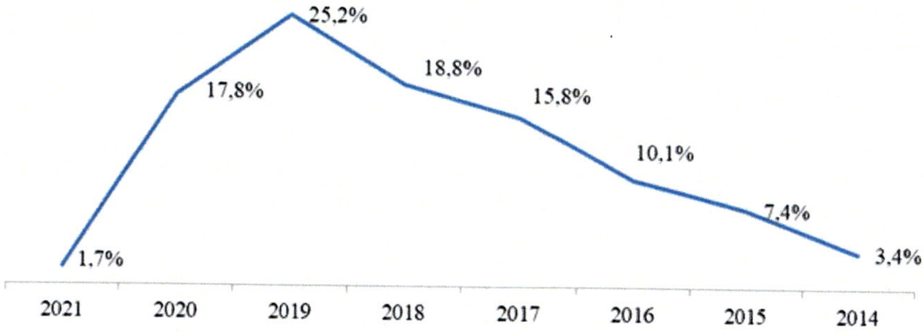

Fig. 67 Distribution of picture fuzzy sets papers with respect to years

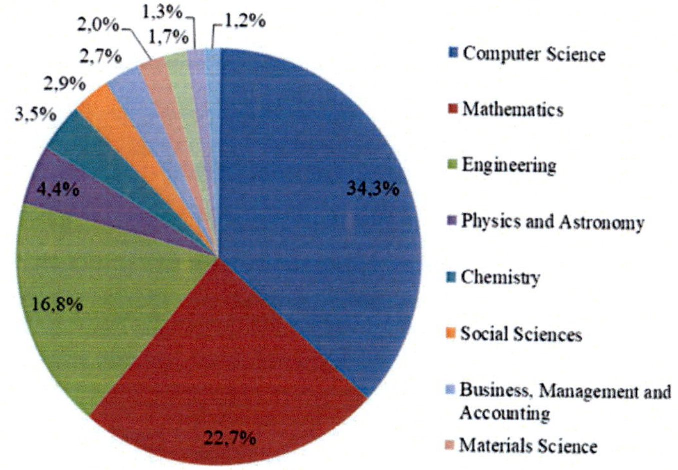

Fig. 68 Distribution of picture fuzzy sets papers with respect to subject areas

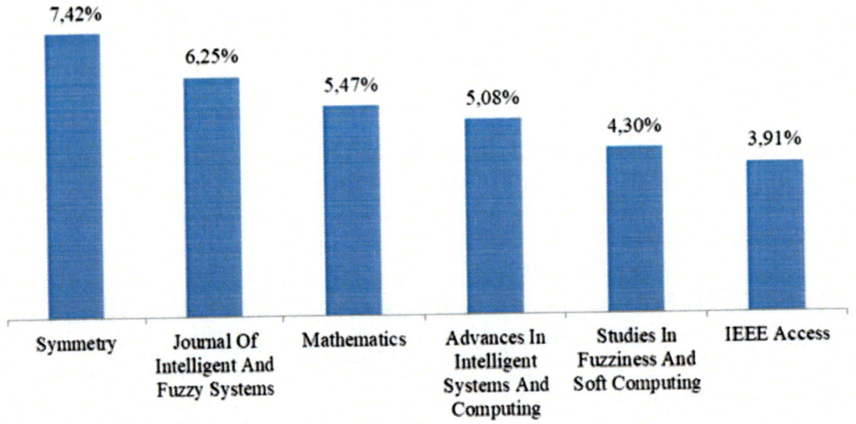

Fig. 69 Picture fuzzy sets papers by their sources

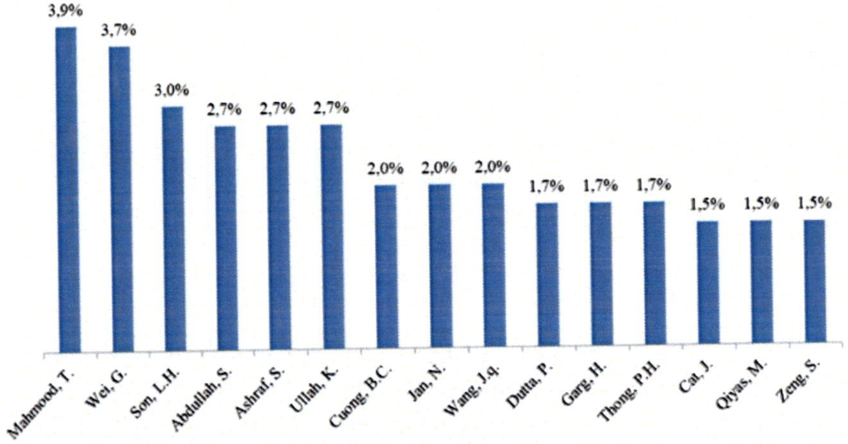

Fig. 70 Distribution of publication percentages of authors on picture fuzzy sets

Science Energy Economics, Econometrics and Finance Multidisciplinary Agricultural and Biological Sciences Arts and Humanities Biochemistry, Genetics and Molecular Biology Medicine Chemical Engineering Neuroscience Earth and Planetary Sciences Health Professions Immunology and Microbiology Pharmacology, Toxicology and Pharmaceutics Psychology.

Figure 69 shows the sources of picture fuzzy set publications. Most of the publications on picture fuzzy sets have been published in Symmetry and International Journal of Intelligent Systems.

In Fig. 70, the publication percentages and the corresponding numbers of authors on picture fuzzy sets are presented. Mahmood and Wei are the leader among these authors.

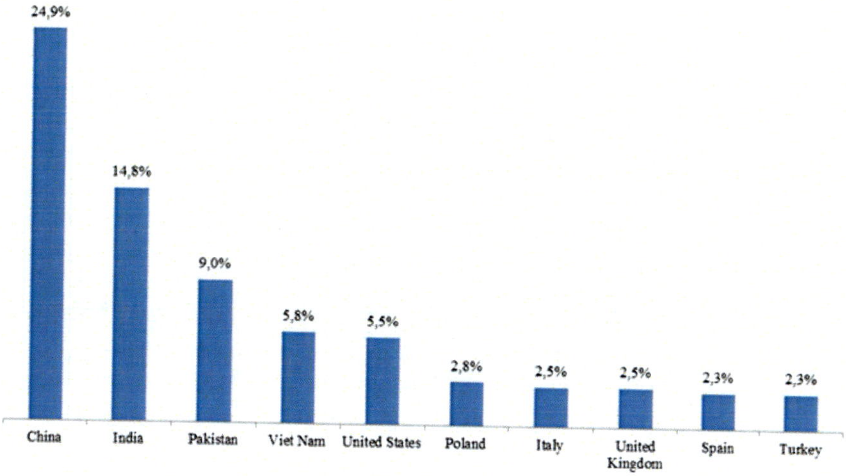

Fig. 71 Distribution of picture fuzzy sets papers by their countries

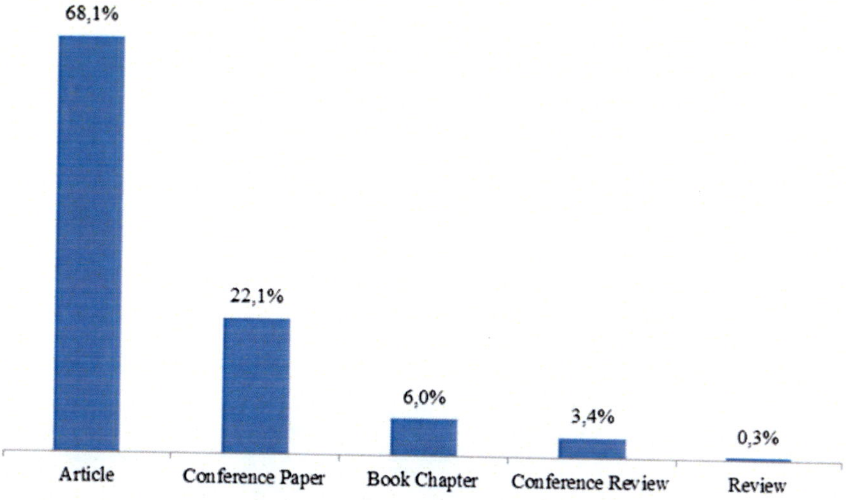

Fig. 72 Distribution of picture fuzzy sets papers by their document types

In Fig. 71, the distribution of publications on picture fuzzy sets with respect to their source countries is illustrated. China and India are the leading countries on picture fuzzy publications. Pakistan and Vietnam are the next two countries after India.

The percentages and the corresponding numbers of picture fuzzy papers are illustrated in Fig. 72. The document types on Picture fuzzy sets are of five types: articles with a percentage of 68.1%, book chapters with a percentage of 28, conference papers with a percentage of 10% and conference review with a percentage of 3.

1.2.12 Literature Review on q-rung Orthopair Fuzzy Sets

We search the *q-rung Orthopair Fuzzy Sets* term in Scopus and find 126 results. The analyses of q-rung Orthopair fuzzy sets are given in Figs. 73 and 74. Figure 73 shows that the distribution of "q-rung Orthopair fuzzy" sets papers with respect to years. It is seen that most of the studies have been published in 2019 with a rate of 49.2%.

In Fig. 74, q-rung Orthopair fuzzy set papers are summarized by their subject areas. The other subject areas are Economics, Econometrics and Finance Energy Environmental Science Multidisciplinary Agricultural and Biological Sciences Biochemistry, Genetics and Molecular Biology Social Sciences Arts and Humanities Business, Management and Accounting Chemical Engineering Medicine Neuroscience. In Fig. 75, it is seen that most of the publications on q-rung Orthopair fuzzy sets have been published in International Journal of Intelligent Systems and Symmetry.

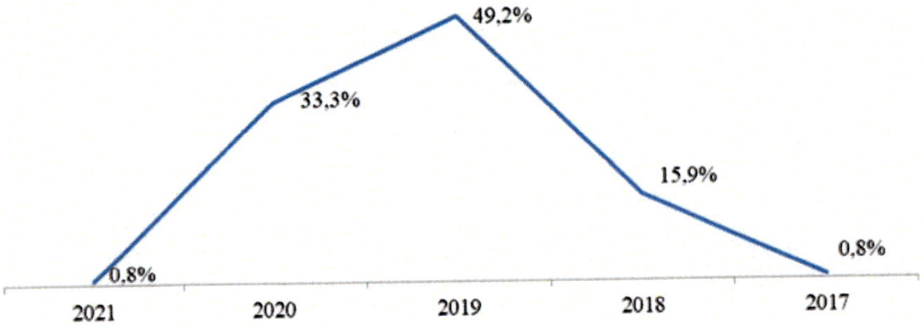

Fig. 73 Distribution of q-rung orthopair fuzzy sets papers with respect to years

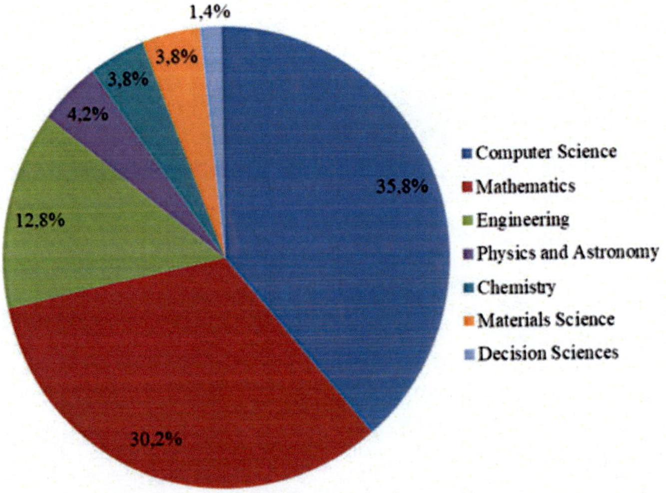

Fig. 74 Distribution of q-rung Orthopair fuzzy sets papers with respect to subject areas

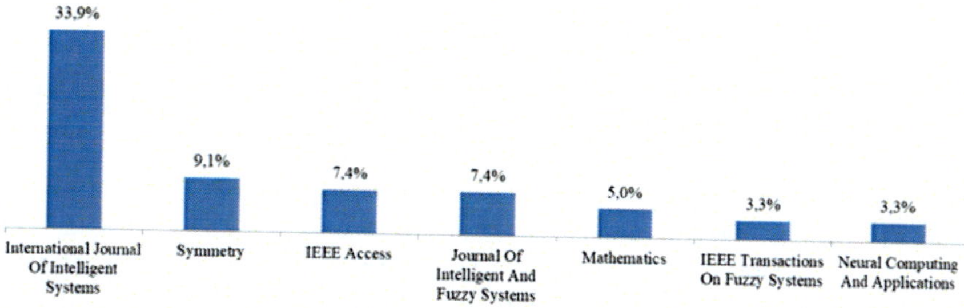

Fig. 75. q-rung Orthopair fuzzy sets papers by their sources

Most of the publications on q-rung Orthopair fuzzy sets have been published in International Journal of Intelligent Systems and Symmetry.

In Fig. 76, the publication percentages and the corresponding numbers of authors on q-rung Orthopair fuzzy sets are presented. Liu and Wei are the leader among these authors.

In Fig. 78, the distribution of publications on q-rung Orthopair fuzzy sets with respect to their source countries is illustrated. China and Pakistan are the leading countries on q-rung orthopair fuzzy publications. India and Saudi Arabia are the next two countries after Pakistan. In Fig. 77, the distribution of publications on q-rung Orthopair fuzzy sets with respect to ther source countries is illustrated. China and Pakistan are the leading countries on q-rung Orthopair fuzzy publications. India and Saudi Arabia are the next two countries after Pakistan.

The percentages and the corresponding numbers of q-rung orthopair fuzzy papers are illustrated in Fig. 78. The document types on q-rung Orthopair fuzzy sets are of four types: articles with a percentage of 92.1, book chapters with a percentage of 1.6,

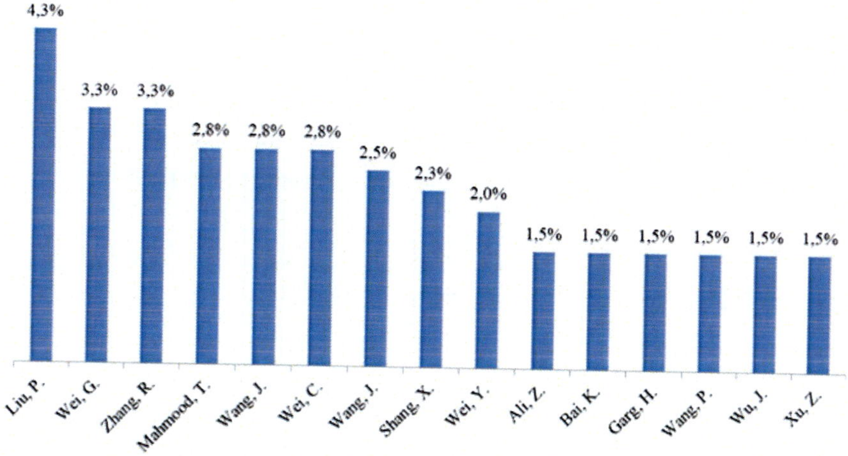

Fig. 76 Publication percentages and numbers of authors on q-rung Orthopair fuzzy sets

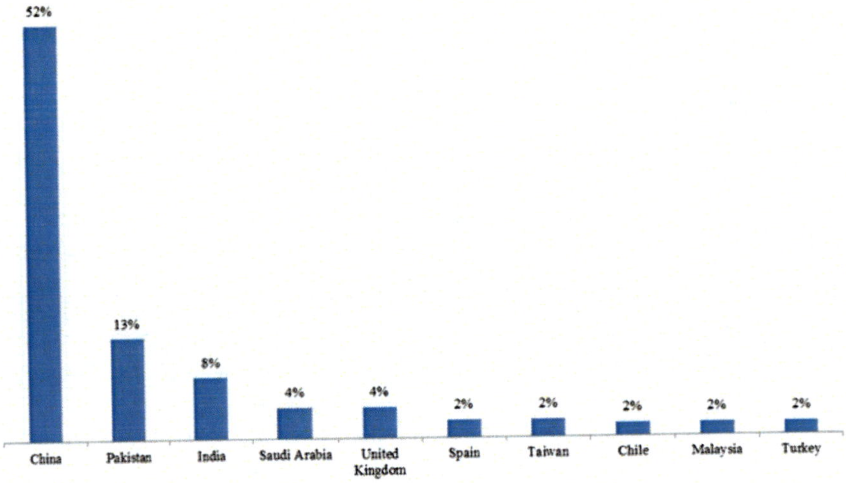

Fig. 77 Distribution of q-rung orthopair fuzzy sets papers by their countries

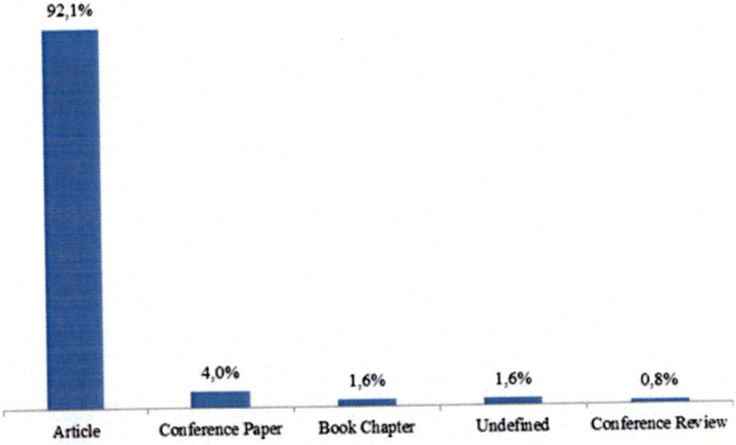

Fig. 78 Distribution of q-rung Orthopair fuzzy sets papers by their types

conference papers with a percentage of 4 and conference review with a percentage of 0.8.

1.2.13 Literature Review on Fermatean Fuzzy Sets

We search the *Fermatean Fuzzy Sets* term in Scopus and find 8 results. The analysis of fermatean fuzzy sets are given in Figs. 79, 80, 81, 82, 83 and 84. Figure 79 shows

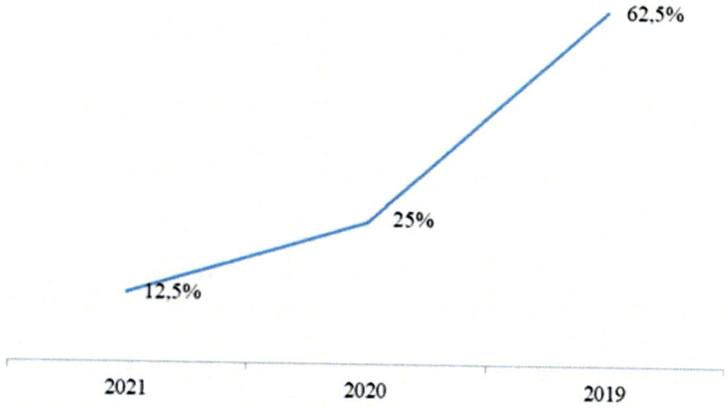

Fig. 79 Distribution of Fermatean fuzzy sets papers with respect to years

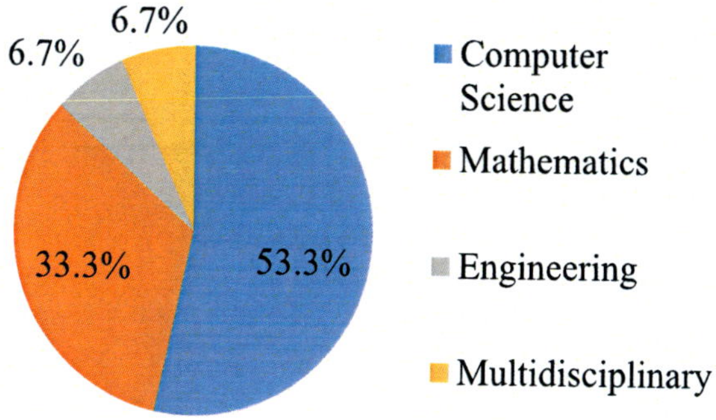

Fig. 80 Distribution of Fermatean fuzzy sets papers with respect to subject areas

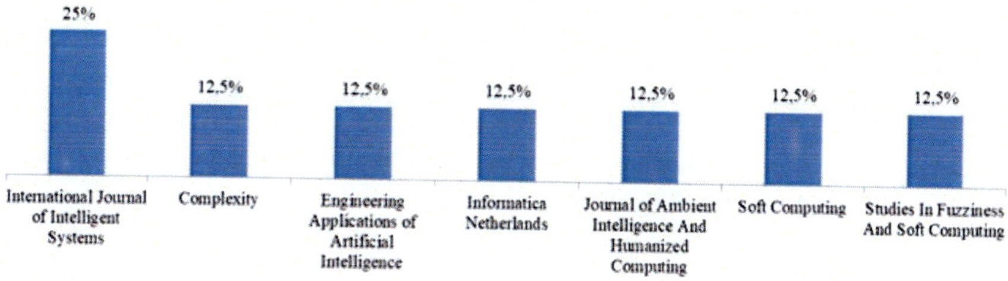

Fig. 81 Fermatean fuzzy sets papers by their sources

the distribution of fermatean fuzzy set papers with respect to years. It is seen that most of the studies have been published in 2019 with a rate of 62.5%.

In Fig. 80, fermatean fuzzy set papers are summarized by their subject areas.

Figure 81 illustrates the sources of fermatean fuzzy sets. Most of the publications on fermatean fuzzy sets have been published in International Journal of Intelligent Systems.

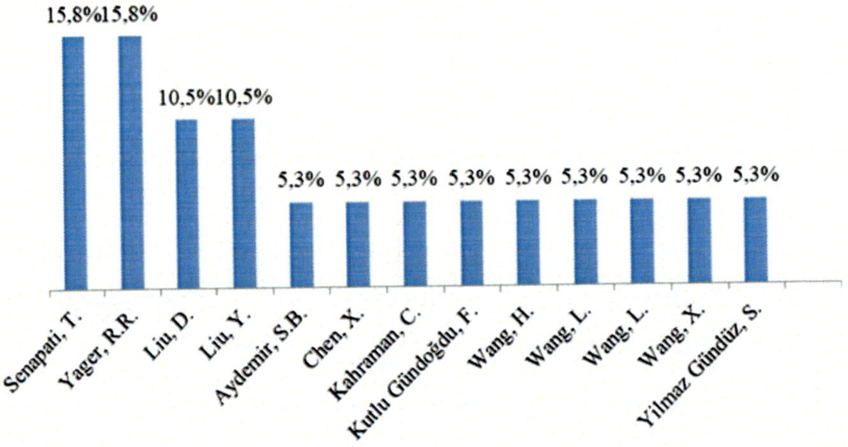

Fig. 82 Publication percentages and numbers of authors on Fermatean fuzzy sets

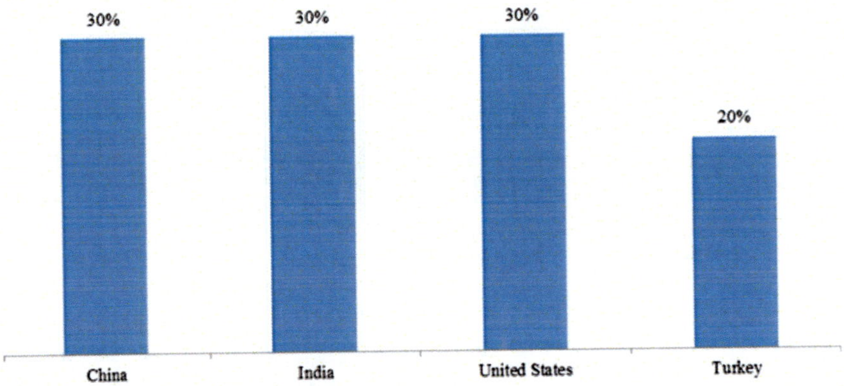

Fig. 83 Distribution of fermatean fuzzy sets papers by their countries

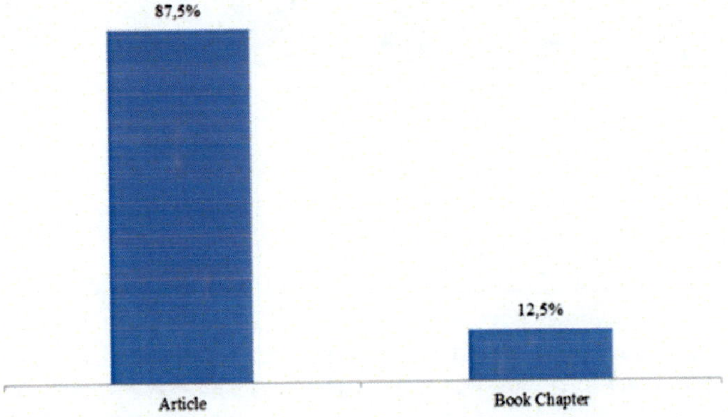

Fig. 84 Distribution of Fermatean fuzzy sets papers by their types

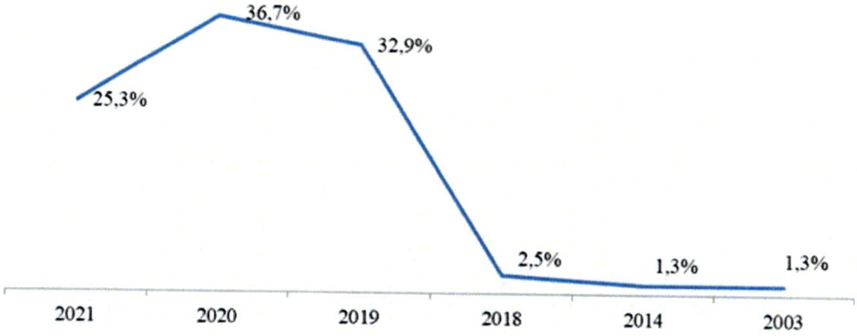

Fig. 85 Distribution of spherical fuzzy sets papers with respect to years

In Fig. 82, the publication percentages and the corresponding numbers of authors on fermatean fuzzy sets are presented. Senapati and Yager are the leader among these authors.

In Fig. 83, the distribution of publications on fermatean fuzzy sets with respect to their source countries is illustrated. China, India and USA are the leading countries on fermatean fuzzy publications.

The percentages and the corresponding numbers of fermatean fuzzy set papers by their document types are illustrated in Fig. 85. The document types on fermatean fuzzy sets are of four types: articles with a percentage of 59%, book chapters with a percentage of 28, conference papers with a percentage of 10% and conference review with a percentage of 3.

1.2.14 Literature Review on Spherical Fuzzy Sets

We search the *Spherical Fuzzy Sets* term in Scopus and find 79 results. The analyses of spherical fuzzy sets are given in Figs. 85, 86, 87, 88, 89 and 90.

Figure 85 shows the distribution of spherical fuzzy set papers with respect to years. It is seen that most of the studies have been published in 2020 with a rate of 37%.

In Fig. 86, spherical fuzzy sets papers are summarized by their subjects areas.

Most of the publications on SFSs have been published in Studies in Fuziness and Soft Computing and Journal of Intelligent and Fuzzy Systems.

In Fig. 88, the publication percentages and the corresponding numbers of authors on spherical fuzzy sets are presented. Cengiz Kahraman and Fatma Kutlu Gündoğdu are the leader among these authors.

In Fig. 89, the distribution of publications on spherical fuzzy sets with respect to their source countries is illustrated. Turkey is the leading country on Spherical fuzzy publications. Pakistan and India are the next two countries after Turkey.

The percentages and the corresponding numbers of spherical fuzzy papers by their document types are illustrated in Fig. 90. The document types on spherical

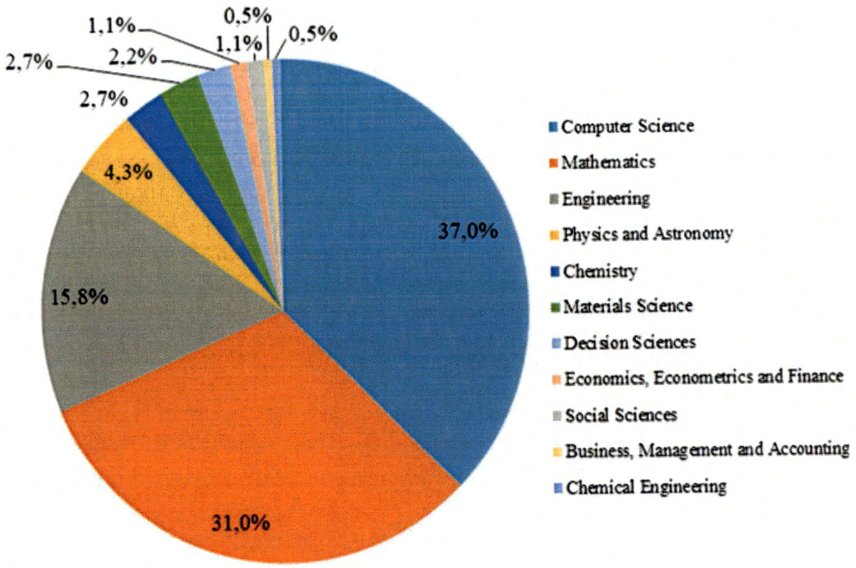

Fig. 86 Distribution of spherical fuzzy sets papers with respect to subject areas

fuzzy sets are of four types: articles with a percentage of 59, book chapters with a percentage of 28, conference papers with a percentage of 10 and conference review with a percentage of 3.

1.3 Conclusion

The overall aim of the fuzzy set extensions is to define membership functions with more parameters for more flexible and more realistic modeling. Intuitionistic fuzzy sets, Pythagorean fuzzy sets, and q-rung orthopair fuzzy sets are the members of the same class, each having two parameters, namely membership and non-membership degrees whereas neutrosophic sets, picture fuzzy sets, fermatean fuzzy sets, and spherical fuzzy sets are the members of another class, each having three independent parameters, namely membership, nonmembership, and indeterminacy degrees. Human behaviour has numerous features that are very complex, difficult to model, nonrigid and difficult to predict. There is a big difference between the happiness of a person having a success in his/her business life and the happiness of having a lunch with his/her best friend. Modeling these two kinds of happiness completely requires different membership functions of happiness. While the happiness in business success increases one's self-confidence, the happiness he/she will have at lunch with his/her best friend increases his/her emotions. It is possible to model these features using fuzzy logic rather than rigid classical logic models. If the behavior of

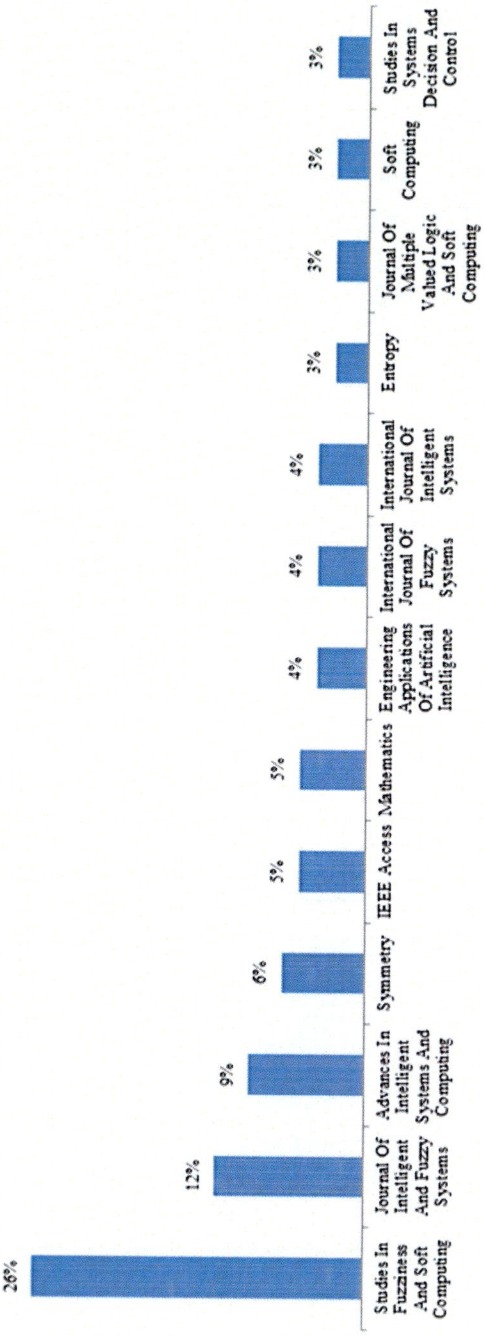

Fig. 87 Spherical fuzzy sets papers by their sources

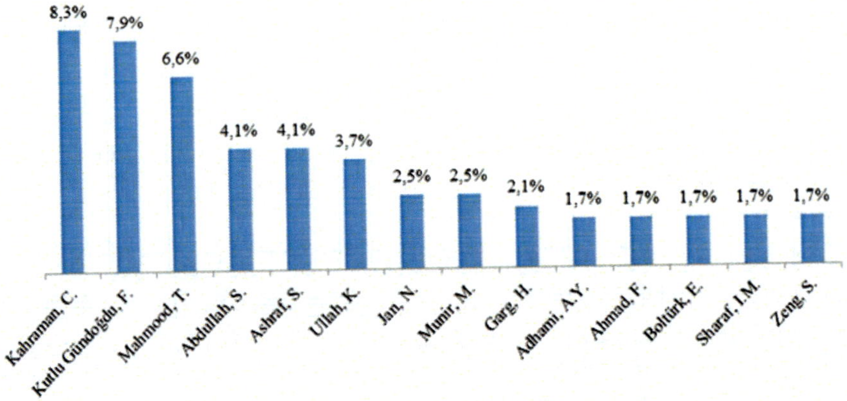

Fig. 88 Publication percentages and numbers of authors on Spherical fuzzy sets

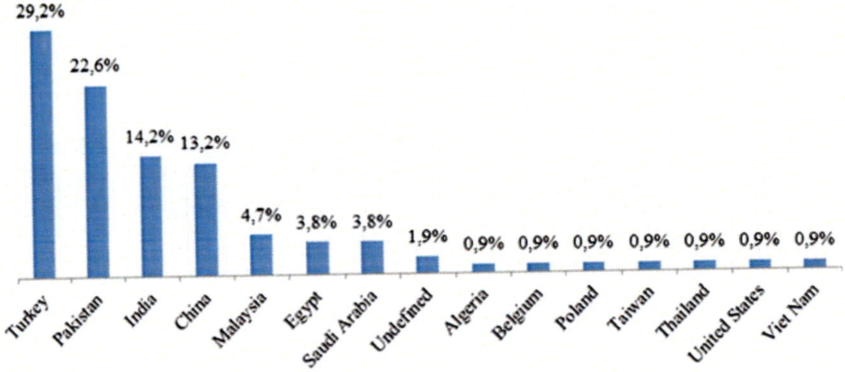

Fig. 89 Distribution of Spherical fuzzy sets papers by their countries

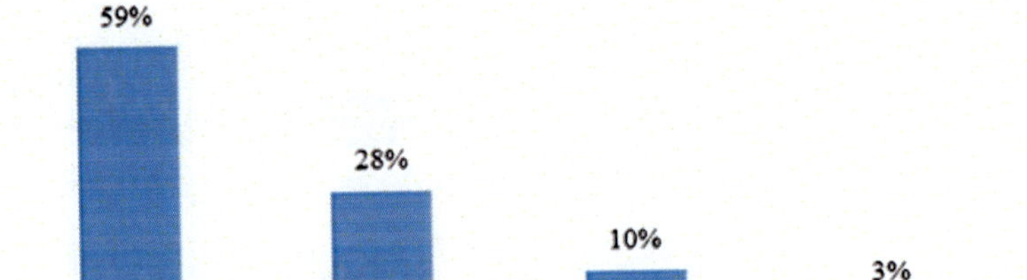

Fig. 90 Distribution of spherical fuzzy set papers by their document types

humanoid robots is intended to resemble that of humans, it is inevitable to use fuzzy set extensions in modeling.

The classical control methods cannot handle the differences between these kinds of happiness because of their rigid modelling techniques. Hence, the functions of humanoid robots' mouth, face, eyes, and other body mimics must be modelled based on fuzzy logic rather than classical logic. Fuzzy sets composed of more parameters clearly present more opportunities to model human behaviours.

References

1. Chen, C.H., Wang, C.C., Wang, Y.T., Wang, P.T.: Fuzzy logic controller design for intelligent robots. Math. Probl. Eng. (2017)
2. Zadeh, L.A.: Information and control. Fuzzy sets **8**(3), 338–53 (1965)
3. Zadeh, L.A.: The concept of a linguistic variable and its application to approximate reasoning. Inf. Sci. **8**(3), 199–249 (1975)
4. Atanassov, K.T.: Intuitionistic fuzzy sets. Fuzzy Sets Syst. **20**(1), 87–96 (1986)
5. Sari, I.U., Kahraman, C.: Intuitionistic Fuzzy Z-numbers. In: Kahraman, C., Cevik Onar, S., Oztaysi, B., Sari, I., Cebi, S., Tolga, A. (eds.) Intelligent and Fuzzy Techniques: Smart and Innovative Solutions. INFUS 2020. Advances in Intelligent Systems and Computing, vol. 1197. Springer, Cham (2021)
6. Xu, Z.: Methods for aggregating interval-valued intuitionistic fuzzy information and their application to decision making. Kongzhi Yu Juece/Control Decis. **22**(2), 215–219 (2007)
7. Wei, G., Wang, X.: Some geometric aggregation operators based on interval-valued intuitionistic fuzzy sets and their application to group decision making. In: 2007 International Conference on Computational Intelligence and Security (CIS 2007), pp. 495–499. Harbin (2007)
8. Abdullah, L., Ismail, W.K.: Hamming distance in intuitionistic fuzzy sets and interval- valued intuitionistic fuzzy sets: a comparative analysis. Adv. Comput. Math. Appl. **1**(1), 7–11 (2012)
9. Karasan, A., Kahraman, C.: Interval-valued neutrosophic extension of EDAS method, In: Kacprzyk, J., Szmidt, E., Zadrożny, S., Atanassov, K., Krawczak, M. (eds.) Advances in Fuzzy Logic and Technology 2017. EUSFLAT 2017, IWIFSGN 2017. Advances in Intelligent Systems and Computing, vol. 642, pp. 343–357. Springer, Cham (2017)
10. Yager, R.R.: On the theory of bags. Int. J. Gener. Syst. **13**(1), 23–37 (1986)
11. Smarandache, F.: Neutrosophy, Neutrosophic Probability, Set, and Logic. Amer. Res. Press, Rehoboth, 105 (1998)
12. Bolturk and Kahraman: A novel interval-valued neutrosophic AHP with cosine similarity measure. Soft. Comput. **22**, 4941–4958 (2018)
13. Li, Y., Wang, Y., Liu, P.: Multiple attribute group decisionmaking methods based on trapezoidal fuzzy two-dimension linguistic power generalized aggregation operators. Soft Comput. **20**(7), 2689–2704 (2016)
14. Zhang, H.Y., Wang J.Q., Chen, X.H.: Interval neutrosophic sets and their application in multicriteria decision-making problems. Sci. Worl J. **15** (2014)
15. Karasan, A., Kahraman, C.: A noval ıntuitionistic fuzzy dematel—anp—topsis ıntegrated methodology for freight village location selection. J. Intell. Fuzzy Syst., 1–18 (2017)
16. Kahraman, C., Otay, İ. (eds.).: Fuzzy Multi-criteria Decision-Making Using Neutrosophic Sets, Vol. 16, pp. 382–394. Springer, Berlin, Germany (2019)
17. Boltürk, E., Karasan, A., Kahraman, C.: Simple additive weighting and weighted product methods using neutrosophic sets. In: Kahraman, C., Otay, I. (eds.), Fuzzy Multicriteria Decision Making Using Neutrosophic Sets, Studies in Fuzziness and Soft Computing, vol. 369, pp. 647–676. © Springer Nature Switzerland AG 2019 (2019). https://doi.org/https://doi.org/10.1007/978-3-030-00045-5_25

18. Biswas, P., Pramanik, S., Giri, B.C.: TOPSIS method for multi-attribute group decision-making under single-valued neutrosophic environment. Neural Comput. Appl. **27**(3), 727–737 (2016)
19. Wang, H., Smarandache, F., Sunderraman, R., Zhang, Y.Q.: Interval neutrosophic sets and logic: theory and applications in computing, 1–87 (2005)
20. Garibaldi, J.M., Ozen, T.: Uncertain fuzzy reasoning: a case study in modelling expert decision making. IEEE Trans. Fuzzy Syst. **15**(1), 16–30 (2007)
21. Torra, V.: Hesitant fuzzy sets. Int. J. Intell. Syst. **25**(6), 529–539 (2010)
22. Yager, R.R.: Pythagorean membership grades in multicriteria decision making, IEEE Trans. Fuzzy Syst. **22**, 958–965 (2014)
23. Atanassov, K.T.: More on intuitionistic fuzzy sets. Fuzzy Sets Syst. **33**(1), 37–45 (1989)
24. Zhang, X., Xu, Z.: Extension of TOPSIS to multiple criteria decision making with pythagorean fuzzy sets. Int. J. Intell. Syst. **29**(12), 1061–1078 (2014)
25. Peng, X., Yang, Y.: Some results for Pythagorean fuzzy sets. Int. J. Intell. Syst. **30**(11), 1133–1160 (2015)
26. Peng, X., Yang, Y.: Fundamental properties of interval-valued Pythagorean fuzzy aggregation operators. Int. J. Intell. Syst. **31**(5), 444–487 (2016)
27. Kahraman, C., Oztaysi, B., Onar, S.c.: Multicriteria Scoring Methods using Pythagorean Fuzzy Sets. Adv. Intell. Syst. Comput. 642 (2018)
28. Kahraman, C., Onar, S.C., Oztaysi, B: Present worth analysis using pythagorean fuzzy sets. In: Kacprzyk, J., Szmidt, E., Zadrożny, S., Atanassov, K., Krawczak, M. (eds.) Advances in Fuzzy Logic and Technology 2017. IWIFSGN 2017, EUSFLAT 2017. Advances in Intelligent Systems and Computing, vol 642. Springer (2018)
29 Karasan, A., Kahraman, C.: A novel intuitionistic fuzzy DEMATELANPTOPSIS integrated methodology for freight village location selection. J. Intell. Fuzzy Syst. Preprint, pp. 1–18 (2017)
30. Karasan, A., Ilbahar, E., Kahraman, C.: A novel pythagorean fuzzy AHP and its application to landfill site selection problem. Soft. Comput. (2018). https://doi.org/10.1007/s00500-018-3649-0
31. Cường, B.C.: Picture fuzzy sets. J. Comput. Sci. Cybernet. **30**(4), 409 (2014)
32. Cuong B.C., Kreinovich, V.: Picture fuzzy sets-a new concept for computational intelligence problems. In: 2013 Third World Congress on Information and Communication Technologies (WICT 2013), pp. 1–6 (2013)
33. Wei, G.W.: Picture fuzzy aggregation operators and their application to multiple attribute decision making. J. Intell. Fuzzy Syst. **33**, 1–12 (2017). https://doi.org/10.3233/JIFS-161798
34. Meksavang, P., Shi, H., Lin, S.M., Li, H.C.: An extended picture fuzzy VIKOR approach for sustainable supplier management and its application in the beef industry. Symmetry **11**(4), 468 (2019)
35. Chen, S.M., Tan, J.M.: Handling multicriteria fuzzy decision-making problems based on vague set theory. Fuzzy Sets Syst. **67**(2), 163–172 (1994)
36. Dutta, P.: Medical diagnosis based on distance measures between picture fuzzy sets. Int. J. Fuzzy Syst. Appl. (IJFSA) **7**(4), 15–36 (2018)
37. Yager, R.R.: Generalized orthopair fuzzy sets. IEEE Trans. Fuzzy Syst. **25**, 1222–1230 (2017)
38. Ju, Y., Luo, C., Ma, J., Gao, H., Santibanez Gonzalez, E.D., Wang, A.: Some interval-valued q-rung orthopair weighted averaging operators and their applications to multiple-attribute decision making. Int. J. Intell. Syst. **34**(10), 2584–260 (2019)
39. Senapati, T., Yager, R.R.: Fermatean fuzzy weighted averaging/geometric operators and its application in multi-criteria decision-making methods. Eng. Appl. Artif. Intell. **85**, 112–121 (2019)
40. Senapati, T., Yager, R.: Fermatean fuzzy sets. J. Amb. Intell. Humanized Comput. 11 (2019) https://doi.org/10.1007/s12652-019-01377-0
41. Kahraman, C., Gündoğdu, F.K., Onar, S.C., Oztaysi, B.: Hospital Location Selection Using Spherical Fuzzy TOPSIS. In: 2019 Conference of the İnternational Fuzzy Systems Association and the European Society for Fuzzy Logic and Technology (EUSFLAT 2019). Atlantis Press (2019)

42. Kutlu Gündoğdu, F., Kahraman, C.: A novel spherical fuzzy analytic hierarchy process and its renewable energy application. Soft Comput. (2019)
43. Kutlu Gündoğdu, F., Kahraman, C.: Extension of WASPAS with spherical fuzzy sets. Informatica **30**(2), 269–292 (2019)
44. Kutlu Gündoğdu, F., Kahraman, C.: A novel spherical fuzzy analytic hierarchy process and its renewable energy application. Soft Comput. (2019)
45. Kutlu Gündoğdu, F., Kahraman, C.: A novel VIKOR method using spherical fuzzy sets and its application to warehouse site selection. J. Intell. Fuzzy Syst. **37**(1), 1197–1211 (2019)
46. Biswas, P., Pramanik, S., Giri, B.C.: A new methodology for neutrosophic multi-attribute decisionmaking with unknown weight information. Neutrosophic Sets Syst. **3**, 42–50 (2014)
47. Gündoğdu, FK., Kahraman, C.: Spherical Fuzzy Sets and Decision Making Applications. In International Conference on Intelligent and Fuzzy Systems, pp. 979–987. Springer, Cham
48. Karasan, A., Kahraman, C.: Interval-valued neutrosophic extension of EDAS method. Adv. Intell. Syst. Comput. **642**, 343–357 (2018)
49. Kutlu Gündoğdu, F., Kahraman, C.: Spherical fuzzy sets and spherical fuzzy TOPSIS method. J. Intell. Fuzzy Syst. **36**(1), 337–352 (2019)
50. Kutlu Gündoğdu, F., Kahraman, C.: A novel fuzzy TOPSIS method using emerging interval-valued spherical fuzzy sets. Eng. Appl. Artif. Intell. **85**, 307–323 (2019)
51. Kutlu Gündoğdu, F., Kahraman, C.: A novel spherical fuzzy QFD method and its application to the linear delta robot technology development. Eng. Appl. Artif. Intell. **87**, 103348 (2020)
52. Kutlu Gündoğdu, F., Kahraman, C.: Spherical fuzzy sets and spherical fuzzy topsis method. J. Intell. Fuzzy Syst. **36**(1), 337–52 (2019)
53. Smarandache, F.: A unifying field in logics. neutrosophy: neutrosophic probability, set and logic (1999)
54. Ju, Y., Luo, C., Ma, J., Gao, H., Santibanez Gonzalez, E. D., Wang, A.: A Some interval-valued q-rung orthopair weighted averaging operators and their applications to multiple-attribute decision making. Int. J. Intell. Syst. **34**(10), 2584-2606 (2019)
55. Wang, H., Smarandache, F., Zhang, Y.Q., Sunderraman, R.: Single valued neutrosophic sets. Multispace Multistruct. **4**, 410–413 (2010)
56. Wang, J., Gao, H., Wei, G.W., Wei, Y.: Methods for multiple-attribute group decision making with q-Rung interval-valued orthopair fuzzy information and their applications to the selection of green suppliers. Symmetry **11**, 56 (2019)

Fuzzy Logic in Humanoid Biomechanics

Modeling Humanoid Robots Using Fuzzy Set Extensions

Cengiz Kahraman, Eda Bolturk, Sezi Cevik Onar, and Basar Oztaysi

Abstract Humanoid robots will be very usable in human society to improve the quality of human life in the future. They can be used for assisting the sick and elderly people, and dirty or dangerous jobs. Human behaviors are mostly emotional and based on the degrees of feelings such as being strongly sad or being slightly happy. Fuzzy sets theory can be very useful in modeling humanoid robots' emotions and behaviors. Pythagorean fuzzy sets (PFS) are an extension of intuitionistic fuzzy sets introduced by Atanassov (Atanassov in Fuzzy Sets Syst. 20:87–96, [1]). PFSs have the advantage of providing larger domains for assigning membership and non-membership degrees satisfying that their squared sum is at most equal to one. PFS have been often used in modeling the problems under vagueness and impreciseness in order to better define the problems together with the hesitancy of decision makers. Different human emotions and behaviors can be modeled in humanoid robots (HR) by fuzzy sets. In this paper, facial expressions of a humanoid robot are modeled depending on the degrees of the emotions. Larger degree of emotion causes a stronger indicator of the facial mimic.

1 Introduction

A humanoid robot (HR) is a robot having a body shape similar to the human body in order to resemble the human behaviors. Some humanoid robots have heads designed to replicate human facial features such as eyes and mouths. Androids are humanoid robots built to aesthetically resemble humans. Humanoid robots imitate human mechanisms of decision making and information processing. In the literature, humanoid robots have been studied for biped walk control [22], cooperative object transformation [11], gender representation [5], emotion and sociable humanoid robots [4], healt assist [19], customer acceptance [3] speaker recognition [7].

C. Kahraman (✉) · E. Bolturk · S. C. Onar · B. Oztaysi
Department of Industrial Engineering, Istanbul Technical University, 34367, Macka Istanbul, Turkey
e-mail: kahramanc@itu.edu.tr

C. Kahraman and E. Bolturk (eds.), *Toward Humanoid Robots: The Role of Fuzzy Sets*, Studies in Systems, Decision and Control 344,
https://doi.org/10.1007/978-3-030-67163-1_3

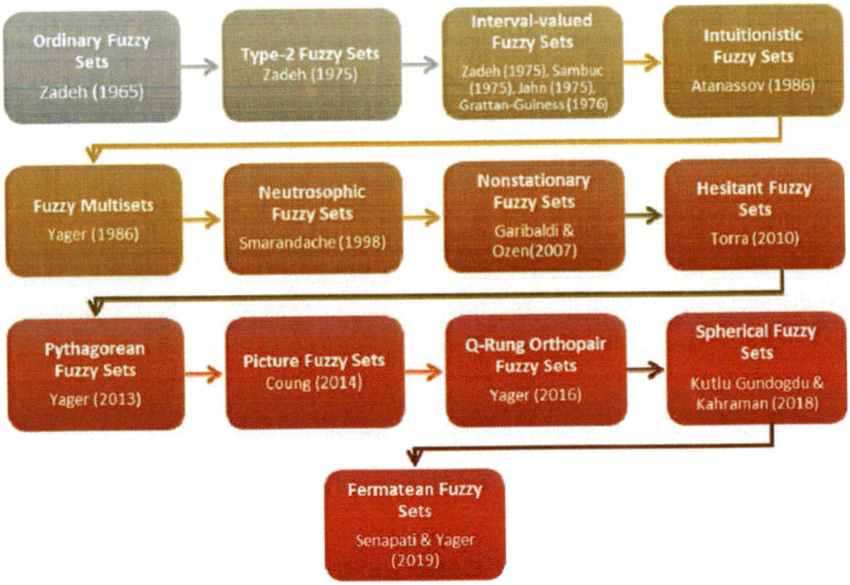

Fig. 1 Extensions of fuzzy sets

Many new extensions since first introduction Ordinary fuzzy sets theory by Zadeh [30] to the present. New extensions try to represent the thoughts of humans more correctly with more details. It is clear that classical logic is not sufficient to handle so complex human emotions. Figure 1 illustrates the flow chart of fuzzy logic history with new extensions each as a milestone.

Extensions of ordinary fuzzy sets can be classified into two main categories: (1). Intuitionistic fuzzy sets and their extensions, (2) Picture fuzzy sets and their extensions. Intuitionistic fuzzy sets (IFS) have been introduced by Atanassov [1] which differ from ordinary fuzzy sets by an independent membership degree that is not necessarily the complement of membership degree. The secondary extensions of IFS, which are Pythagorean fuzzy sets, fermatean fuzzy sets and ultimately q-rung orthopair fuzzy sets aim at providing a larger domain to experts than it is in IFS. Picture fuzzy sets have been introduced by Coung [6]. Then, the extensions of picture fuzzy sets have appeared as spherical fuzzy sets and t-spherical fuzzy sets.

Sophia has been developed by Hong Kong based company Hanson Robotics in 2016. Sophia has only 60 different facial expressions whereas the number of real human facial expressions might be infinite. Kodomoroid developed by Hiroshi Ishiguro Laboratories in 2014 is a teleoperated humanoid robot resembling a human child. It can recite news reports gathered from around the world 24 h a day in a variety of voices and languages. Jia Jia has been developed by a Chinese team at the University of Science and Technology of China in Hefei. Jia Jia can move her arms, create different facial expressions, and answer to human conversations. Mirai Madoka, which is similar to Toshiba robots, has been developed by a Japanese tech

Orix car rental agency in 2016. Junko Chihira is a humanoid developed at Toshiba. She has the ability to make human-like facial expressions. Junko Chihira currently works at a tourist information centre on Tokyo's waterfront, greeting visitors in Japanese, English, and Chinese.

In this chapter, we try to model the face expressions of HRs using Pythagorean fuzzy sets and fuzzy functions in order to express the emotional feelings of HRs. In the literature, there are very few works on fuzzy or intelligent modeling of humanoid robots. Katic and Vukobratovi [14] surveyed intelligent control techniques for humanoid robots. Wong et al. [26] studied fuzzy control of humanoid robots. Fang et al. [8] studied fuzzy brain emotional learning models.

The rest of the paper is organized as follows. Section 2 summarizes the basic definitions of IFS and PFS together with the face expressions of HRs. Section 3 concludes the paper with future directions and suggestions.

2 Extensions of Fuzzy Sets in Modeling HRs

2.1 Intuitionistic Fuzzy Sets

Intuitionistic fuzzy sets have been proposed to take into account the hesitancy of decision makers in defining membership functions [1]. These sets let the sum of membership and non-membership degrees be less than or equal to 1.0 whereas it has to be equal to 1.0 in ordinary fuzzy sets.

2.1.1 Single-Valued and Interval-Valued Intuitionistic Fuzzy Sets

Definition 1 Intuitionistic fuzzy set: Let $X \neq \emptyset$ be a given set. An intuitionistic fuzzy set in X is an object $\tilde{A}$ given by

$$\tilde{A} = \left\{x, \mu_{\tilde{A}}(x), v_{\tilde{A}}(x); xX\right\}, \tag{1}$$

where $\mu_{\tilde{A}} : X \to [0, 1]$ and $v_{\tilde{A}} : X \to [0, 1]$ satisfy the condition $0 \leq \mu_{\tilde{A}}(x) + v_{\tilde{A}}(x) \leq 1$, for every $x \varepsilon X$. Hesitancy of an expert is calculated as $\pi_{\tilde{A}} = 1 - \mu_{\tilde{A}}(x) - v_{\tilde{A}}(x)$.

Definition 2 Let $D \subseteq [0, 1]$ be the set of all closed subintervals of the interval and X be a universe of discourse. An interval-valued intuitionistic fuzzy set (IVIFS) in $\tilde{A}$ over X is an object having the form [13, 16].

$$\tilde{A} = \left\{< x, \mu_{\tilde{A}}(x), v_{\tilde{A}}(x) > | xX\right\}, \tag{2}$$

where $\mu_{\tilde{A}} \to D \subseteq [0, 1]$, $v_{\tilde{A}}(x) \to D \subseteq [0, 1]$ with the condition $0 \leq sup\mu_{\tilde{A}}(x) + supv_{\tilde{A}}(x) \leq 1, \forall x \in X$.

The intervals $\mu_{\tilde{A}}(x)$ and $\upsilon_{\tilde{A}}(x)$ denote the membership function and the non-membership function of the element x to the set $\tilde{A}$, respectively. Thus for each $x \in X$, $\mu_{\tilde{A}}(x)$ and $\upsilon_{\tilde{A}}(x)$ are closed intervals and their starting and ending points are denoted by $\mu^-_{\tilde{A}}(x)$, $\mu^+_{\tilde{A}}(x)$, $\upsilon^-_{\tilde{A}}(x)$ and $\upsilon^+_{\tilde{A}}(x)$, respectively. IVIFS $\tilde{A}$ is then denoted by

$$\tilde{A} = \left\{ < x, \left[\mu^-_{\tilde{A}}(x), \mu^+_{\tilde{A}}(x)\right], \left[\upsilon^-_{\tilde{A}}(x), \upsilon^+_{\tilde{A}}(x)\right] > | x \in X \right\} \quad (3)$$

where $0 \le \mu^+_{\tilde{A}}(x) + \upsilon^+_{\tilde{A}}(x) \le 1$, $\mu^-_{\tilde{A}}(x) \ge 0$, $\upsilon^-_{\tilde{A}}(x) \ge 0$.

For each element x, the hesitancy (indeterminacy) degree of an IVIFS of $x \in X$ in $\tilde{A}$ defined as follows;

$$\pi_{\tilde{A}}(x) = 1 - \mu_{\tilde{A}}(x) - \upsilon_{\tilde{A}}(x) = \left[1 - \mu^+_{\tilde{A}}(x) - \upsilon^+_{\tilde{A}}(x), 1 - \mu^-_{\tilde{A}}(x) - \upsilon^-_{\tilde{A}}(x)\right]. \quad (4)$$

Definition 3 Let $\tilde{I} = \left(\left[\mu^-_{\tilde{A}}(x), \mu^+_{\tilde{A}}(x)\right], \left[\upsilon^-_{\tilde{A}}(x), \upsilon^+_{\tilde{A}}(x)\right]\right)$ be a IVIF number. Equation (5) can be used for the defuzzification of $\tilde{I}$:

$$Defuzz\left(\tilde{I}\right) = \frac{\mu^-_{\tilde{A}}(x) + \mu^+_{\tilde{A}}(x) + \left(1 - \upsilon^-_{\tilde{A}}(x)\right) + \left(1 - \upsilon^+_{\tilde{A}}(x)\right) + \mu^-_{\tilde{A}}(x) \times \mu^+_{\tilde{A}}(x) - \sqrt{\left(1 - \upsilon^-_{\tilde{A}}(x)\right) \times \left(1 - \upsilon^+_{\tilde{A}}(x)\right)}}{4} \quad (5)$$

Definition 4 A Triangular Intuitionistic Fuzzy Number (TIFN) $\tilde{A}$ is a subset of IFS in $\mathbb{R}$ with the following membership and non-membership functions as follows:

$$\mu_{\tilde{A}}(x) = \begin{cases} \frac{x-a_1}{a_2-a_1}, \; for a_1 \le x \le a_2 \\ \frac{a_3-x}{a_3-a_2}, \; for a_2 \le x \le a_3 \\ 0, \; otherwise \end{cases} \quad (6)$$

$$\upsilon_{\tilde{A}}(x) = \begin{cases} \frac{a_2-x}{a_2-a_1'}, \; for a_1' \le x \le a_2 \\ \frac{x-a_2}{a_3'-a_2}, \; for a_2 \le x \le a_3' \\ 1, \; otherwise \end{cases} \quad (7)$$

where $a_1' \le a_1 \le a_2 \le a_3 \le a_3'$,$0 \le \mu_{\tilde{A}}(x) + \upsilon_{\tilde{A}}(x) \le 1$and TIFN (Fig. 2) is denoted by $\tilde{A}_{TIFN} = \left(a_1, a_2, a_3; a_1', a_2, a_3'\right)$.

Facial Expressions of HRs Using IFS

The facial expression of a HR must be appropriate to the event it meets. Even the scientists discovered that humans have 21 different facial expressions, which are

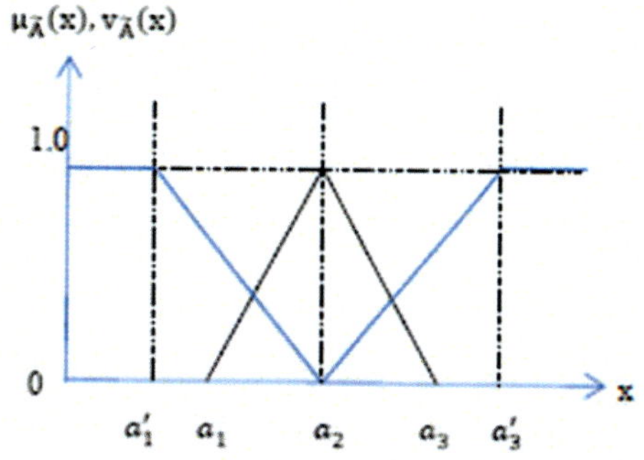

Fig. 2 Triangular intuitionistic fuzzy number

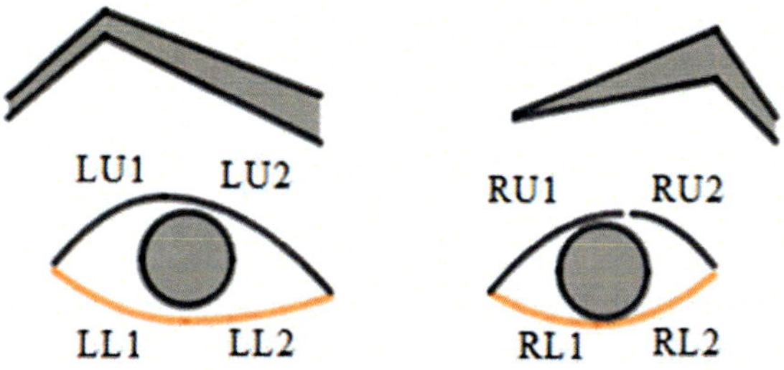

Fig. 3 Eye circles indicating a slightly surprised person

Fig. 4 Eye circles indicating a very strongly surprised person

happy, sad, fearful angry, surprised, disgusted, appalled, happily surprised, happily disgusted, sadly fearful, sadly angry, sadly surprised, sadly disgusted, fearfully angry, fearfully surprised, fearfully disgusted, angrily surprised, angrily disgusted, disgustedly surprised, hatred, and awed, you cannot classify the facial expressions by discrete sets. Besides, there are discussions on if the facial expressions are universal or vary from a culture to another. A HR's facial expression is a matter of degree that the HR should feel when it meets an event (Web-1).

Mathematical relations for some of the most common facial expressions can be in the form of simple mathematical functions such as translations and sinusoidal functions which can model expressions such as smiling, frowning and eyebrow raising. Using classical functions, the face expressions in Figs. 3, 4 and 5 are tried to model as follows (Web-2). Left eye upper and lower curve functions are represented by LU1, LU2, and LL1, LL2 and Right eye upper curve functions are represented by RU1, RU2 and RL1, RL2.

Fig. 5 Eye circles indicating a quite strongly surprised person

In Fig. 3, the left eye circle and right eye circle are modeled by the following group of equations:

Left eye circle functions	Right eye circle functions
$7(y - 4.3) = (x + 2.2)^2, -2.2 < x < -1$	$3.2(y - 4.3) = (x - 2)^2, 1.1 < x < 2$
$3(y - 4.3) = (x + 2.2)^2, -3.3 < x < -2.2$	$2.4(y - 4.3) = (x - 2)^2, 2 < x < 3$
$1.5(y - 5.3) = -(x + 2.4)^2, -3.3 < x < -2.4$	$1.9(y - 5.3) = -(x - 2.2)^2, 1.1 < x < 2.2$
$2.6(y - 5.3) = -(x + 2.4)^2, -2.4 < x < -1$	$1.2(y - 5.3) = -(x - 2.3)^2, 2.3 < x < 3$

In Fig. 4, the left eye circle and right eye circle are modeled by the following group of equations:

Left eye circle functions	Right eye circle functions
$7(y - 4.0) = (x + 2.2)^2, -2.2 < x <$	$3.2(y - 4.0) = (x - 2)^2, 1.1 < x < 2$
$-13(y - 4.0) = (x + 2.2)^2, -3.3 < x <$	$2.4(y - 4.0) = (x - 2)^2, 2 < x < 3$
$-2.21.5(y - 5.8) = -(x + 2.4)^2, -3.3 < x < -2.4$	$1.9(y - 5.8) = -(x - 2.2)^2, 1.1 < x < 2.2$
$2.6(y - 5.8) = -(x + 2.4)^2, -2.4 < x < -1$	$1.2(y - 5.8) = -(x - 2.3)^2, 2.3 < x < 3$

In Fig. 5, the left eye circle and right eye circle are modeled by the following group of equations. This is a case between Figs. 3 and 4.

Left eye circle functions	Right eye circle functions
$7(y - 4.1) = (x + 2.2)^2, -2.2 < x <$	$3.2(y - 4.1) = (x - 2)^2, 1.1 < x < 2$
$-13(y - 4.1) = (x + 2.2)^2, -3.3 < x < -2.2$	$2.4(y - 4.1) = (x - 2)^2, 2 < x < 3$
$1.5(y - 5.5) = -(x + 2.4)^2, -3.3 < x < -2.4$	$1.9(y - 5.5) = -(x - 2.2)^2, 1.1 < x < 2.2$
$2.6(y - 5.5) = -(x + 2.4)^2, -2.4 < x < -1$	$1.2(y - 5.5) = -(x - 2.3)^2, 2.3 < x < 3$

We can express the given equations above depending on two variables k and p. Thus, the facial expression affected by the surprise can be expressed by a continuous linguistic variable. It will be a matter of degree in this way. We define the intervals of these variables as $0.1 \leq p \leq 0.8$ and $0 \leq k \leq 0.3$

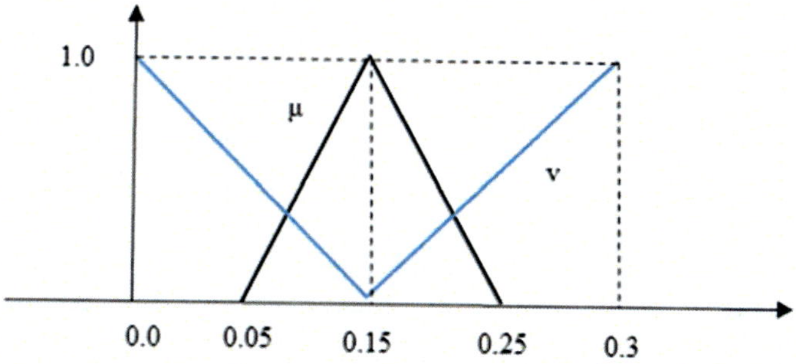

Fig. 6 Triangular intuitionistic fuzzy number for the parameter *k*

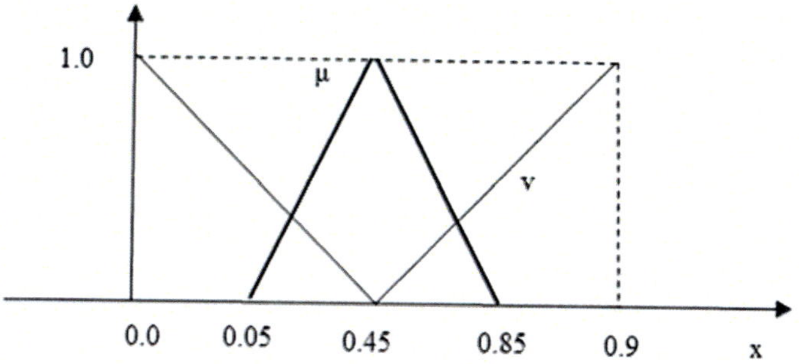

Fig. 7 Triangular intuitionistic fuzzy number for the parameter *p*

Left eye circle functions	Right eye circle functions
$7(y-4-k)=(x+2.2)^2, -2.2<x<-1$	$3.2(y-4-k)=(x-2)^2, 1.1<x<2$
$3(y-4-k)=(x+2.2)^2, -3.3<x<-2.2$	$2.4(y-4-k)=(x-2)^2, 2<x<3$
$1.5(y-5-p)=-(x+2.4)^2, -3.3<x<-2.4$	$1.9(y-5-p)=-(x-2.2)^2, 1.1<x<2.2$
$2.6(y-5-p)=-(x+2.4)^2, -2.4<x<-1$	$1.2(y-5-p)=-(x-2.3)^2, 2.3<x<3$

Now let the intuitionistic fuzzy sets for *k* and *p* be defined as follows.

For a slightly surprised person, interval valued intuitionistic fuzzy number may be defined as ([0.5, 0.6], [0.3, 0.4]) for k = 0.3 and p = 0.3

For a quite strongly surprised person, interval valued intuitionistic fuzzy number may be defined as ([0.7, 0.9], [0, 0.1]) for k = 0.1 and p = 0.5.

For an extremely surprised person, interval valued intuitionistic fuzzy number may be defined as ([0.85, 0.95], [0, 0.05]) for k = 0 and p = 0.8.

Alternatively, we can define the parameters *k* and *p* for a surprised person's facial expression as triangular intuitionistic fuzzy numbers as in Figs. 6 and 7, respectively.

The function in Fig. 6 can be represented by the following α-cuts:

For membership function: $x_L = 0.05 + 0.10\alpha$, $x_R = 0.25 - 0.10\alpha$,

For non-membership function: $x_L = 0.15 - 0.15\alpha$, $x_R = 0.15 + 0.15\alpha$.
The function in Fig. 7 can be represented by the following α-cuts:
For membership function:$x_L = 0.05 + 0.40\alpha$, $x_R = 0.85 - 0.40\alpha$,
For non-membership function:$x_L = 0.45 - 0.45\alpha$, $x_R = 0.45 + 0.45\alpha$.
Thus, the equations with k and p converted to their α-cuts equivalents as follows:

Lefl eye circle functions	Right eye circle functions
$3(y - 4 - ([0.05 + 0.10\alpha, 0.25 - 0.10\alpha], [0.15 - 0.15\alpha, 0.15 + 0.15\alpha])) - 0 + 2.2) - 3.3 < j < -2.2$	$3.2(y - 4 - ([0.05 + 0.10\alpha, 0.25 - 0.10\alpha], [0.15 - 0.15\alpha, 0.15 + 0.15\alpha])) -(x - 2)^2, 1.1 < x < 2$
$7(y - 4 - ([0.05 + 0.10\alpha, 0.25 - 0.10\alpha], [0.15 - 0.15\alpha, 0.15 + 0.15\alpha])) = (x + 2.2): -2.2 < x < -1$	$2.4(y - 4 - ([0.05 + 0.10\alpha, 0.25 - 0.10\alpha], [0.15 - 0.15\alpha, 0.15 + 0.15\alpha])) = (x-2)^2, 2 < x < 3$
$1.5(y - 5 - ([0.05 + 0.40\alpha, 0.85 - 0.40\alpha], [0.45 - 0.4S\alpha, 0.45 - 0.45\alpha]) = -(x + 2.4)^z, -3.3 < J < -2.4$	$1.9(y - 5 - ([0.05 + 0.40\alpha, 0.85 - 0.40\alpha], [0.45 - 0.45\alpha, 0.45 + 0.45\alpha])) = -(x - 2.2)^2, 1.1 < x < 2.2$
$2.6(y - 5 - ([0.05 + 0.40\alpha, 0.85 - 0.40\alpha], [0.45 - 0.45\alpha, 0.45 - 0.45\alpha]) = -\{x + 2.4)^z, -2.4 < x < -1$	$1.2(y - 5 - ([0.05 + 0.40\alpha, 0.85 - 0.40\alpha], [0.45 - 0.45\alpha, 0.45 - 0.45\alpha]) = -(x - 2.3)^2, 2.3 < x < 3$

Table 1 gives us the defuzzified values of the IVIF numbers depending on the various α values in left eye upper and lower curve functions LU1, LU2, and LL1, LL2 and right eye upper and lower curve functions RU1, RU2 and RL1, RL2.

Table 2 presents the defuzzified results of left eye upper and lower curve functions LU1, LU2, and LL1, LL2 and right eye upper and lower curve functions RU1, RU2 and RL1, RL2.

Figures 8, 9 and 10 illustrate the LU1 functions for the given α values.

2.1.2 Pythagorean Fuzzy Sets (PFS)

The main idea of PFS comes from intuitionistic fuzzy sets. PFS have been proposed by Atanassov [2] first time as type-2 IFS. Later, Yager [29] developed type-2 IFS and called it PFS. In the following, preliminaries of PFS are briefly presented.

Single-Valued and Interval-Valued Pythagorean Fuzzy Sets

Yager [29] introduced Pythagorean fuzzy set (PFS) characterized by a membership degree and a non-membership degree satisfying the condition that the squared sum of its membership degree and non-membership degree is equal to or less than 1, which is a generalization of IFS.

Definition 5 A single-valued PFS is defined as follows:

$$\tilde{P} = \{< x, P\,(\mu_P(x), v_P(x)) > | x \in X\} \tag{8}$$

Table 1 Defuzzified values of IVIF numbers in upper and lower functions

α	Defuzzified IVIF in left and right lower functions	Defuzzified IVIF in left and right upper functions
0	0.290625	0.373125
0.05	0.290877	0.37714
0.1	0.291133	0.381186
0.15	0.291393	0.385264
0.2	0.291657	0.389378
0.25	0.291926	0.393532
0.3	0.292198	0.397731
0.35	0.292474	0.401983
0.4	0.292755	0.406297
0.45	0.29304	0.410684
0.5	0.293329	0.415157
0.55	0.293622	0.419734
0.6	0.29392	0.424434
0.65	0.294221	0.429282
0.7	0.294528	0.43431
0.75	0.294838	0.439557
0.8	0.295153	0.445072
0.85	0.295473	0.450922
0.9	0.295797	0.457194
0.95	0.296126	0.464015
1	0.29646	0.471568

where $\mu_P : X \to [0, 1]$ is the membership degree and $v_P : X \to [0, 1]$ is the nonmembership degree. Then, Eq. (9) is valid:

$$0 \leq (\mu_P(x))^2 + (v_P(x))^2 \leq 1 \tag{9}$$

The degree of indeterminacy is defined as follows:

$$\pi_P(x) = \sqrt{1 - (\mu_P(x))^2 - (v_P(x))^2} \tag{10}$$

For two PFSs, the following arithmetic operations are valid:

$$\tilde{\beta}_1 \oplus \tilde{\beta}_2 = P\left(\sqrt{\mu_{\beta_1}^2 + \mu_{\beta_2}^2 - \mu_{\beta_1}^2 \mu_{\beta_2}^2}, v_{\beta_1} v_{\beta_2}\right) \tag{11}$$

Table 2 Defuzzified values of LU1, LU2, LL1, LL2, RU1, RU2, RL1, and RL2

α	LU1	LU2	LL1	LL2	RU1	RU2	RL1	RL2
0	4.658	4.291	5.772	5.292	4.655	4.373	5.998	5.373
0.05	4.624	4.291	5.728	5.296	4.636	4.379	5.946	5.378
0.1	4.592	4.291	5.686	5.302	4.618	4.386	5.897	5.385
0.15	4.561	4.293	5.647	5.312	4.601	4.396	5.850	5.394
0.2	4.532	4.295	5.609	5.324	4.584	4.408	5.806	5.403
0.25	4.505	4.298	5.574	5.339	4.569	4.422	5.765	5.415
0.3	4.480	4.301	5.540	5.357	4.555	4.438	5.726	5.428
0.35	4.456	4.305	5.509	5.377	4.542	4.456	5.690	5.442
0.4	4.434	4.310	5.480	5.401	4.530	4.476	5.657	5.458
0.45	4.413	4.316	5.453	5.427	4.519	4.499	5.626	5.476
0.5	4.394	4.322	5.428	5.456	4.510	4.524	5.598	5.495
0.55	4.377	4.329	5.406	5.488	4.501	4.550	5.573	5.516
0.6	4.361	4.337	5.385	5.522	4.493	4.579	5.551	5.539
0.65	4.348	4.346	5.367	5.559	4.487	4.611	5.531	5.563
0.7	4.335	4.355	5.351	5.599	4.482	4.644	5.514	5.588
0.75	4.325	4.365	5.337	5.642	4.478	4.680	5.500	5.616
0.8	4.316	4.376	5.325	5.687	4.475	4.718	5.489	5.645
0.85	4.309	4.387	5.315	5.736	4.474	4.759	5.481	5.676
0.9	4.303	4.399	5.307	5.787	4.474	4.802	5.476	5.709
0.95	4.299	4.412	5.302	5.841	4.475	4.848	5.474	5.744
1.0	4.297	4.425	5.298	5.897	4.479	4.897	5.476	5.782

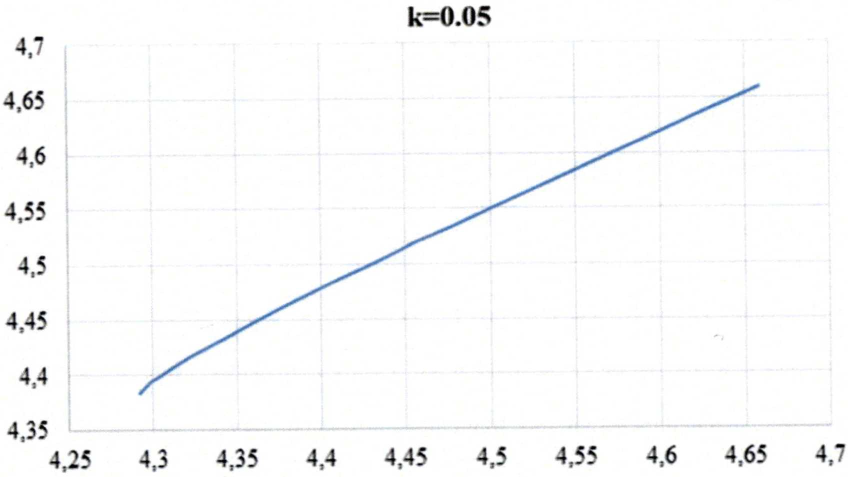

Fig. 8 LU1 function indicating very slightly surprised person

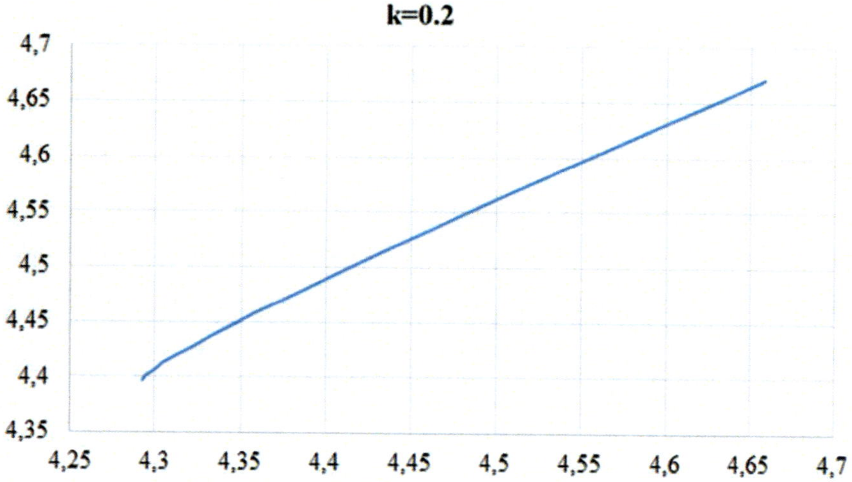

Fig. 9 LU1 function indicating slightly surprised person

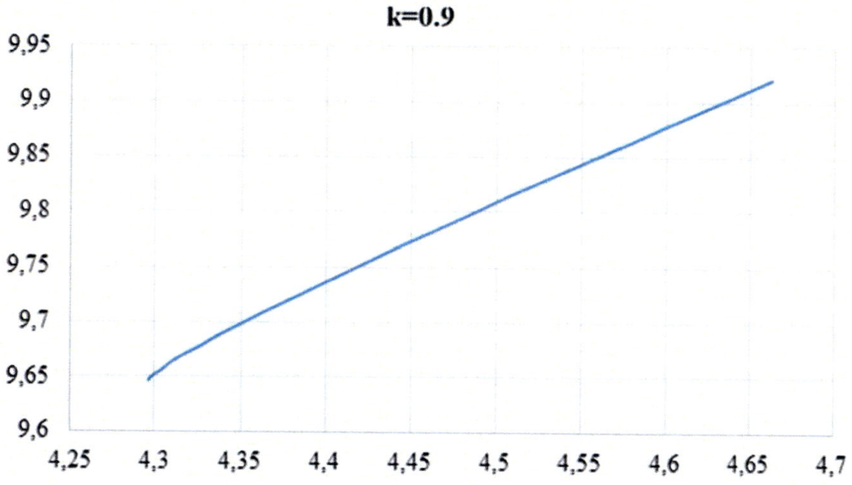

Fig. 10 LU1 function indicating strongly surprised person

$$\tilde{\beta}_1 \otimes \tilde{\beta}_2 = P\left(\mu_{\beta_1}\mu_{\beta_2}, \sqrt{v_{\beta_1}^2 + v_{\beta_2}^2 - v_{\beta_1}^2 v_{\beta_2}^2}\right) \quad (12)$$

$$\lambda\tilde{\beta} = P\left(\sqrt{1 - \left(1 - \mu_{\beta}^2\right)^{\lambda}}, (v_{\beta})^{\lambda}\right), \lambda > 0 \quad (13)$$

$$\tilde{\beta}^{\lambda} = P\left((\mu_{\beta})^{\lambda}, \sqrt{1 - \left(1 - v_{\beta}^2\right)^{\lambda}}\right), \lambda > 0 \quad (14)$$

Definition 6 Let Int([0, 1]) denote the set of all closed subintervals of [0, 1], and X be a universe of discourse. An interval-valued PFS (IVPFS) $\tilde{P}$ in X is given by Eq. (15) [18].

$$\tilde{P} = \left\{x, \mu_p(x), v_p(x) \middle| x \in X\right\} \tag{15}$$

where the functions $\mu_p: X \to \text{Int}([0, 1])(x \in X \to \mu_p(x) \subseteq [0, 1])$ and $v_p: X \to \text{Int}([0, 1])$ $(x \in X \to v_p(x) \subseteq [0, 1])$ denote the membership degree and non-membership degree of the element $x \in X$ to the set $\tilde{P}$, respectively, and for every $x \in X$, $0 \leq \sup\left\{(\mu_p(x))^2\right\} + \sup\left\{(v_p(x))^2\right\} \leq 1$. Also, for each $x \in X$, $\mu_p(x)$ and $v_p(x)$ are closed intervals and their lower and upper bounds are denoted by $\mu_p^L(x)$, $\mu_p^U(x)$, $v_p^L(x)$, $v_p^U(x)$, respectively. Therefore, $\tilde{P}$ can also be expressed as follows:

$$\tilde{P} = \left\{x, \left[\mu_p^L(x), \mu_p^U(x)\right], \left[v_p^L(x), v_p^U(x)\right] \middle| x \in X\right\} \tag{16}$$

The degree of indeterminacy is given by Eq. (17).

$$\pi_P(x) = \begin{bmatrix} \sqrt{1 - \left(\mu_p^U(x)\right)^2 - \left(v_p^U(x)\right)^2}, \\ \sqrt{1 - \left(\mu_p^L(x)\right)^2 - \left(v_p^L(x)\right)^2} \end{bmatrix} \tag{17}$$

Definition 7 Let $\tilde{A} = \left[\mu_A^L, \mu_A^U\right], \left[v_A^L, v_A^U\right]$, $\tilde{B} = \left[\mu_B^L, \mu_B^U\right], \left[v_B^L, v_B^U\right]$ be two IVPF numbers, and $\lambda > 0$, then some operations of IVPF numbers are defined as follows [18].

$$\tilde{A} \oplus \tilde{B} = \left(\begin{matrix} \begin{bmatrix} \sqrt{(\mu_A^L)^2 + \left(\mu_B^L\right)^2 - \left(\mu_A^L\right)^2\left(\mu_B^L\right)^2}, \\ \sqrt{(\mu_A^U)^2 + \left(\mu_B^U\right)^2 - \left(\mu_A^U\right)^2\left(\mu_B^U\right)^2} \end{bmatrix}, \\ \left[v_A^L v_B^L, v_A^U v_B^U\right] \end{matrix} \right) \tag{18}$$

$$\tilde{A} \otimes \tilde{B} = \left(\begin{matrix} \left[\mu_A^L \mu_B^L, \mu_A^U \mu_B^U\right], \\ \begin{bmatrix} \sqrt{(v_A^L)^2 + (v_B^L)^2 - (v_A^L)^2(v_B^L)^2}, \\ \sqrt{(v_A^U)^2 + (v_B^U)^2 - (v_A^U)^2(v_B^U)^2} \end{bmatrix} \end{matrix} \right) \tag{19}$$

$$\lambda\tilde{A} = \left(\begin{bmatrix} \sqrt{1 - \left(1 - (\mu_A^L)^2\right)^\lambda}, \\ \sqrt{1 - \left(1 - (\mu_A^U)^2\right)^\lambda} \end{bmatrix}, \left[\left(v_A^L\right)^\lambda, \left(v_A^U\right)^\lambda\right] \right) \tag{20}$$

$$\left(\tilde{A}\right)^{\lambda} = \left(\left[\left(\mu_{A}^{L}\right)^{\lambda}, \left(\mu_{A}^{U}\right)^{\lambda}\right], \left[\begin{array}{c}\sqrt{1-\left(1-\left(\nu_{A}^{L}\right)^{2}\right)^{\lambda}}, \\ \sqrt{1-\left(1-\left(\nu_{A}^{U}\right)^{2}\right)^{\lambda}}\end{array}\right]\right) \tag{21}$$

Definition 8 An IVIF number Ã = ([a, b], [c, d]) can be defuzzified by using Eq. (22).

$$Def\left(\tilde{A}\right) = \frac{a + b + (1 - c) + (1 - d) + a \times b - \sqrt{(1 - c) \times (1 - d)}}{4} \tag{22}$$

In Eq. (22), the terms (1 − c) and (1 − d) convert non-membership degrees to membership degrees while the term $\sqrt{(1 - c) \times (1 - d)}$ decreases the defuzzified value.

Definition 9 Score function for an IVPF number Ã = ([a, b], [c, d]) is given by Eq. (23) [17]:

$$S\left(\tilde{A}\right) = \frac{1}{2}\left(a^{2} + b^{2} - c^{2} - d^{2}\right) \tag{23}$$

where

$$S\left(\tilde{A}\right) \in [-1, +1].$$

Facial Expressions of HRs Using PFS

The facial expression of a HR must be appropriate to the event it meets. Even the scientists discovered that humans have 21 different facial expressions, which are Happy, Sad, Fearful Angry, Surprised, Disgusted, Appalled, Happily surprised, Happily disgusted, Sadly fearful, Sadly angry, Sadly surprised, Sadly disgusted, Fearfully angry, Fearfully surprised, Fearfully disgusted, Angrily surprised, Angrily disgusted, Disgustedly surprised, Hatred, and Awed, you cannot classify the facial expressions by discrete sets. Besides, there are discussions on if the facial expressions are universal or vary from a culture to another. A HR's facial expression is a matter of degree that the HR should feel when it meets an event.

Mathematical relations for some of the most common facial expressions can be in the form of simple mathematical functions such as translations and sinusoidal functions which can model expressions such as smiling, frowning and eyebrow raising. Using classical functions, the face expressions are tried to model as illustrated in Fig. 11 (Web-2).

The crisp smile functions can be modeled by using IFS and PFS as follows. Assume that the humanoid robot's smile upon an event should be between medium smile ($k = 0.6$) and very slight smile ($k = 0.1$) by considering the lower lip function

$f_{strong\ smile\ lower\ lip} = -0.8cosx - 3.5,$
$-2\pi/3 < x < 2\pi/3$
$f_{strong\ smile\ upper\ lip} = 0.4cosx - 3,$
$-2\pi/3 < x < 2\pi/3$

$f_{slight\ smile\ lower\ lip} = -0.5cosx - 3.3, -2\pi/3 \leq x$
$\leq 2\pi/3$
$f_{slight\ smile\ upper\ lip} = 0.5cosx - 3.3, -\pi/2 \leq x \leq \pi/2$

Fig. 11 Smile using mathematical functions

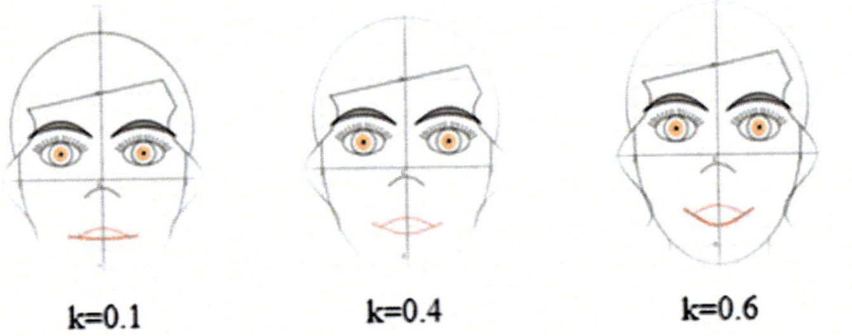

Fig. 12 Smiles between very slight smile to medium smile for lower lip

and upper lip function where the coefficient k is as in Eqs. (24) and (25):

$$f_{smilelowerlip} = -kcosx - 3.3, -2\pi/3 \leq x \leq 2\pi/3 \quad (24)$$

$$f_{smileupperlip} = kcosx - 3.3, -\pi/2 \leq x \leq \pi/2 \quad (25)$$

The smile function of Eq. (24) may vary as in Fig. 12 with different k values between 0.1 and 0.6 for very slight smile to medium smile. We also examine the smile function of Eq. (25) with different values of k between 0.7 and 1.0 for strong smile to very strong smile as in Fig. 13.

Let us assume the membership function be defined between 0.20 and 0.50 and the non-membership function be defined between 0.10 and 0.60 for very slight smile to medium smile in Eq. (24). This can be represented by a Pythagorean fuzzy number whose squared sum of upper limits of membership and non-membership is $0.50^2 + 0.60^2 = 0.61$ as given in Fig. 14.

For Eq. (25), the membership function can be defined between 0.70 and 0.90 and the non-membership function be defined between 0.10 and 0.20 for strong smile to very strong smile. This can be represented by a Pythagorean fuzzy number whose

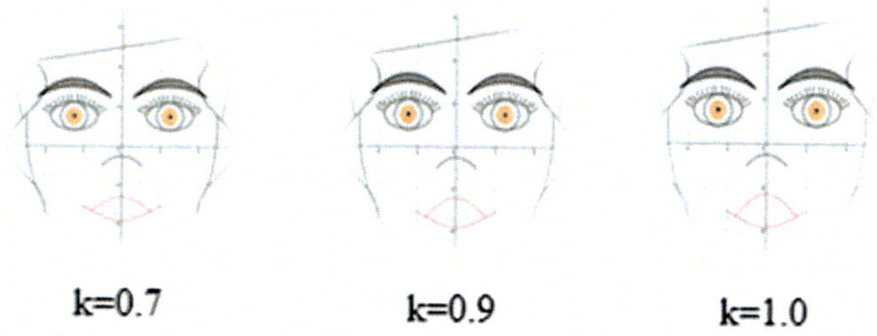

Fig. 13 Smiles between strong smile to very strong smile for upper lip

Fig. 14 Pythagorean fuzzy number for the coefficient k in Eq. (23)

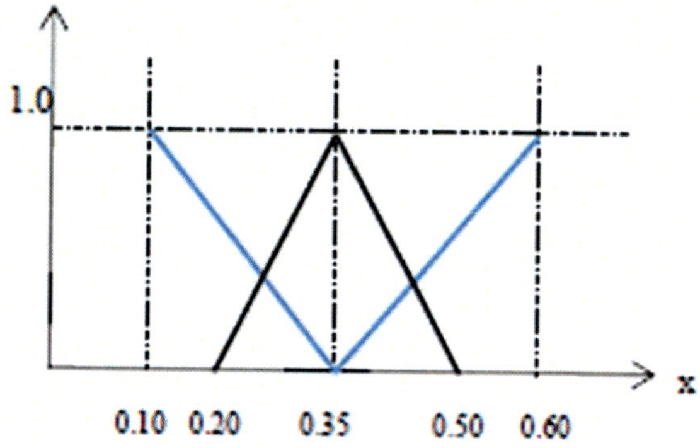

Fig. 15 Pythagorean fuzzy number for the coefficient k in Eq. (24)

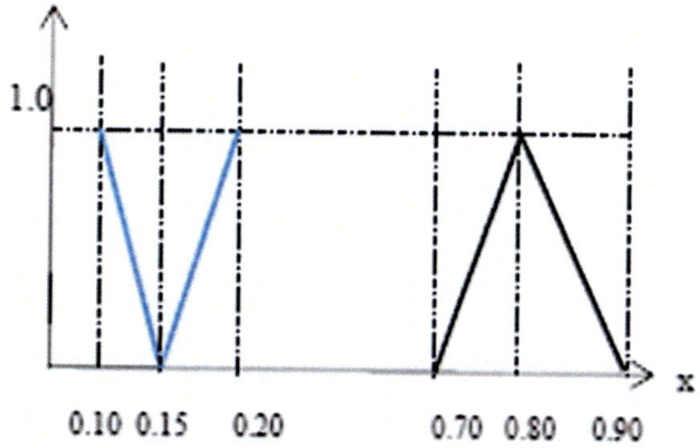

squared sum of upper limits of membership and non-membership is $0.90^2 + 0.20^2 = 0.85$ as given in Fig. 15.

The function in Fig. 14 can be represented by the following α-cuts:

$$\mu_L = 0.20 + 0.15\alpha, \quad \mu_R = 0.50 - 0.15\alpha,$$
$$v_L = 0.35 - 0.25\alpha, \quad v_R = 0.35 + 0.25\alpha.$$

The function in Fig. 15 can be represented by the following α-cuts:

$$\mu_L = 0.70 + 0.10\alpha, \quad \mu_R = 0.90 - 0.10\alpha$$
$$v_L = 0.15 - 0.05\alpha, \quad v_R = 0.15 + 0.05\alpha.$$

The fuzzy function *between very slight smile and medium smile (vss&ms)* then becomes

$$f_{vss\&ms} = -([0.20 + 0.15\alpha, 0.50 - 0.15\alpha], [0.35 - 0.25\alpha, 0.35 + 0.25\alpha]) \cos x - 3.3, -2\pi/3 \le x \le 2\pi/3$$

Using Eq. (21), we obtain

$$f_{vss\&ms} = -\left(\frac{2 + (0.20 + 0.15\alpha) \times (0.50 - 0.15\alpha) - \sqrt{(0.65 + 0.25\alpha) \times (0.65 - 0.25\alpha)}}{4}\right) \cos x - 3.3,$$

where $-2\pi/3 \le x \le 2\pi/3$.

The fuzzy function *between strong smile and very strong smile (ss&vss)* becomes

$$f_{ss\&vss} = ([0.70 + 0.10\alpha, 0.90 - 0.10\alpha], [0.15 - 0.05\alpha, 0.15 + 0.05\alpha]) \cos x - 3.3, -\pi/2 \le x \le \pi/2$$

Using Eq. (21), we obtain

$$f_{ss\&vss} = \left(\frac{3.3 + (0.70 + 0.10\alpha) \times (0.90 - 0.10\alpha) - \sqrt{(0.85 + 0.05\alpha) \times (0.85 - 0.05\alpha)}}{4}\right) \cos x - 3.3,$$

where $-\pi/2 \le x \le \pi/2$.

Table 3 gives us the defuzzified values of the IVPF numbers depending on the various α values for lower lip. Table 4 presents the results of the $f_{vss\&ms}$ function with respect to the defuzzified values given in Table 3.

Using the values in Table 4, the graph in Fig. 16 is obtained for α = [0, 1]. For larger membership degrees in IVPF numbers, the smile gets stronger. Table 5 gives us the defuzzified values of the IVPF numbers depending on the various α values for upper lip. Table 6 presents the results of the $f_{ss\&vss}$ function with respect to the defuzzified values given in Table 5. Using the values in Table 6, the graph in Fig. 17 is obtained for α = [0, 1].

The similar continuous membership functions can be defined for the other elements of a face such as eyebrow and eye orb. Figure 18 illustrates the behavior of

Table 3 Defuzzified values of IVPF number in $f_{vss\&ms}$

α	Deffuzzified value	α	Deffuzzified value
0	0.3625	0.55	0.3707
0.05	0.3631	0.6	0.3716
0.1	0.3637	0.65	0.3726
0.15	0.3643	0.7	0.3736
0.2	0.3650	0.75	0.3747
0.25	0.3657	0.8	0.3758
0.3	0.3665	0.85	0.3769
0.35	0.3672	0.9	0.3781
0.4	0.3680	0.95	0.3793
0.45	0.3689	1	0.3806
0.5	0.3698		

Table 4 Results of the function $f_{vss\&ms}$ with respect to the defuzzified values

α	DV	x								
		−20	−90	−60	−30	0	30	60	90	120
0.0h00	0.363	−3.005	−3.462	−3.645	−3.244	−2.938	−3.244	−3.645	−3.462	−3.005
0.050	0.363	−3.004	−3.463	−3.646	−3.244	−2.937	−3.244	−3.646	−3.463	−3.004
0.100	0.364	−3.004	−3.463	−3.646	−3.244	−−2.936	−3.244	−3.646	−3.463	−3.004
0.150	0.364	−3.003	−3.463	−3.647	−3.244	−2.936	−3.244	−3.647	−3.463	−3.003
0.200	0.365	−3.003	−3.464	−3.648	−3.244	−2.935	−3.244	−3.648	−3.464	−3.003
0.250	0.366	−3.002	−3.464	−3.648	−3.244	−2.934	−3.244	−3.648	−3.464	−3.002
0.300	0.366	−3.002	−3.464	−3.649	−3.243	−2.934	−3.243	−3.649	−3.464	−3.002
0.350	0.367	−3.001	−3.465	−3.650	−3.243	−2.933	−3.243	−3.650	−3.465	−3.001
0.400	0.368	−3.000	−3.465	−3.651	−3.243	−2.932	−3.243	−3.651	−3.465	−3.000
0.450	0.369	−3.000	−3.465	−3.651	−3.243	−2.931	−3.243	−3.651	−3.465	−3.000
0.500	0.370	−2.999	−3.466	−3.652	−3.243	−2.930	−3.243	−3.652	−3.466	−2.999
0.550	0.371	−2.998	−3.466	−3.653	−3.243	−2.929	−3.243	−3.653	−3.466	−2.998
0.600	0.372	−2.997	−3.467	−3.654	−3.243	−2.928	−3.243	−3.654	−3.467	−2.997
0.650	0.373	−2.997	−3.467	−3.655	−3.243	−2.927	−3.243	−3.655	−3.467	−2.997
0.700	0.374	−2.996	−3.467	−3.656	−3.242	−2.926	−3.242	−3.656	−3.467	−2.996
0.750	0.375	−2.995	−3.468	−3.657	−3.242	−2.925	−3.242	−3.657	−3.468	−2.995
0.800	0.376	−2.994	−3.468	−3.658	−3.242	−2.924	−3.242	−3.658	−3.468	−2.994
0.850	0.377	−2.993	−3.469	−3.659	−3.242	−2.923	−3.242	−3.659	−3.469	−2.993
0.900	0.378	−2.992	−3.469	−3.660	−3.242	−2.922	−3.242	−3.660	−3.469	−2.992
0.950	0.379	−2.991	−3.470	−3.661	−3.241	−2.921	−3.241	−3.661	−3.470	−2.991
1.000	0.381	−2.990	−3.471	−3.663	−3.241	−2.919	−3.241	−3.663	−3.471	−2.990

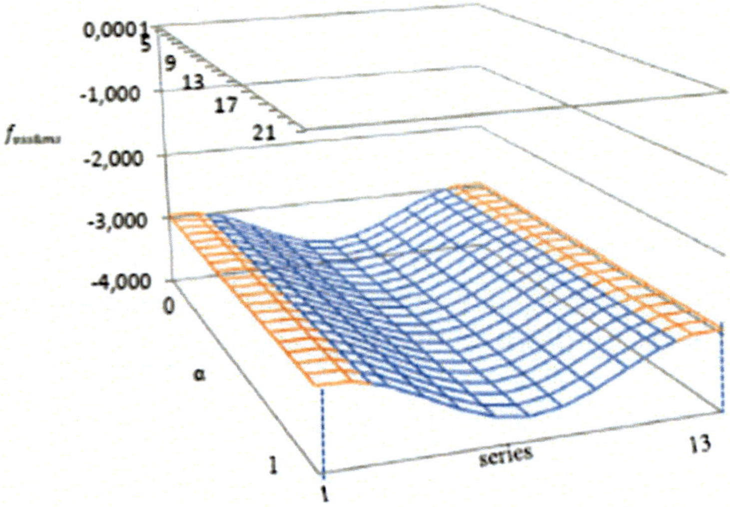

Fig. 16 Functions of lower lip for α = [0, 1]

Table 5 Defuzzified values o IVPF number in $f_{ss\&vss}$

α	Deffuzzified value	α	Deffuzzified value
0	0.7700	0.55	0.7721
0.05	0.7702	0.6	0.7722
0.1	0.7705	0.65	0.7723
0.15	0.7707	0.7	0.7725
0.2	0.7709	0.75	0.7726
0.25	0.7711	0.8	0.7726
0.3	0.7713	0.85	0.7727
0.35	0.7715	0.9	0.7728
0.4	0.7717	0.95	0.7728
0.45	0.7718	1	0.7729
0.5	0.7720		

eye orb for the different values of k in Eq. (26).

$$(y - 1.5)^2 + (x - 2.5)^2 = k \tag{26}$$

Table 6 Results of the $f_{ss\&vss}$ function with respect to the defuzzified values

α	DV	x						
		−90	−60	−30	0	30	60	90
0.000	0.77000	−3.64502	−4.03336	−3.18123	−2.53000	−3.18123	−4.03336	−3.64502
0.050	0.77024	−3.64513	−4.03359	−3.18119	−2.52976	−3.18119	−4.03359	−3.64513
0.100	0.77048	−3.64523	−4.03381	−3.18115	−2.52952	−3.18115	−4.03381	−3.64523
0.150	0.77070	−3.64533	−4.03403	−3.18112	−2.52930	−3.18112	−4.03403	−3.64533
0.200	0.77091	−3.64543	−4.03423	−3.18109	−2.52909	−3.18109	−4.03423	−3.64543
0.250	0.77112	−3.64552	−4.03442	−3.18105	−2.52888	−3.18105	−4.03442	−3.64552
0.300	0.77131	−3.64560	−4.03460	−3.18102	−2.52869	−3.18102	−4.03460	−3.64560
0.350	0.77149	−3.64568	−4.03478	−3.18100	−2.52851	−3.18100	−4.03478	−3.64568
0.400	0.77166	−3.64576	−4.03494	−3.18097	−2.52834	−3.18097	−4.03494	−3.64576
0.450	0.77182	−3.64583	−4.03509	−3.18095	−2.52818	−3.18095	−4.03509	−3.64583
0.500	0.77197	−3.64590	−4.03523	−3.18092	−2.52803	−3.18092	−4.03523	−3.64590
0.550	0.77210	−3.64596	−4.03536	−3.18090	−2.52790	−3.18090	−4.03536	−3.64596
0.600	0.77223	−3.64602	−4.03548	−3.18088	−2.52777	−3.18088	−4.03548	−3.64602
0.650	0.77235	−3.64607	−4.03560	−3.18086	−2.52765	−3.18086	−4.03560	−3.64607
0.700	0.77246	−3.64612	−4.03570	−3.18085	−2.52754	−3.18085	−4.03570	−3.64612
0.750	0.77255	−3.64616	−4.03579	−3.18083	−2.52745	−3.18083	−4.03579	−3.64616
0.800	0.77264	−3.64620	−4.03587	−3.18082	−2.52736	−3.18082	−4.03587	−3.64620
0.850	0.77271	−3.64623	−4.03594	−3.18081	−2.52729	−3.18081	−4.03594	−3.64623
0.900	0.77277	−3.64626	−4.03600	−3.18080	−2.52723	−3.18080	−4.03600	−3.64626
0.950	0.77283	−3.64628	−4.03605	−3.18079	−2.52717	−3.18079	−4.03605	−3.64628
1.000	0.77287	−3.64630	−4.03609	−3.18078	−2.52713	−3.18078	−4.03609	−3.64630

3 Conclusions

New extensions of ordinary fuzzy sets provide a more detailed membership function defining additional new parameters. These parameters can be successfully used in modeling the human-like behaviors of humanoid robots. The fuzzy extensions such as intuitionistic fuzzy sets, Pythagorean fuzzy sets or spherical fuzzy sets can present larger domains for humanoid robots to make judgments on the decision making problems they face. Facial expressions of humanoid robots can be directed by these detailed membership functions much easier than classical control mechanisms. The limitation of this work is its two dimensional approach to face expressions while a real HR requires three dimensional modeling. For further research, we suggest these new extensions to be used in modeling the humanoid robots' other body parts such as hands, arms, and legs.

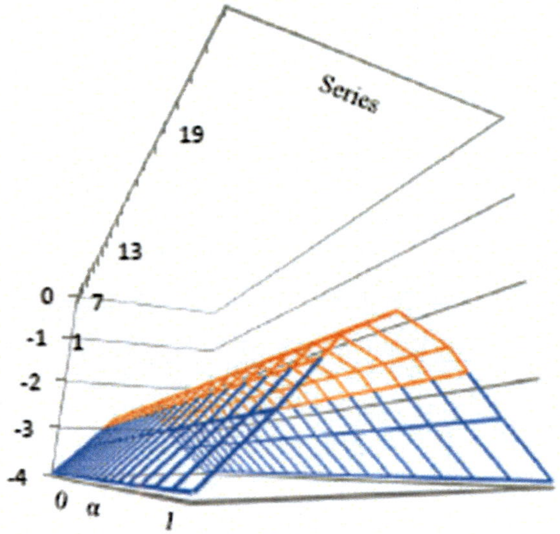

Fig. 1.17 $f_{ss\&vss}$ functions of upper lip for α = [0. 1]

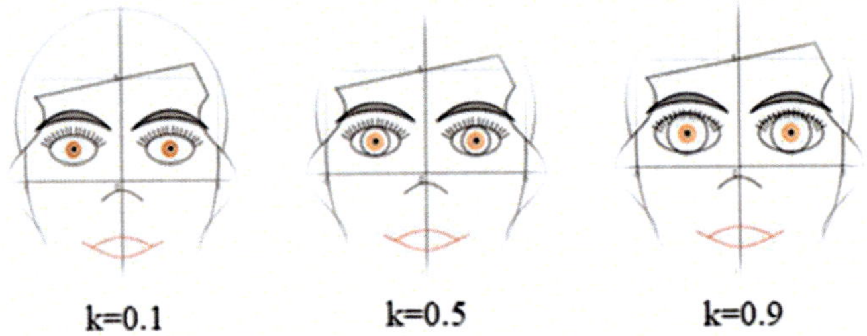

Fig. 18 Eye orb for different k values

References

1. Atanassov, K.T.: Intuitionistic fuzzy sets. Fuzzy Sets Syst. **20**(1), 87–96 (1986)
2. Atanassov, K.T.: Intuitionistic Fuzzy Sets: Theory and Applications. Studies in Fuzziness and Soft Computing. Springer (1999)
3. Belanche, D., Casalo, L.V., Flavian, C.: Customers acceptance of humanoid robots in services: the moderating role of risk aversion. In: Marketing and Smart Technologies, pp. 449–458. Springer (2020)
4. Breazeal, C.: Emotion and sociable humanoid robots. Int. J. Hum. Comput. Stud. **59**(1–2), 119–155 (2003)
5. Carpenter, J., Davis, J.M., Erwin-Stewart, N., Lee, T.R., Bransford, J.D., Vye, N.: Gender representation and humanoid robots designed for domestic use. Int. J. Soc. Robot. **1**(3), 261 (2009)
6. Cuong, B.C.: Picture fuzzy sets. J. Comput. Sci. Cybern. **30**, 409–420 (2014)
7. Ding, J., Shi, J.Y.: Kinect microphone array-based speech and speaker recognition for the exhibition control of humanoid robots. Comput. Electr. Eng. **62**, 719–729 (2017)

8. Fang, W., Chao, F., Lin, C.M., Yang, L., Shang, C., Zhou, C.: An improved fuzzy brain emotional learning model network controller for humanoid robots. Front. Neurorobot. **13**, 2 (2019)
9. Garibaldi, J.M., Ozen, T.: Uncertain fuzzy reasoning: A case study in modelling expert decision making. IEEE Trans. Fuzzy Syst. **15**(1), 16–30 (2007)
10. Grattan-Guinness, I.: Fuzzy membership mapped onto intervals and many-valued quantities. Zeitsehr Math. Logik Und Grundlagen Math **22**, 149–160 (1976)
11. Hawley, L., Suleiman, W.: Control framework for cooperative object transportation by two humanoid robots. Robot. Auton. Syst. **115**, 1–16 (2019)
12. Jahn, K.U.: Intervall-Wertige Mengen. Math. Nach **68**, 115–132 (1975)
13. Kahraman, C., Cebi, S., Cevik Onar, S., Oztaysi, B.: A novel trapezoidal intuitionistic fuzzy information axiom approach: An application to multicriteria landfill site selection. Eng. Appl. Artif. Intell. **67**, 157–172 (2018)
14. Katic, D., Vukobratovic, M.: Survey of intelligent control techniques for humanoid robots. J. Intell. Rob. Syst. **37**(2), 117–141 (2003)
15. Kutlu Gündoğdu, F., Kahraman, C.: Spherical fuzzy sets and spherical fuzzy TOPSIS method. J. Intell. Fuzzy Syst. **36**(1), 337–352 (2019)
16. Otay, I., Oztaysi, B., Cevik Onar, S., Kahraman, C.: Multi-expert performance evaluation of healthcare institutions using an integrated intuitionistic fuzzy AHP&DEA methodology. Knowl.-Based Syst. **133**, 90–106 (2017)
17. Peng, X.: New operations for interval-valued Pythagorean fuzzy Set. Scientia Iranica E **26**(2), 1049–1076 (2019)
18. Peng, X., Yang, Y.: Fundamental properties of interval-valued Pythagorean fuzzy aggregation operators. Int. J. Intell. Syst. **31**(5), 444–487 (2016)
19. Robins, B., Dautenhahn, K., Te Boekhorst, R., Billard, A.: Robotic assistants in therapy and education of children with autism: can a small humanoid robot help encourage social interaction skills? Univ. Access Inf. Soc. **4**(2), 105–120 (2005)
20. Sambuc, R.: Function Φ-Flous These de Doctorat en Medicine, University of Marseille Application a l'aide au Diagnostic en Pathologie Thyroidienne (1975)
21. Smarandache, F.: A Unifying Field in Logics. Neutrosophy: Neutrosophic Probability, Set and Logic. American Research Press, Rehoboth (1999)
22. Tawara, T., Okumura, Y., Shimizu, M., Tomiyama, K.: Design and construction of a series of compact humanoid robots and development of biped walk control strategies. Robot. Auton. Syst. **37**(2–3), 81–100 (2001)
23. Torra, V.: Hesitant fuzzy sets. Int. J. Intell. Syst. **25**(6), 529–539 (2010)
24. Web-1: https://www.dailymail.co.uk/sciencetech/article-2593459/Which-face-happily-disgusted-Scientists-discover-humans-21-different-facial-expressions.html
25. Web-2: https://www.desmos.com/calculator/25oi4snhve
26. Wong, C.C., Cheng, C.T., Huang, K.H., Yang, Y.T.: Fuzzy control of humanoid robot for obstacle avoidance. Int. J. Fuzzy Syst. 10(1) (2008)
27. Xu, Z.: Methods for aggregating interval-valued intuitionistic fuzzy information and their application to decision making. Control Decis. **22**(2), 1179–1187 (2007)
28. Yager, R.R.: On the theory of bags. Int. J. General Syst. **13**(1), 23–37 (1986)
29. Yager, R.R.: Pythagorean fuzzy subsets. In: Joint IFSA World Congress and NAFIPS Annual Meeting (IFSA/NAFIPS). 24–28 June, pp. 57–61. Edmonton, AB, Canada (2013)
30. Zadeh, L.A.: Fuzzy sets. . Inf. Control **8**(3), 338–353 (1965)
31. Zadeh, L.A.: The concept of a linguistic variable and its application to approximate reasoning. Inf. Sci. **8**, 199–249 (1975)

We Need Fuzzy Techniques to Design Successful Human-Like Robots

Vladik Kreinovich, Olga Kosheleva, and Laxman Bokati

Abstract In this chapter, we argue that to make sure that human-like robots exhibit human-like behavior, we need to use fuzzy techniques—and we also provide details of this usage. The chapter is intended both for researchers and practitioners who are very familiar with fuzzy techniques and also for researchers and practitioners who do not know these techniques—but who are interested in designing human-like robots.

Keywords Fuzzy · Human-like robots · Human-like robot behavior

1 What Is the Main Objective of Designing Human-Like Robots

Most successful robots do not look like (and do not behave like) humans. For decades, robots have been successfully used in industrial applications to perform routine and/or dangerous tasks. Let us give a few examples.

- Robots have been used to perform manipulations in dangerous environments—e.g., after a disaster and building collapse, when the situation is too dangerous for human rescuers to enter.
- Robots have been used to help human astronauts to perform important repairs in space.

In all these example of successful applications, the design of a robot was determined by its desired functionality.

V. Kreinovich (✉) · O. Kosheleva · L. Bokati
University of Texas at El Paso, 500 W. University,El Paso, TX 79968, USA
e-mail: vladik@utep.edu

O. Kosheleva
e-mail: olgak@utep.edu

L. Bokati
e-mail: lbokati@miners.utep.edu

C. Kahraman and E. Bolturk (eds.), *Toward Humanoid Robots: The Role of Fuzzy Sets*, Studies in Systems, Decision and Control 344,
https://doi.org/10.1007/978-3-030-67163-1_4

- Sometimes, some part of the robot somewhat resembles the corresponding part of a human being—e.g., there is some similarity between the robotic arm—which is used to grab, move, and manipulate objects—and a human arm.
- However, it is not that the designers wanted to simulate a human arm—if a more efficient but less similar design becomes available, engineers would gladly abandon the current similarity and switch to a new design.

The only reason why this similarity happens in actual robots is that some of the robotic tasks are similar to tasks that have been performed by humans and their ancestors for millions of years, and, as a result of millions of years of improving biological evolution, we humans developed (almost) optimal ways to implement these tasks. Naturally, designers—who also look for optimal ways to perform these tasks—come up with solutions resembling what nature has found by evolutionary trial-and-error.

Not all human features are optimal from this viewpoint: e.g., while we move on two feet, most robots use either wheels or more than two legs.

Since many possible human-like features are usually *not* implemented—they would make a robot less efficient—then why do we need to design human-like robots in the first place?

Why do we need to design human-like robots in the first place? If a robot works on its own—e.g., in assembling a car or in investigating the state of the collapsed building—functionality takes priority, we do not care whether this robot looks like a human being or not. But many robots are intended to collaborate with and communicate with people; for example:

- It is desirable to have robots that help medical doctors and nurses take care of patients.
- It is desirable to have robots that help elderly people and people with disabilities.
- It is desirable to have robots that would help astronauts explore distant planets.

In all these tasks, the success of using these robots comes not only from how well they perform their tasks, but also from how well they collaborate with and communicate with humans—and how convenient it is for humans to collaborate and communicate with these robots.

Humans are very skilled in collaborating and communicating with each other. We are much less skilled in collaborating with and communicating with objects which are different from us, be it animals or machines. So, to make it easier for people to collaborate with and communicate with robots, it is desirable to make these robots look like us—i.e., to design human-like robots.

In other words, there is only one major reason why we want to design human-like robots: to make it more convenient for humans to collaborate with and communicate with robots.

Robots should also behave human-like. For human-robot collaboration and communication to be successful, we need to make sure that robots not only *look* like us, but also that they *behave* like us.

This is a much more difficult task than creating outside resemblance.

What we do in this chapter. In this chapter, we provide arguments that the desire to make robotic behavior human-like naturally leads to the use of fuzzy techniques.

To explain why, let us briefly recall what are fuzzy techniques and how they have been used in control so far.

Comment. Readers who want to know more about fuzzy techniques are referred to [1, 3, 5, 10, 11, 14].

2 Fuzzy Techniques: A Brief Reminder

The origin of fuzzy techniques. Fuzzy techniques originated when Lotfi Zadeh, a renowned specialist in control and a co-author of the most successful textbook on control—so successful that the word z-transform is now used by all control students, practitioners, and researchers, most of whom have no idea that z comes from Zadeh – started thinking about how to make control more efficient.

Traditional control methods—in particular, techniques of optimal control—led to many successful applications, many successful designs of automatic and semi-automatic controllers. Somewhat puzzling was the fact that sometimes, the results of the supposedly optimal controllers were worse than the control provided by expert human controllers.

Of course, in reality, there was no puzzle, no contradiction. There is a straightforward explanation for this phenomenon: models of the corresponding systems used in designing an optimal controllers are approximate—since all models are approximate. And, of course, a control strategy that optimizes an approximate model may not be absolutely optimal for the actual system.

From this viewpoint, it looks like experts have some additional knowledge of the system, knowledge which has not yet been incorporated into the model. So, a natural way to improve the control is to elicit this knowledge from the experts.

Most experts are willing and eager to share their knowledge, but there is often a problem—the experts describe this knowledge not in precise terms (which would be feasible to add to the model), but rather by using imprecise ("fuzzy") words from natural language such as "small".

This fact is natural. For example, many folks know how to drive a car. However, if someone asks you what do you do if you are traveling on a highway at the speed of 100 km/h and a car 10 meters in front of you slows down to 95, a natural answer will be: I break a little bit. Very few people will able to explain with what exactly force they press the break pedal and for exactly how long. And this is typical.

Lotfi Zadeh came up with a technique to describe this knowledge, technique that he called *fuzzy*.

What are fuzzy techniques: a brief reminder. In contrast to precise statement like "speed is greater than 90 km/h" which are always either true or false, a typical natural-language statement like "the car is going fast" is not precise. For some values of the car speed, this is absolutely true, for some it is absolutely false, but for intermediate values, the expert him/herself is not 100% sure whether this statement is true or not.

To describe this uncertainty, a natural idea is to ask the expert to describe his/her degree of confidence—that the speed is fast—by selecting a number on some numerical scale: e.g., on a scale from 0 to 1, in which:

- 0 means absolutely false,
- 1 means absolutely true, and
- intermediate values correspond to uncertainty.

This is something we all do when we answer surveys, this is something students do when they evaluate their instructors, etc.

This way, for different values of the corresponding quantity x, we get a degree $\mu(x)$ to which the corresponding statement (like "x is fast") is true. Of course, there are infinitely many possible values of the quantity, and we can only ask finitely many questions. So, we need to use some interpolation—e.g., linear interpolation—to get the values $\mu(x)$ for all x. The resulting function $\mu(x)$ is called a *membership function*.

This is only the beginning: expert rules usually have several conditions. For example, we can have a rule "if a car in front is close and it slows down a little bit, then press the break pedal a little bit". To describe the consequences of this rule, it is not sufficient to find the degree to which, for the given distance d, the car is close, and the degree to which, for a given change Δv in speed, the car in front slowed down a little bit – we also need to know the degree to which the entire "and"-statement "the car in front is close and it slows down a little bit" is true.

Theoretically, we can get this degree by asking the same expert to mark his/her degree of confidence in this "and"-statement for all possible pairs $(d, \Delta v)$—and if there are three or four conditions, for all possible triples, quadruples, etc. However, in practice, this is not possible. Even if we consider 10 values for each quantity, for quadruples, there are 10^4 combinations of these statements—and it is not possible to ask 10,000 questions just to find the meaning of each of the expert's rules.

Since we cannot elicit the desired degrees of such "and"-statements $A \,\&\, B$ directly from the expert, we need to estimate these degrees based on whatever information we have: namely, on the expert's degrees of confidence a and b in statements A and B. The algorithm that estimates the expert's degree of confidence in the composite statement $A \,\&\, B$ based on the values a and b is called an *"and"-operation* (or, for historical reason, a t-norm). We will denote its result by $f_\&(a, b)$.

Similarly, an algorithm that estimates the expert's degree of confidence in the composite statement $A \vee B$ based on the values a and b is called an *"or"-operation* (or, for historical reason, a t-conorm). We will denote its result by $f_\vee(a, b)$.

Once we selected the "and"- and "or"-operations, we can translate the expert rules into a precise control strategy that transforms the inputs $x_1, \ldots, x_n$ into a control value u. Indeed, suppose that the expert has the following r rules containing imprecise terms A_{ij} and B_i:

- If x_1 is A_{11}, ..., and x_n is A_{1n}, then u is B_1.
- ...

- If x_1 is A_{r1}, ..., and x_n is A_{rn}, then u is B_r.

These rules are usually called *fuzzy rules.*

These rules mean that u is a reasonable control for given values $x_1, \ldots, x_n$ if one of these rules is applicable, i.e.:

- either x_1 is A_{11}, ..., x_n is A_{1n}, and u is B_1;
- ...
- or x_1 is A_{r1}, ..., x_n is A_{rn}, and u is B_r.

We can elicit, from the experts, the membership functions $\mu_{ij}(x_j)$ and $\mu_i(u)$ corresponding to the terms A_{ij} and B_i. If we use an "and"-operation $f_\&(a, b)$ and an "or"-operation $f_\vee(a, b)$, then the degree $\mu(u)$ to which u is reasonable is equal to

$$\mu(u) =$$

$$f_\vee(f_\&(\mu_{11}(x_1), \ldots, \mu_{1n}(x_n), \mu_1(u)), \ldots, f_\&(\mu_{r1}(x_1), \ldots, \mu_{rn}(x_n), \mu_r(u))).$$

For manual control or decision making, this formula allows us to provide recommendations for the decision maker.

In the case of automatic control, we need to select a single value $\bar{u}$. A reasonable idea is to select a value which is close to all possible value of control. The degree of confidence μ can be naturally interpreted as saying that out of N cases, we would make this conclusion $N \cdot \mu$ times.

So, in the simplified case when we have finitely many possible control values $u_1, \ldots, u_m$, then we have the following system of approximate equations:

$$\bar{u} \approx u_1 \quad (N \cdot \mu(u_1) \text{ times});$$

$$\ldots$$

$$\bar{u} \approx u_m \quad (N \cdot \mu(u_m) \text{ times}).$$

In other words, we want to find a single value $\bar{u}$ for which the tuple $(\bar{u}, \ldots, \bar{u})$ is as close as possible to the tuple

$$(\mu(u_1), \mu(u_1), \ldots, \mu(u_m), \mu(u_m)).$$

Minimizing the distance between the two couples is equivalent to minimizing the square of this distance, i.e., the sum

$$\sum_{i=1}^{m} N \cdot \mu(u_i) \cdot (\bar{u} - u_i)^2.$$

Minimizing this quantity is, in its turn, equivalent to minimize the same quantity divided by a constant N, i.e., the sum

$$\sum_{i=1}^{m} \mu(u_i) \cdot (\bar{u} - u_i)^2.$$

In the continuous limit, we get $\int \mu(u) \cdot (\bar{u} - u)^2 \, du$. Differentiating this expression with respect to the unknown $\bar{u}$ and equating the derivative to 0, we conclude that

$$2 \int \mu(u) \cdot (\bar{u} - u) \, du = 2 \left(\bar{u} \cdot \int \mu(u) \, du - \int u \cdot \mu(u) \, du \right) = 0,$$

hence

$$\bar{u} = \frac{\int u \cdot \mu(u) \, du}{\int \mu(u) \, du}.$$

This formula is known as *centroid defuzzification*.

Comments.

- It is known that to describe possible values of each quantity, people select between 5 and 9 different terms like "small" or "big"; this corresponds to the well-known "seven plus minus two rule"; see, e.g., [6, 12]. How many values we use depends on the person—it is 5 for some folks, 9 for others, in between values for the rest.
- What we described is the basic way of translating experts' natural-language statements into an exact control strategy. In practice, there are more sophisticated techniques that lead to a more accurate description of expert rules.
  Let us give just one example. It is reasonable to take into account that an expert often cannot describe his/her degree of confidence in a statement by a single number—no one can distinguish between degree of confidence 0.509 and degree of confidence 0.510. It is more reasonable to ask an expert to provide the whole interval of possible values of this degree—which leads to interval-valued control, etc.
  We can also use machine learning and/or genetic algorithms to come up with appropriate interpolation techniques.
  In general, all these techniques are known as *fuzzy techniques*, not only the above-described basic ones—as long as we understand fuzzy techniques as techniques that allow us to translate natural-language rules into a precise control strategy.

3 Fuzzy Techniques: Successes and Limitations

Successes. Fuzzy techniques have been very successful in many applications, especially in applications where the objective function itself is fuzzy; e.g.:

- how to design a train with the maximally smooth ride,
- how to design a rice cooker that cooks the most tasty rice, etc.

Fuzzy techniques have also been very useful in more traditional engineering applications, where a straightforward formalization of available imprecise expert knowledge leads to reasonably good automatic controls.

Limitations. The main limitation of fuzzy techniques is that the resulting control is usually not optimal.

Indeed, if we, crudely speaking, only consider 5 to 9 different possible values of each input, we then cannot compete with more thorough optimization techniques in which we can find the optimal control value for each of the numerous possible values of the inputs.

How is fuzzy control used. Because of this limitation, fuzzy control is usually used:

- either on the initial stage, where we need to design *a* controller,
- or—on the later stages—in combination with more traditional techniques: the more traditional techniques are used to take care of precise defined objective functions, while fuzzy techniques deal with subjective difficult-to-formalize objective functions like smoothness or taste.

4 So How Should We Make Robotic Behavior More Human-Like: General Idea

How do we describe our own behavior. What does it mean for a behavior to be human-like? It means that robots behave the same way we do.

How do we ourselves behave? A natural way to answer this question is to ask us how we behave. And this is exactly what we do when we try to translate human behavior into a precise control strategy: we ask experts how they do it. As we have mentioned, what experts do as a reply is provide us with imprecise rules—corresponding to combinations of 5-9 words for each input. In other words, when we describe our own behavior—we use what is called *fuzzy rules*.

Maybe this is a biased description and our actual behavior is different? Not really: it turns out that many sub-optimal features of our behavior—features well-studied by psychologists (see, e.g., [2]) can practically all be explained if we take into account the fuzzy-rule model of our behavior; see, e.g., [4].

So let us use fuzzy techniques in designing human-like robots. The fact that our own behavior is well described by fuzzy rules leads to a natural conclusion:

- *if we want robots to show human-like behavior,*
- *we need to use fuzzy control!*

This goes contrary to the usual robotic design, where a lot of optimization is implemented—but this is the whole point: we need understandable, human-like behavior, *not* necessarily optimal control.

- Explainability is what many of us want from AI-based bank systems deciding whether to give us a loan or not.
- Explainability is what we expect from human-like robots.

In situations when we do not need explainability, we do not need to make robots human-like at all, neither in the way they look nor in the way they behave.

Let us illustrate on an example of planning a spaceflight. If we want an automatic space mission to go to several planets, a natural idea is to go to the nearest planet, they move to the next one, etc. This is reasonable and understandable, but the optimal trajectory is often quite different: to save fuel, the optimal way is to follow some weird non-intuitive trajectory that uses gravitational pull from nearby planets to perform needed changes in the spaceship's trajectory without spending too much of the limited amount of fuel. In these terms, what we want from a human-like robot is *not* such an optimal trajectory, we want a not-so-efficient trajectory that we will be able to understand.

Similarly, when students learn new material, they are much more productive when they can understand *why* they are studying, e.g., this particular math class—and not as productive if they are simply told that this leads to the best possible knowledge at the end, without any additional explanations.

Our hope. Our hope is that fuzzy control will be intensively used in designing human-like robots.

5 How to Make Robotic Behavior More Human-Like: Technical Details

Now that we have described the main idea, let us provide some details about how exactly this idea should be implemented.

Human-like robots should be individualized. We want human-like robots' behavior to be explainable—but at the same time, under this constraint, we want this behavior to be as optimal as possible.

Of course, the more values (5 to 9) of each quantity we use, the more parameters we have in our control strategy and thus, the better control we can achieve. From this viewpoint, we need to use as many values as possible.

However, as we have mentioned, understandability depends on the user:

- for some users, understandability is limited to 5 levels,
- others users can gain understandability even if we have 9 levels.

Thus, for each user, we need to select, for each input, the number of levels corresponding to this particular user:

- 5 levels for some users,
- 6 levels for other users,
- ..., all the way to 9 levels for some users.

In other words, human-like robots must be individualized.

- It is possible—but not optimal—to use a 5-level robot with a higher-level (e.g., 9-level) patients.
- However, if we try to do the opposite—e.g., use a 9-level robot with a 5-level patient, this will make the robot's behavior less explainable and thus, defeat the whole purpose of human-like robots.

Which "and"- and "or"-operations should we use when designing human-like robots. When we elicit knowledge from experts, as we have mentioned, the corresponding degrees of certainty come with some uncertainty:

- what can be one time described as 0.6,
- the same expert next time can describe as 0.7,

since in reality, there is a whole interval of possible values describing the expert's opinion.

It is desirable to minimize the effect of this uncertainty on the resulting robot's behavior. Since an important step in fuzzy control is estimating the degrees $f_{\&}(a, b)$ and $f_{\vee}(a, b)$ based on the known values a and b, we therefore need to minimize the changes in the values of these functions based on changes in a and b. How can we do it?

Since the robot needs to perform many tasks, a natural idea is to minimize the overall effect of this uncertainty in all these tasks, i.e., to minimize the average effect of these changes. In mathematical terms, we need to select the most robust operations.

A natural way to describe these changes is to add random deviations Δa and Δb to the original values and consider the mean square value of the difference

$$f_{\&}(a + \Delta a, b + \Delta b) - f_{\&}(a, b),$$

i.e., equivalently, the smallest possible value of the integral

$$\int_0^1 \int_0^1 E\left[(f_{\&}(a + \Delta a, b + \Delta b) - f_{\&}(a, b))^2\right] da\, db,$$

where $E[\cdot]$ stands for expected value – and similarly for the "or"-operation.

It turns out that the smallest possible value of this mean square difference is attained when $f_{\&}(a, b) = a \cdot b$ and $f_{\vee}(a, b) = a + b - a \cdot b$, i.e., when we use what in fuzzy techniques is called algebraic product and algebraic sum; see, e.g., [9, 10].

Alternatively, instead of aiming for overall robustness—and allowing a few actions to be possible less robust—we can aim for making each action robust. In this case, instead of minimizing the *average* difference, we should minimize the *worst-possible* difference

$$\max_{a,b} |f_{\&}(a + \Delta a, b + \Delta b) - f_{\&}(a, b)|.$$

In this case, the optimal "and"- and "or"-operations are $f_{\&}(a, b) = \min(a, b)$ and $f_{\vee}(a, b) = \max(a, b)$ [7, 8, 10].

Comment. These may be not always the operations leading to the optimal control—see, e.g., [13] for optimal operations—but, as we have mentioned several times, the main objective of controlling a human-like robot is *not* to come up with the optimal control, but to come up with an *understandable* control (to be more precise, a control which is optimal among understandable controls).

Control of human-like robots should not always be limited to fuzzy control. In everyday behavior, robots should behave like humans—in particular, they should show human-like behavior—and, as we have argued, for this, we need to use fuzzy rules and fuzzy techniques.

However, in many cases, these robots have an additional functionality that goes beyond user comfort. For example, a robot taking care of elderly patients should bring them medicine and food, should help them feel better—but in the case of an emergency, this robot should be able to react as fast and as efficiently as possible. If a robot needs to move to bring the needed emergency help to a patient at risk of dying, it should not use a suboptimal explainable control strategy, it should get there as fast as possible.

For this purpose, ideal human-like robots should have *two* control strategies:

- a control strategy based on fuzzy rules for everyday situations and
- an optimal (or at least as-optimal-as-possible) strategy for emergency situations.

Acknowledgements This work was supported in part by the National Science Foundation grants 1623190 (A Model of Change for Preparing a New Generation for Professional Practice in Computer Science), and HRD-1834620 and HRD-2034030 (CAHSI Includes).
The authors are greatly thankful to the editors for their encouragement.

References

1. Belohlavek, R., Dauben, J.W., Klir, G.J.: Fuzzy Logic and Mathematics: A Historical Perspective. Oxford University Press, New York (2017)
2. Kahneman, D.: Thinking, Fast and Slow. Farrar, Straus, and Giroux, New York (2013)
3. Klir, G., Yuan, B.: Fuzzy Sets and Fuzzy Logic. Prentice Hall, Upper Saddle River, New Jersey (1995)
4. Lorkowski, J., Kreinovich, V.: Bounded Rationality in Decision Making Under Uncertainty: Towards Optimal Granularity. Springer, Cham, Switzerland (2018)
5. Mendel, J.M.: Uncertain Rule-Based Fuzzy Systems: Introduction and New Directions. Springer, Cham, Switzerland (2017)
6. Miller, G.A.: The magical number seven plus or minus two: some limits on our capacity for processing information. Psychol. Rev. **63**(2), 81–97 (1956)

7. Nguyen, H.T., Kreinovich, V., Lea, B., Tolbert, D.: How to control if even experts are not sure: robust fuzzy control. In: Proceedings of the Second International Workshop on Industrial Applications of Fuzzy Control and Intelligent Systems, College Station, 2–4 Dec 1992, pp. 153–162
8. Nguyen, H.T., Kreinovich, V., Tolbert, D.: On robustness of fuzzy logics. Proceedings of the 1993 IEEE International Conference on Fuzzy Systems FUZZ-IEEE'93, San Francisco, California, March 1993, vol. 1, pp. 543–547
9. Nguyen, H.T., Kreinovich, V., Tolbert, D.: A measure of average sensitivity for fuzzy logics. Int. J. Uncertain. Fuzziness Knowl. Based Syst. **2**(4), 361–375 (1994)
10. Nguyen, H.T., Walker, C.L., Walker, E.A.: A First Course in Fuzzy Logic. Chapman and Hall/CRC, Boca Raton, Florida (2019)
11. Novák, V., Perfilieva, I., Močkoř, J.: Mathematical Principles of Fuzzy Logic. Kluwer, Boston, Dordrecht (1999)
12. Reed, S.K.: Cognition: Theories and Application. Wadsworth Cengage Learning, Belmont, California (2010)
13. Smith, M.H., Kreinovich, V.: Optimal strategy of switching reasoning methods in fuzzy control, Chapter 6. In: Nguyen, H.T., Sugeno, M., Tong, R., Yager, R. (eds.) Theoretical aspects of fuzzy control, pp. 117–146. Wiley, NY (1995)
14. Zadeh, L.A.: Fuzzy sets. Inf. Control **8**, 338–353 (1965)

Humanoid Robots and Metaheuristics

Metaheuristics in Modeling Humanoid Robots: A Literature Review

Cengiz Kahraman and Eda Bolturk

Abstract Metaheuristics are designed to find, generate, or select a heuristic that can provide a sufficiently good solution to a complex optimization problem, especially with incomplete, imperfect, vague and imprecise information. Fuzzy set theory is an excellent tool to capture this kind of information. Metaheuristics can be used as important building blocks in humanoid robots together with fuzzy set theory. In this chapter, we present a literature review on metaheuristics used in modeling robots.

Keywords Metaheuristics · Humanoid robots · Literature review · Particle swarm optimization · Ant colony optimization · Artificial bee colony

1 Introduction

Heuristics are flexible problem-solving techniques that try to produce sufficient solutions for too much complex problems that can not be solved by classical modeling methods. Metaheuristics are iterative master processes that guide and modify the operations of subordinate heuristics to efficiently produce high quality solutions. Meta-heuristic algorithms imitate natural phenomena such as behaviors of ants or bees in ant colony optimization or artificial bee colony optimization, physical annealing in simulated annealing, human memory in a tabu search, evolution in evolutionary algorithms, and wolves hunting animals bigger or faster than themselves in grey wolf optimizer. Metaheuristics can be classified as in Table 1.

Kahraman et al. [14] addressed the Hybrid Flow Shop (HFS) scheduling problems to minimize the makespan value. The proposed algorithm is tested by Carlier and Neron's [3] benchmark problem from the literature. The computational results indicate that the proposed efficient genetic algorithm approach is effective in terms of reduced total completion time or makespan (C_{max}) for HFS problems. Fuzzy sets that better express uncertainties and reduce complexity can be successfully used with metaheuristic algorithms to achieve more concrete and realistic results. Cevik Onar

C. Kahraman (✉) · E. Bolturk
Department of Industrial Enginheering, Istanbul Technical University, 34367 Macka, Istanbul, Turkey
e-mail: kahramanc@itu.edu.tr

C. Kahraman and E. Bolturk (eds.), *Toward Humanoid Robots: The Role of Fuzzy Sets*, Studies in Systems, Decision and Control 344,
https://doi.org/10.1007/978-3-030-67163-1_5

Table 1 Classification of Metaheuristics

Metaheuristics			
Under certainty conditions	Under probabilistic conditions		
Tabu search	Single solution based methods	Population based methods	
	Simulated annealing	Ant colony optimization	Evoluationary algorithm
		Artificial bee colony	Biogeography-based optimization
		Particla swarm optimization	Water flow-like algorithm

et al. [5] presented a literature survey on metaheuristics in production systems. Senvar et al. [30] presented the usage of metaheuristics in engineering through a literature review. Alkan and Kahraman [1] presented a literature review on fuzzy metaheuristics. Kılıç and Kahraman [20] dealt with a permutation flowshop problem with fuzzy processing times. First they explained how to compute start and finish time of each operation on related machines for a given sequence of jobs using fuzzy arithmetic. Next they used a fuzzy ranking method in order to select the best schedule with minimum fuzzy makespan. They proposed an ant colony optimization algorithm for generating and finding near optimal schedules.

Metaheuristics gained significant popularity in many application areas. Robotics is a wide research discipline that embraces artificial intelligence in a complex individually-thinking robot and distributed robots. Metaheuristics made a significant impact on the application areas of collaborating robotics. This offers the possibility of enhanced task performance, high reliability, low unit complexity and decreased cost over traditional robotic systems. Collaborating robots however are more than just networks of independent agents; they are potentially reconfigurable networks of communicating agents capable of coordinated sensing and interaction with the environment [9].

The rest of the chapter is organized as follows. Section 2 presents a brief introduction to most used metaheuristics. Section 3 gives a literature review on the metaheuristics used in modeling robots. Section 4 concludes the chapter.

2 Metaheuristics

In this section, we summarize the most used metaheuristic techniques briefly [15].

Particle swarm optimization method is inspired from the social behaviour of biological swarm systems such as the movement of organisms in a bird flock or fish school. PSO method was developed originally by Kennedy and Eberhart [18].

It is a population-based computational method which achieves optimization by iteratively improving the candidate solutions. Particles, which are candidate solutions, form a population. The particles of the population are located in the search space according to the particle's position and velocity and the current optimum particles. The particles communicate either directly or indirectly with one another for search directions. As a result, the swarm is directed to the best solution. PSO has been used as an effective metaheuristic technique for various problem types of different applications.

Ant Colony Optimization (ACO) is a metaheuristic approach for solving hard combinatorial optimization problems. Ant colony optimization (ACO) algorithm based on the foraging behaviour of ants has been first introduced by Dorigo and Gambardella [7]. The basic idea of ACO is to imitate the cooperative behaviour of ant colonies. When searching for food, ants initially explore the area surrounding their nest in a random manner. As soon as an ant finds a food source, it evaluates it and carries some food back to the nest. During the return trip, the ant deposits a pheromone trail on the ground. The pheromone deposited, the amount of which may depend on the quantity and quality of the food, guides other ants to the food source [31]. Quantity of pheromone on the arc is decreased in time due to evaporating. Each ant decides to a path or way according to the quantity of pheromone which has been leaved by other ants. More pheromone trail consists in short path than long path. Because the ants drop pheromones every time they bring food, shorter paths are more likely to be stronger, hence optimizing the solution. The first ACO algorithm developed was the ant system (AS) [8], and since then several improvement of the AS have been devised [11] [10, 33].

Artificial bee colony (ABC) algorithm was proposed by Karaboğa [17]. Bee Colony Optimization (BCO) algorithm imitates the procedure of collective food search of honeybees. The initial search for the food is executed by a group of bees which inform their remaining bees in the hive about the location quantity and the quality of the food they have explored. A bee carrying out random search is called a scout. Moreover, the scout bees which will lead the followers also try to attract follower bees from the hive by a dance behaviour named as waggle dance. During the waggle dance, the quantity of the food is also given to the followers. Besides, it is known that the quality food is an important factor for strong commitment among the bees. The foraging bees under the lead of the explorer bee leave the hive and collect the food in the explored area. The collected food is returned back to the hive. As the bees collect the food, they return back to the hive to store the food. Then, those bees may choose one of the following options to go through: (1) it may continue to collect food at the same location under its previous leader; (2) it may choose to build up its own team and try to attract followers to join its team; or (3) they may separate from the leader bee and become an uncommitted bee. The exploration of new areas and food collection processes continuously take place.

Genetic Algorithms (GAs) are heuristic procedures that use the principles of evolutionary algorithms. The methodology of Genetic algorithms have been developed by Holland [12] and applied extensively to various types of optimization problems. GAs are inspired from the biological process of natural selection and the survival of the fittest. A pool of solutions defined as a population of chromosomes and a search process is achieved by generations of crossovers. Improvement is aimed to be obtained by selecting the competitive chromosomes that weed out poor solutions and carry over the genetic material to the offspring. At each iteration, the competitive solutions are recombined with other solutions to obtain hopefully better solutions in terms of objective function value or the "fitness" value. The resulting better solutions are then used to replace inferior solutions in the population. For further details on Genetic Algorithms, the interested reader is referred to the study by Reeves [29].

Simulated annealing (SA) methods are the methods proposed for the problem of finding, numerically, a point of the global minimum of a function defined on a subset of a k-dimensional Euclidean space. The motivation of the methods lies in the physical process of annealing, in which a solid is heated to a liquid state and, when cooled sufficiently slowly, takes up the configuration with minimal inner energy. Metropolis et al. [25] described this process mathematically. SA uses this mathematical description for the minimization of other functions than the energy. SA algorithm is a technique to find a good solution of an optimization problem using a random variation of the current solution. A worse variation is accepted as the new solution with a probability that decreases as the computation proceeds. The slower the cooling schedule, or rate of decrease, the more likely the algorithm is to find an optimal or near-optimal solution [37].

The word tabu (or taboo) comes from Tongan, a language of Polynesia, where it was used by the aborigines of Tonga island to indicate things that cannot be touched because they are sacred. According to Webster's Dictionary, the word now also means "a prohibition imposed by social custom as a protective measure" or of something "banned as constituting a risk." Difficulty in optimization problems encountered in practical settings such as telecommunications, logistics, financial planning, transportation and production has motivated in development of optimization techniques. Tabu search (TS) is a higher level heuristic algorithm for solving combinatorial optimization problems. It is an iterative improvement procedure that starts form an initial solution and attempts to determine a better solution.

Differential evolution (DE) is introduced by Storn and Price in [32]. DE is known as population-based optimization algorithm similar to GAs using similar operators; crossover, mutation and selection. According to Karaboğa and Ökem [16], the main difference in constructing better solutions is that genetic algorithms rely on crossover while DE relies on mutation operation. This main operation is based on the differences of randomly sampled pairs of solutions in the population. DE algorithm uses mutation operation as a search mechanism and selection operation to direct the search toward the prospective regions in the search space. In addition to this, the DE algorithm uses a non-uniform crossover which can take child vector parameters from one parent more often than it does from others. By using the components of the existing population members to construct trial vectors, the recombination (crossover) operator efficiently shuffles information about successful combinations, enabling the search for a better solution space. An optimization task consisting of D parameters can be represented by a D-dimensional vector. In DE, a population of NP solution vectors is randomly created at the start. This population is successfully improved by applying mutation, crossover and selection operators.

3 Graphical Analyses

In Scopus database, you find around 40 publications on robot applications using metaheuristics. In the following, we present most recent publications.

Umar et al. [35] presented a novel algorithm for robot kinematic analysis with enhanced parameters. The algorithm is capable of analyzing all the known robot configurations. This was achieved by studying the convergence behavior of PSO under various robot configurations, with a view of determining new PSO parameters for robot analysis and a suitable adaptive technique for parameter identification. Most of the parameters tested stagnated in the vicinity of strong local minimizers. A few parameters escaped stagnation but were incapable of finding the global minimum solution, which is undesirable because accuracy is an important criterion for robot analysis and control. The algorithm was trained to identify stagnating solutions.

Wahab et al. [36] launched a mobile robot to find its path from the starting point to the destination in three different simulated environments using a proposed hybrid metaheuristic algorithm between Particle Swarm Optimization and Fringe Search Algorithm, named PSOFS, and its simultaneous localization and mapping (SLAM) capability. During runtime, the path is optimized by considering the path length. The performance of PSOFS for local path planning is compared against two existing algorithms by evaluating the path smoothness as well as robot safety.

Azar et al. [2] presented a theoretical and practical implementation of a drawing robot using BA to tune the PID controller governing the robotic arm which is a nonlinear system difficult to be controlled using classical control. In order to achieve this aim and meet high performance feedback and robust dynamic stability of the system, the PID controller is designed considering the realistic constraints. In the proposed design, MATLAB was used for trajectory reckoning. Afterwards, the value of coordinate position of the shape to be drawn is translated into a joint angle by applying the inverse kinematics to control the two DC motors through the ATMEGA 2560 microcontroller.

Palmieri et al. [27] focus on the application of different bio-inspired metaheuristics for the coordination of a swarm of mobile robots that have to explore an unknown area in order to rescue and handle cooperatively some distributed targets. This problem is formulated by first defining an optimization model and then considering two sub-problems: exploration and recruiting. Firstly, the environment is incrementally explored by robots using a modified version of ant colony optimization. Then, when a robot detects a target, a recruiting mechanism is carried out to recruit a certain number of robots to deal with the found target together. For this latter purpose, they proposed and compared three approaches based on three different bio-inspired algorithms (Firefly Algorithm, Particle Swarm Optimization, and Artificial Bee Algorithm).

Janardhanan et al. [13] presented a study on robotic assembly line balancing, with the aim of minimizing cycle time by considering sequence-dependent setup times. A mathematical model for the problem is formulated and CPLEX solver is utilized to solve small-sized problems. A recently developed metaheuristic Migrating Birds Optimization (MBO) algorithm and set of metaheuristics were implemented to solve the problem. Three different scenarios were tested (with no setup time, and low and high setup times).

Zaldivar et al. [39] presented an educational platform to assist the learning of the principles of classical and metaheuristic optimization algorithms at undergraduate level, by providing a simple and easy-to-follow teaching setup. The proposed study aims to accompany students through the learning of optimization fundamentals by building hands-on robotic experiments.

Mahanta et al. [21] presented the application of soft computing techniques to obtain the inverse kinematics of Kawasaki RS06L 6-DOF robotic manipulator for a pick and place operation. For validating and checking the efficiency of the proposed approaches, a comprehensive study was conducted among the techniques such as artificial bee colony (ABC), firefly algorithm (FA), invasive weed optimization (IWO), and particle swarm optimization (PSO).

Carvajal et al. [4] described the optimization of a Fuzzy Logic Controller (FLC) for an autonomous mobile robot that needs to follow a desired path. The FLC is for the simulation of its trajectory, the parameters of the membership functions of the FLC had not been previously optimized. They consider the flower pollination algorithm (FPA) as a method for optimizing the FLC. For this reason, they use the FPA to find the best parameters with the objective of minimizing the error between the trajectory of the robot and the reference. A comparative study of results with different metaheuristics is also presented in this work.

Pierezan et al. [28] proposed the static modeling of a humanoid robot and the optimization of its static force capability through a modified self-adaptive differential evolution (MSaDE) approach. Unlike the original SaDE, MSaDE employs a new combination of strategies and an adaptive scaling factor mechanism. In order to verify the effectiveness of the proposed MSaDE, a series of controlled experiments are performed. Moreover, some statistical tests are applied, an analysis of the results is carried out, and a comparative study of the MSaDE performance with other metaheuristics is presented.

Merabti et al. [23] presented a comparison between the use of a simple and multi objective MBPC in robots control for tracking trajectories and obstacle avoidance. Two cases were considered, in the first each robot has its own MPC controller where in the second a single two- objectives MPC controller is used for both robots. In the second case; two approaches were proposed to solve the multi objective optimization problem arising in the MOMPC: the multi objective Particle Swarm Optimization (MOPSO) and weighted sum method.

Nouri et al. [26] proposed hybrid metaheuristics based on clustered holonic multi-agent model for the FJSPT–MR. Firstly, a scheduler agent applies a Neighborhood-based Genetic Algorithm (NGA) for a global exploration of the search space. Secondly, a set of cluster agents uses a tabu search technique to guide the research in promising regions. Computational results are presented using three sets of benchmark literature instances.

Merabti et al. [24] presented a comparison between the uses of three different heuristics, namely particle swarm optimization (PSO), ant colony optimization, and gravitational search algorithm for the solution of the nonlinear MBPC for a mobile robot tracking trajectory with dynamic obstacle avoidance. The computation times obtained show that PSO is a feasible alternative for real-time applications. The MBPC based on the PSO is applied to controlling a LEGO mobile robot with encouraged results.

Kouzehgar et al. [19] considered a hide-and-seek process in a maze imaginable in any form between two robots, one of which is basically trying to deceive the other. The fuzzy definition of behavioral strategies based on past experience for both the deceiver and the competitor robot makes them act like human beings in conflict with each other. Combining the fuzzy reasoning with ant-inspired metaheuristics is another aspect of novelty: Fulfilling the deception, the deceiver is supposed to produce two deceptive signals (track and pheromone) using a fuzzy inference system in order to arrange the environment as desired. After the deceiver decides where to go, the robot under deception is to decide which path to choose based on a utility function calculated within a hierarchical fuzzy inference system whose direct inputs are the value of deception signals and also his behavioral strategy.

Xing [38] gave a vision of what heterogeneous ambient assistive living (AAL) robots is supposed to look like and how a human is to act, navigate and function in it. The author investigated the effect of artificial neural network (ANN) based control techniques for AAL robots. To enhance the accuracy and convergence rate of ANN, a new method of neural network training is explored, i.e., grey wolf optimization (GWO). Moreover, the author provided an overview of applying emerging meta-heuristic approaches to various smart robot control scenarios which, have a great influence on various AAL robot related activities, such as location identification, manipulation, communication, vision, learning, and docking capabilities.

Masehian and Amin-Nesari [22] uses a tabu search approach for sensor-based robot motion planning. Chatterjee et al. [6] showed the possible development of particle swarm optimization (PSO)-based fuzzy-neural networks (FNNs) that can be employed as an important building block in real robot systems, controlled by voice-based commands. The PSO is employed to train the FNNs that can accurately output the crisp control signals for the robot systems, based on fuzzy linguistic spoken language commands, issued by a user. The FNN is also trained to capture the user-spoken directive in the context of the present performance of the robot system. Hidden Markov model (HMM)-based automatic speech recognizers (ASRs) are developed, as part of the entire system, so that the system can identify important user directives from the running utterances.

In Fig. 1, frequencies of metaheuristic papers on robots by years are presented. In 2016, it has the largest frequency. There is an increasing positive trend for this research area starting in 2014.

In Fig. 2, publication frequencies of researchers working on robots using meta-heuristics are illustrated. Dr. Ghedira and Dr Belarbi are the most productive researchers on this area.

Figure 3 shows affiliations of the researchers studying robots using metaheuristics. Tunisia University and Manouba University in Tunisia are the most productive universities on robots using metaheuristics.

Figure 4 presents the countries publishing robot studies using metaheuristics.

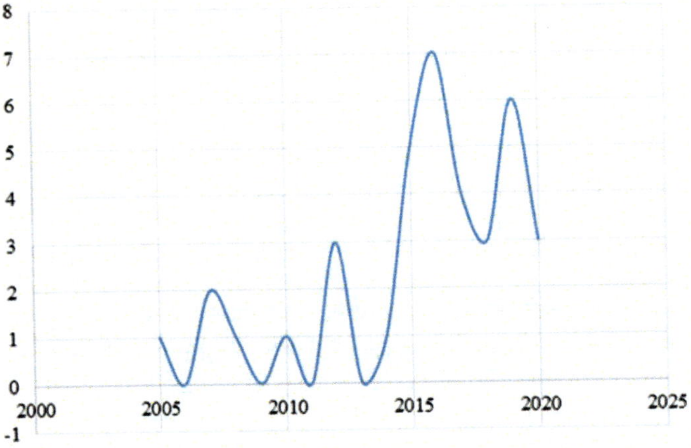

Fig. 1 Frequencies of metaheuristic papers on robots by years

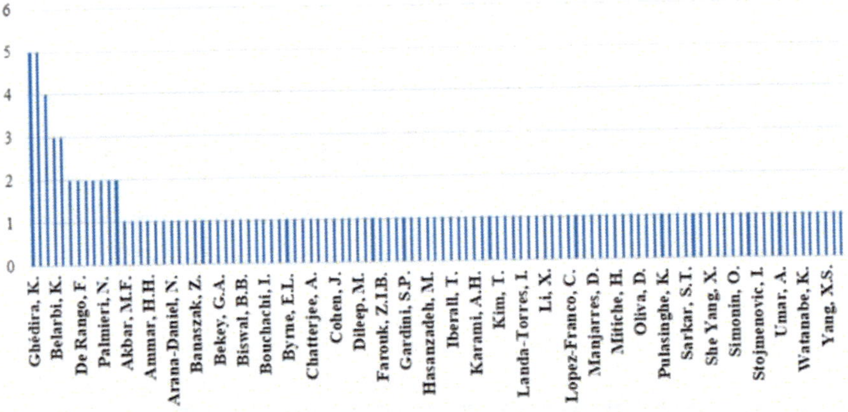

Fig. 2 Publication frequencies of researchers working on robots using metaheuristics

Tunisia, Algeria and United Kingdom are the most productive countries on robots modeled by using metaheuristics.

Figure 5 illustrates the types of publications on robots using metaheuristics. Conference papers and articles have almost equal frequencies.

Figure 6 shows the subject areas of the publications on robots modeled by using metaheuristics. Computer science and engineering are the two top subject areas on this topic.

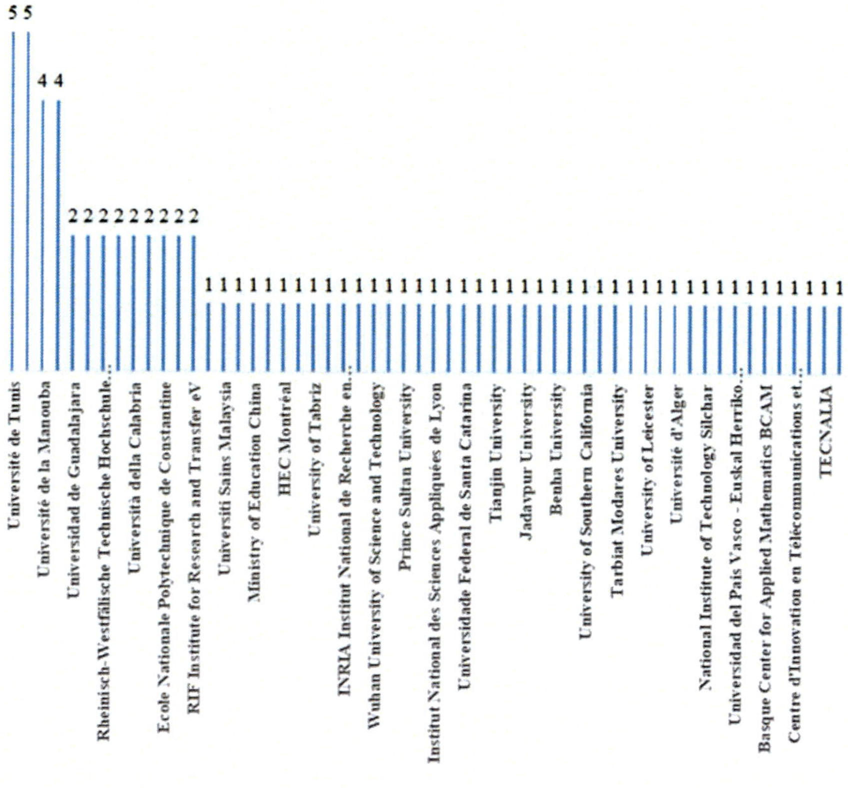

Fig. 3 Affiliations of the researchers studying robots using metaheuristics

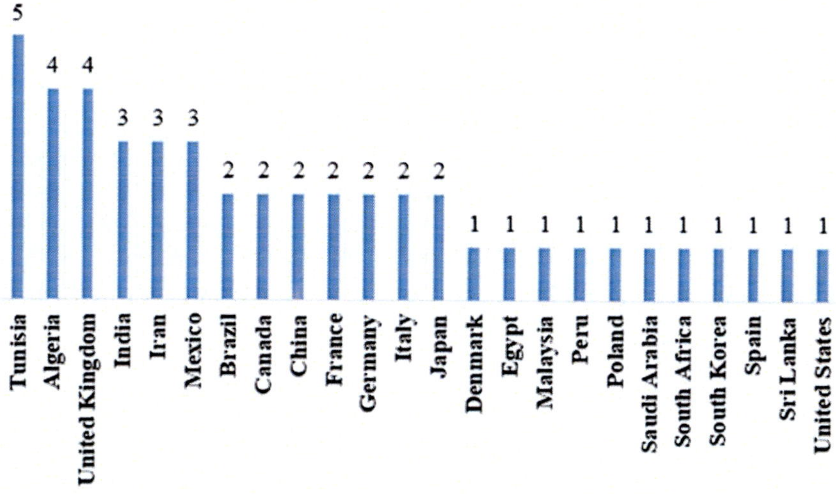

Fig. 4 Countries publishing robot studies using metaheuristics

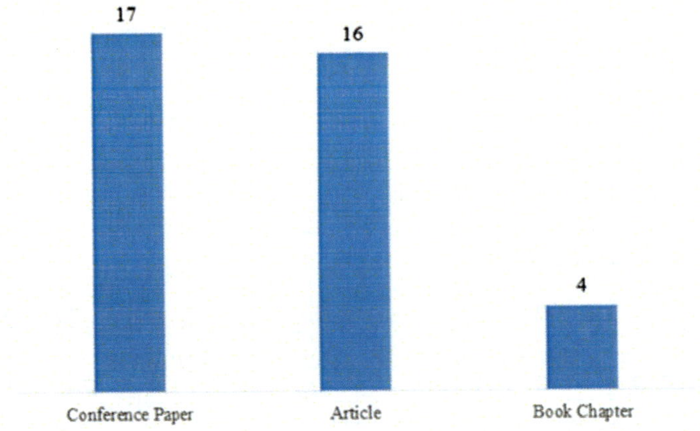

Fig. 5 Types of publications on robots using metaheuristics

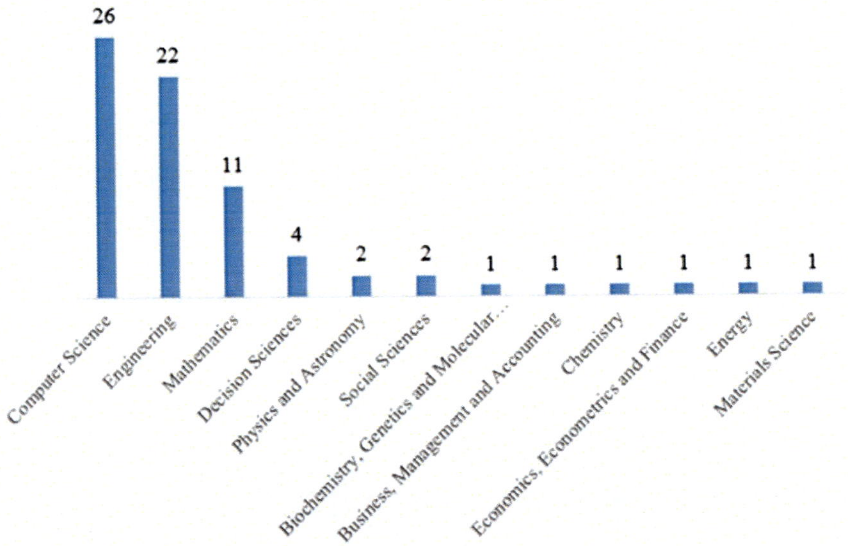

Fig. 6 Subject areas of the publications on robots using metaheuristics

4 Conclusions

A metaheuristic is a higher-level procedure or heuristic designed to find, generate, or select a heuristic that may provide a sufficiently good solution to an optimization problem, especially with incomplete or imperfect information or limited computation capacity. Metaheuristics can be used as important building blocks in humanoid robots. For instance, particle swarm optimization based fuzzy-neural networks that can be employed as an important building block controlled by voice-based commands.

Fuzzy set theory is an important compenent in metaheuristic modeling. Metaheuristics in modeling robots have a significant positive trend after 2013 in the literature. They are usually used for controlling robots actions. Integration of fuzzy logic to metaheuristics will help to obtain a humanoid robots control system more similar to human thoughts and decisions.

References

1. Alkan, N., Kahraman, C.: Fuzzy metaheuristics: a state-of-the-art review. In: Proceedings of International Conference on Intelligent and Fuzzy Systems (INFUS, 2020), pp. 1447–1455. Springer (2020)
2. Azar, A.T., Ammar, H.H., Beb, M.Y., Garces, S.R., Boubakari, A.: Optimal design of PID controller for 2-DOF drawing robot using bat-inspired algorithm. Adv. Intell. Syst. Comput. **1058**, 175–186 (2020)
3. Carlier, J., Néron, E.: An exact method for solving the multi-processor flow-shop. RAIRO-Oper. Res. -Recherche Opérationnelle **34**(1), 1–25 (2016)
4. Carvajal, O.R., Castillo, O., Soria, J.: Optimization of membership function parameters for fuzzy controllers of an autonomous mobile robot using the flower pollination algorithm. J. Autom. Mobile Robot. Intell. Syst. **12**(1), 44–49 (2018)
5. Cevik Onar, S., Öztayşi, B., Kahraman, C., Yanık, S., Şenvar, Ö.: A literature survey on metaheuristics in production systems. In: Metaheuristics for Production Systems, pp. 1–16. Springer (2016)
6. Chatterjee, A., Pulasinghe, K., Watanabe, K., Izumi, K.: A particle-swarm-optimized fuzzy-neural network for voice-controlled robot systems. IEEE Trans. Industr. Electron. **52**(6), 1478–1489 (2005)
7. Dorigo, M., Gambardella, L.M.: Ant colony system: a cooperative learning approach to the traveling salesman problem. IEEE Trans. Evol. Comput. **1**, 53–66 (1997)
8. Dorigo, M.: Optimization, learning and natural algorithms. Unpublished Doctoral Dissertation. University of Politecnico di Milano, Italy (1992)
9. Fonga, S., Debb, S., Chaudhary, A.: A review of metaheuristics in robotics. Comput. Electr. Eng. **43**, 278–291 (2015)
10. Gambardella, L.M., Dorigo, M.: Ant-Q: a reinforcement learning approach to the travelling salesman problem. In: Proceedings of the Twelfth International Conference on Machine Learning. California, USA (1995)
11. Gambardella, L.M., Dorigo, M.: Solving symmetric and asymmetric TSPs by ant colonies: In Proceedings of the IEEE conference on Evolutionary Computation. ICEC96, Nagoya, Japan. 622–627 (1996)
12. Holland, J.H. (ed.): Adaptation in Natural and Artificial Systems: An Introductory Analysis with Applications to Biology, Control, and Artificial Intelligence. University of Michigan Press, Ann Arbor, MI (1975)
13. Janardhanan, M.N., Li, Z., Bocewicz, G., Banaszak, Z., Nielsen, P.: Metaheuristic algorithms for balancing robotic assembly lines with sequence-dependent robot setup times. Appl. Math. Model. **65**, 256–270 (2019)
14. Kahraman, C., Engin, O., Kaya, I., Yilmaz, M.K.: An application of effective genetic algorithms for solving hybrid flow shop scheduling problems. Int. J. Comput. Intell. Syst. **1**(2), 134–147 (2008)
15. Kahraman, C., Yanık, S., Intelligent decision making techniques in quality management: a literature review. In: Kahramanc, Yanık, S. (eds.) Intelligent Decision Making in Quality Management Theory and Applications. Springer, Switzerland (2016)

16. Karaboğa, D., Ökdem, S.: A simple and global optimization algorithm for engineering problems: differential evolution algorithm. Turk. J. Electron. Eng. **12**(1) (2004)
17. Karaboğa, D.: An idea based on honeybee swarm for numerical optimization. Technical Report TR06, Erciyes University (2005)
18. Kennedy, J., Eberhart, R.C.: Particle swarm optimization. In: Proceedings of the IEEE international conference on neural networks IV, pp. 1942–1948 (1995)
19. Kouzehgar, M., Badamchizadeh, M., Feizi-Derakhshi, M.-R.: Ant-inspired fuzzily deceptive robots. IEEE Trans. Fuzzy Syst. Vol. **24**(2), 374–387 (2016)
20. Kılıç, S., Kahraman, C.: Scheduling a flowshop problem with fuzzy processing times using ant colony optimization. In: Applied Artificial Intelligence, Proceedings of the 7th International FLINS Conference, Genova, Italy, pp. 449–456 (2006)
21. Mahanta, G.B., Deepak, B.B.V.L., Dileep, M., Biswal, B.B., Pattanayak, S.K.: Prediction of inverse kinematics for a 6-DoF industrial robot arm using soft computing techniques. Adv. Intell. Syst. Comput. **817**, 519–530 (2019)
22. Masehian, E., Amin-Naseri, M.R.: Sensor-based robot motion planning—A Tabu search approach. IEEE Robot. Autom. Mag. **15**(2), 48–57 (2008)
23. Merabti, H., Belarbi, K., Bouchachi, I.: Single and multi objective predictive control of mobile robots. Lect. Notes Electr. Eng. **411**, 70–79 (2017)
24. Merabti, H., Belarbi, K., Bouchemal, B.: Nonlinear predictive control of a mobile robot: a solution using metaheuristcs. J. Chin. Inst. Eng. Trans. Chin. Inst. Eng. Ser. A **39**(3), 282–290 (2016)
25. Metropolis, N., Rosenbluth, A., Rosenbluth, M., Teller, A., Teller, E.: Equation of state calculations by fast computing machines. J. Chem. Phys. **21**(6), 1087–1092 (1953)
26. Nouri, H.E., Belkahla Driss, O., Ghédira, K.: Simultaneous scheduling of machines and transport robots in flexible job shop environment using hybrid metaheuristics based on clustered holonic multiagent model. Comput. Ind. Eng. **102**, 488–501 (2016)
27. Palmieri, N., Yang, X.-S., De Rango, F., Marano, S.: Comparison of bio-inspired algorithms applied to the coordination of mobile robots considering the energy consumption. Neural Comput. Appl. **31**(1), 263–286 (2019)
28. Pierezan, J., Freire, R.Z., Weihmann, L., Reynoso-Meza, G., dos Santos, C.L.: Static force capability optimization of humanoids robots based on modified self-adaptive differential evolution. Comput. Oper. Res. **84**, 205–215 (2017)
29. Reeves, C.R.: Genetic alorithms. In: Glover, F., Kochenberge, G.A. (eds.) Handbook of Metaheuristics, pp. 55–82. Kluwer Academic, Boston (2003)
30. Senvar, O., Turanoglu, E., Kahraman, C.: Usage of metaheuristics in engineering: a literature review. In: Meta-heuristics Optimization Algorithms in Engineering, Business, Economics, and Finance, 484–528. IGI Global (2013)
31. Socha, K., Dorigo, M.: Ant colony optimization for continuous domains. Eur. J. Oper. Res. **185**, 1155–1173 (2008)
32. Storn, R., Price, K.: Minimizing the real functions of the ICEC'96 contest by differential evolution. In: IEEE International Conference on Evolutionary Computation (ICEC'96), pp. 842–844, May 1996 (1996)
33. Stützle, T., Hoos, H.: MAX-MIN ant system. Future Gener. Comput. Syst. **16**(8), 889–904 (2000)
34. Su, C.T., Chiang, T.L.: Optimizing the IC wire bonding process using a neural networks/genetic algorithms approach. J. Intell. Manuf. **14**(2), 229–238 (2003)
35. Umar, A., Shi, Z., Khlil, A., Farouk, Z.I.B.: Developing a new robust swarm-based algorithm for robot analysis. Mathematics **8**(2), Art. no. 158 (2020)
36. Wahab, M.N.A., Lee, C.M., Akbar, M.F., Hassan, F.H.: Path planning for mobile robot navigation in unknown indoor environments using hybrid PSOFS algorithm. IEEE Access **8**(art. no. 9186019), 161805–161815 (2020)
37. Xinchao, Z.: Simulated annealing algorithm with adaptive neighborhood. Appl. Soft Comput. **11**, 1827–1836 (2011)

38. Xing, B.: The spread of innovatory nature originated metaheuristics in robot swarm control for smart living environments. Stud. Syst. Decis. Control **40**, 39–70 (2016)
39. Zaldivar, D., Cuevas, E., Maciel, O., Valdivia, A., Chavolla, E., Oliva, D.: Learning classical and metaheuristic optimization techniques by using an educational platform based on LEGO robots. Int. J. Electr. Eng. Educ. (2019) (Article in Press)

On the Use of Meta-Heuristic Algorithms for Automated Test Suite Generation in Software Testing

Manju Khari, Anunay Sinha, Enrique Herrerra-Viedma, and Rubén González Crespo

Abstract There exists a dire need to automate the process of test suite generation to get the most optimal results as testing accounts for more than 40% of total cost. A solution consists of using meta-heuristic algorithms which iteratively improve the test data to reach the most optimized test suites. The goal of the study is to find the best suited algorithm to narrow down future research in the field of test automation and also provide issues on the design of new proposals. We focus on the performance evaluation of different major Meta-Heuristic Algorithms namely: Hill Climbing Algorithm (HCA), Particle Swarm Optimization (PSO), Firefly Algorithm (FA), Cuckoo Search Algorithm (CA), Bat Algorithm (BA) and Artificial Bee Colony Algorithm (ABC). Each algorithm is implemented to automatically generate test suites based on the program under test. Then, we develop a performance evaluation of each algorithm for five programs written in Java. The algorithms are compared using several process metrics (average time, best time, worst time) and also product metrics (path coverage & objective function values of the generated test suites). Results indicate ABC as the best suited algorithm as it gave the most optimal Test Suites in reasonable time. BA is the fastest one but produced less optimal results. FA is the slowest algorithm while CA, PSO and HCA perform in between. Some issues and strategies to create hybrid algorithms are discusses and pointed out.

M. Khari · A. Sinha
Ambedkar Institute of Advanced Communication Technologies and Research, Delhi, India
e-mail: manjukhari@yahoo.co.in

A. Sinha
e-mail: anunay.sinha212@gmail.com

E. Herrerra-Viedma
Universidad de Granada, Granada, Spain
e-mail: viedma@decsai.ugr.es

R. G. Crespo (✉)
Universidad Internacional de La Rioja, Logroño, Spain
e-mail: ruben.gonzalez@unir.net

C. Kahraman and E. Bolturk (eds.), *Toward Humanoid Robots: The Role of Fuzzy Sets*, Studies in Systems, Decision and Control 344,
https://doi.org/10.1007/978-3-030-67163-1_6

1 Introduction

Software Testing is highly labor intensive and accounts for about 40–50% of the total cost of Software Development. Hence there arises a dire need for automating the process of testing to reduce costs of the project. Software Test Automation also decreases the overall effort in the project while simultaneously increasing the quality of the software [1].

In the realm of software testing, the major contributing factor is the generation of test suites also known as Test Suite Generation (TSG). The study initially focuses on automating the process of test suite generation and optimizing it using six Meta-Heuristic Algorithms namely: Hill Climbing Algorithm (HCA), Bat Algorithm (BA), Cuckoo Search Algorithm (CA), Firefly Algorithm (FA), Particle Swarm Optimization (PSO) and Artificial Bee Colony Algorithm (ABC). Further, the study details a comparative analysis of these algorithms to figure out the best algorithm with respect to automated generation of test suites and their optimization.

The automation process is done by associating the population values in each algorithm as the basis for the Test Suites. Each *i*th member of the population is treated as a potential test suite or more precisely as a collection of test cases. While each collection of test cases is associated with the individual values of the *i*th member. The length of the test cases i.e. the number of test data elements in a single test case is based on the number of inputs in the software program and the number of test cases in each potential test suit (*i*th member of the population) was based on the Cyclomatic complexity of the program. The optimizing function (objective function) used for the approach is a linear scalerization of path coverage and branch coverage. This is done in a fashion similar to other prior work [2]. The population of potential test suites is evaluated iteratively to improve the test suites based on the algorithms evolutionary and meta-heuristic approaches.

The comparative analysis is done using the process and product metrics while generating the test suites for five programs written in java. The process involved in each algorithm and the characteristics of the test suite generated are judged to provide comparative analysis. The process metrics are used to evaluate the performance of the processes involved in each algorithm. This is done by executing each algorithm hundred times to generate test suites for the 5 problems, and the average, best and worst times are noted. For product metrics, the product here being the test suite generated; their path coverage and the value of the objective function are used. A low objective function value along with high path coverage implies a better test case. A brief description of each algorithm is described below.

HCA is a meta-heuristic algorithm that is used for solving np-hard problems. Its most effective implementation is with "the property that the state description itself contains all the information needed for a solution". It is a very efficient algorithm as it doesn't maintain a search tree and only looks over the present state and near future states [3].

CA employs a meta-heuristic approach based on the breeding behavior of cuckoo birds which can primarily be utilized to find the global optimal solution of any mathematical function or problem. It was developed in 2009 by Xin-She Yang and Suash Deb [4].

FA was developed by yang in 2009. It is inspired by the biological nature of fireflies. The FA belongs to class of swarm intelligence algorithms which forms a part of artificial intelligence discipline. FA is used to find solutions for searching and optimization problems. A FA is a search procedure with a goal to find a solution in a multidimensional space [5].

PSO was developed by Eberhart and Kennedy in 1995. It is a meta-heuristic algorithm that imitates the nature of migrating animals such as a group of birds trying to find their destination. This algorithm itself is simple yet very powerful, it has been applied various fields and many researchers have created variations of it [6].

ABC algorithm was created by Dervis Karaboga in 2007 as a self-organizing system based on swarm intelligence and division of labor. It is a Bio-Inspired meta-heuristic algorithm that derives its metaphor from the way a hive of bees work together to find relevant sources of nectar in the nearby environment with the help of segregated tasks for individuals conforming to a division of labor [7].

BA is a very recent meta-heuristic nature inspired algorithm that relies on the echolocation behavior of bats to find obstacles and catch their prey. It works in a fashion similar to PSO but is more efficient as it uses features such as automatic zooming. It was created by Xin-She Yang and published in 2010 [8].

This study will help in narrowing down future research in Test Automation and TSG to only the best performing algorithm and then tweaking that algorithm to create hybrid algorithms that can outperform other algorithms when implemented over various frameworks.

Section 2 describes the relevant past research in the field and gives the overall review of the past literature. Section 3 explains each of the six algorithms in detail, gives their limitations, as well as flow chart and pseudo code. Section 4 describes the testing framework for this study as well as the entire methodology used for comparing the algorithms. Section 5 details the experimental results including average time, best time, worst time, path coverage and objective function value. Section 6 discusses the theoretical and practical contributions for the study as well as the implications that the results indicated. Section 7 gives the conclusion and future works.

2 Literature Review

This section describes the previous work undertaken in the field of test automation and meta-heuristic algorithms.

Karaboga et al. [7] proposed a novel approach for solving np-hard optimization problems with a meta-heuristic approach inspired the behavior of bee colonies. The authors suggested an artificial bee colony where the optimization problem is divided into three phases called the employee bee phase, onlooker bee phase and scout phase.

Ngyuen [9] presented a technique using genetic algorithm and also provided a process of generating test data to perform unit tests over classes for general practice. Test data was represented using chromosomes. On the basis of which - objects were created, methods & inputs were selected. The main objective of proposed algorithm was to mutate them with maximizing a given coverage measure.

Ngo and Tan [10] gave a heuristic base technique to generate test data dynamically by performing unfeasible path discovery. The methodology focuses on the properties of the infeasible path. Experiments indicate that the proposed technique efficiently detected the infeasible paths with a great degree of precision.

Kanmani and Maragathavalli [11] explained evolutionary testing indicates the implementation of evolutionary searching methods that automated test case generation. Meta-heuristic methods lie on few techniques which are- evolutionary algorithms, Simulated Annealing and Tabu Search which are generally used for generating test data. Finally, the strength of a system based on GA was evaluated against a traditional randomized testing schema using parameters like branch, path coverage, accuracy, and fitness. Thus, the research work experimentally proved that the best methodology over certain real time programs was based on genetic algorithm.

Yang et al. [12] proposed a novel approach to solve optimizing problems in a much faster method using automatic zooming and based on the behavior of bats in their natural environment. It was swarm based intelligent algorithm which produced optimized results in a fashion similar to PSO but provided much faster approach with far better results.

Mansouri et al. [13] solved hard fix point problems using a combination of artificial bee colony algorithm and bisection approach. The authors demonstrated that the artificial bee colony algorithm along with bisection method performed effectively in solving hard fix point problems.

Sharma et al. [14] in their study presented a survey on various genetic algorithm methodologies and techniques while implementing them for the test case generation and test data optimization. The studied the functional issues faced while programming test case generation using these genetic algorithm techniques.

Varshney et al. [15] suggested implementation of dominance concepts, elitism and branch distance for generating test data. The authors compared the methodology proposed for data flow testing with random testing and previous works in the field. The findings of the work suggested that the proposed methodology outperformed previous works as well as random testing.

Panichella et al. [16] proposed a new approach to handle automated test case generation using multi objective optimizations and dynamic selection of targets. The authors presented a new algorithm called DynaMOSA which stands for Dynamic Many-Objective Sorting Algorithm. The goal was to provide a new methodology that allows multi objective optimizations which were lacking in previous studies using statement coverage, branch coverage and strong mutation coverage. The authors carried out empirical study over 346 java classes and the results indicated that DynaMOSA outperformed WSA for branch coverage, mutation coverage and statement coverage in 28%, 27% and 51% of the classes respectively.

Kumar et al. (2017) [17] proposed a novel methodology for automated test data generation by combining two major meta-heuristic algorithms namely, PSO and GA. The authors proposed a new hybrid adaptive algorithm called PSO-GA algorithm to produce data flow testing based test cases. The authors also implemented a new fitness function which was more efficient and relied on branch weight, branch distance and dominance relations for guiding the search detections.

Mann et al. [18] presented approaches for automated test case generation using ABC, PSO and GA which was path specific and compared their performances. The authors also presented a new methodology for test case prioritizing method using PSO. The results indicated that ABC outperformed PSO and GA in the path specific automated test case generation implementations. While, the results for test case prioritizing methodology using PSO indicated a better applicability for large and small test suites as opposed to conventional prioritizing methods.

Khari et al. [19] developed an automated testing tool that focused on two major factors namely test suite generation and test suite optimization. The control flow graphs were shown for each software under test. The five main methods focused by this study were random testing, robust worst case testing, worst case testing, robustness testing and boundary value testing. The techniques proposed were able to provide minimal test cases with maximum path coverage when compared with conventional methods. Lastly the optimized test suites that were generated were used to automatically detect faults in the program.

Malhotra and Khari [20] proposed and implemented a novel approach for test suite optimization using Mutated Artificial Bee Colony algorithm. The proposed approach combined aspects of Genetic Algorithm namely mutation to the three phases of artificial bee colony algorithm. The approach was compared on the basis of run time and number of iterations and was found to be suitable for selecting minimal number of test cases for a test suite.

Khari and Kumar [21] reviewed and presented an extensive survey of Search Based Software Testing (SBST). The study focused on meta-heuristic and heuristic approaches to solve the problem of testing. Researchers reviewed earlier studies in the field of SBST from the year 1996 to 2016 along with their various applications and implementations. The study was comprehensive and explored a multitude of search based techniques.

3 Bio Inspired Computational Algorithms

3.1 Hill Climbing

Hill climbing is a methodology for optimization of problems that have been labelled complex or hard in computational terms. It finds its most potent applications in resolving problems with "the property that the state description itself contains all the information needed for a solution" [23].

HCA is a technique for optimization of mathematics-based problems which belong to the category of search classified as "local search". It is an algorithm which is iterative in nature, providing a solution to the specified problem by initially supplying with a capricious solution and then endeavours to find an improved solution by incrementally modifying the composite elements of the solution, focussing on a singular element at a time. In scenarios where the modification is producing a superior solution, the iterative improvement effected on the problem is declared as the new solution. This step is repeated until improvements in the solution come to a standstill. Under the category of local search in Computer Science [22], HCA is a technique for analytical optimization. It starts by proposing a capricious solution to a problem at hand in its inception phase and then iteratively finds an improved version of the solution by making changes to a single element present in the solution, incrementally. This iteration continues till a better solution is being produced and no more development or advancement is observed.

HCA proves to be the best solution in cases where we require local optima, i.e. a solution which has no scope for further improvement, acknowledging the structure of the neighbours. However, this does not ensure of a global optimum (the most effective solution) to the problem. Problems which are convex in nature, like the traditional binary search, linear programming simplex procedures, they serve the best and the most optimal input for a solution to the HCA.

HCA works on a function $f(x)$, which is called as the target function. Here, x denotes a vector. This vector can be of both, discrete and continuous nature values. HCA will start by considering this target function $f(x)$ and working in repetitive steps. At every step, the algorithm will alter one element in the vector x. It will then compute and analyse the development in the target function $f(x)$. If this development is positive, the alteration in the vector is consolidated. This procedure continues till the changes in the vector no more bring any advancement in $f(x)$. At the end of every iteration, the function $f(x)$ is locally optimized. Each value taken for x is anticipated as a separate vertex of a graph, in discrete spaces of vectors. HCA then follows the designed graph from one vertex to another vertex, modifying the value of $f(x)$ always locally (either decreasing or increasing), till a local extrema x_m is attained (local minima or local maxima).

With the development of the algorithm, there were variations proposed to it. The traditional HCA selects the node which is closer in the vector. Then there came a variation, before selecting the neighbouring nodes, there is a comparison done among those nodes, to choose that very node which stands closer to the optimal solution. This variation is called as the Steepest Ascent HCA (SAHCA). This way follows the best-fit procedure, which analyses all possibilities from the current node. However, both these techniques proved to failure in the case when there wasn't any closer node found. This may happen when in the search space there are many local maxima, but none of them is a solution.

Another version of HCA is Stochastic HCA (SHCA) [23], proposed in 2009, and chooses a neighbouring node randomly, unlike SAHCA. It does not do any examination of the nodes. Its decision is governed by the degree of development and growth of the neighbouring node, and then it determines whether it has to proceed with that node or not. Coordinate Descent, an improvement to the HCA, searches in a unidirectional fashion from the current node in the iteration. Variation to this works on a distinguished selection of the direction of the coordinates at every iteration.

Lastly, Random-Restart HCA which builds itself on the conventional hill climbing. It is sometimes also referred to as the Shotgun HCA. It is a meta-algorithm, which works on a new randomly selected inceptive situation x_0. The most optimal x_n is stored. If at any point in the algorithm, a new iteration generates an improved output than x_n, it overrides the previous state which was stored. For example, HCA finds is a very popular application in the Travelling Salesman Problem (TSP). Finding an initial solution which visits every city will be an easy way, however, it will be not as refined in comparison to the optimal solution. In its inception phase, the algorithm begins with such a solution, making gradual little improvements in the solution, like a swap in the turn of visiting two cities. Eventually, this leaves us with a much shorter route.

HCA is best suited for situations where we aim at finding a local optimum. Local Optima are those solutions which can't be further revised by acknowledging neighbouring configuration. However, it doesn't guarantee on finding the finest solution possible, a condition called as the global optimum, from all the possible solutions, known as the search space [24]. Overcoming the specification that singularly local optima are ensured can be achieved by recapitulating (repeating local search), or even more intrinsic arrangement as per the iterations, for instance, step-wise local searches, utilising memory, like TS and reactive search optimization, or non-memory oriented stochastic mutations, like SA.

Relatively, the simplicity in the structure of the algorithm determines it as a familiar prime selection from the other optimizing algorithms. It is used very much in the field of AI, in order to reach a target or desired state, to begin with from a starting node. The process of selection of the next node and beginning node can be diversified to present a group of somewhat related algorithms. Although, improved algorithms, for instance, SA or tabu searching can deliver superior outcomes, while in few cases HCA executes just as good. HCA seldom produces an improved output in comparison with other algorithms, talking in terms of the times when there is a limited span of time in hand, like with real-time systems. It can be called as an 'anytime algorithm', i.e. it is capable of returning a viable solution even if gets interrupted amid its execution flow of HCA is depicted in the figure.

Steps for hill climbing:

```
1. Begin;
2.        Initializing population at t=0;
3.        Select a current node Vc;
4.        Evaluate its value;
5.                WHILE (t<MAX)
6.                Select the string Vnfrom the neighborhood of Vc;
7.                Select Vnwith probability p = 1/1+e^eval(vc) – eval(vn)/t;
8.                t=t+1;
9. END WHILE;
```

Figure 1 explains the HCA in a flow chart depicting all the steps that the algorithm has to follow. The algorithm starts with initialization of the population followed by the selection of node and evaluating that node. After which the loop starts that evaluates the probability of neighboring nodes and selecting the right neighboring node until the termination criteria is met and the loop ends.

Fig. 1 Flow of hill climbing

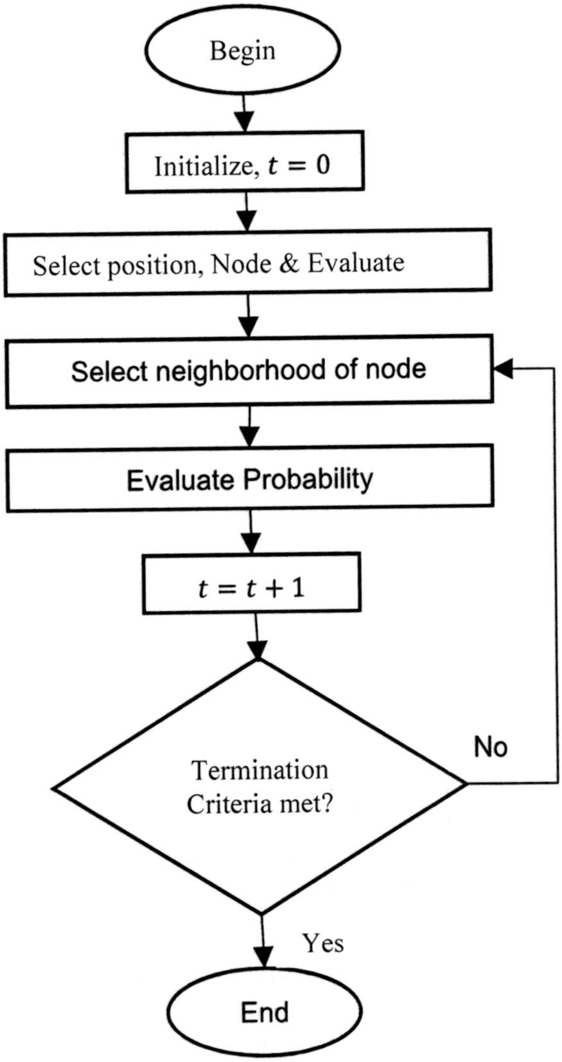

3.2 Bat Algorithm

Bat algorithm (BA) is an algorithm that tries to imitate the way bats find their prey using echolocation. It was created by Yang in 2010.

The algorithm follows the behavior of bats trying to find their prey using echolocation which acts like sonar. Bats send sound waves which get reflected off the prey and bats then estimate the distance and speed at which the prey is moving. This concept is further expanded to fit into optimization problems in a way similar to PSO except here the parameters are not fixed and in fact keep on changing. The parameters involved are r_i the pulse rate which lies between [0, 1] and A_i which is the loudness of the wave sent by the bat to catch the prey and has max and min constraints on it.

Key Features

- Frequency Tuning: In BA frequencies are slowly varied like in other swarm based algorithms. Hence BA has same advantages of other similar swarm based methods.
- Automatic Zooming: This involves zooming into a particular area where a promising solution is present. Hence it automatically switches a explorative mode to a more local mode and hence has a really fast convergence rate.
- Parameter Control: Unlike other NIAs, here the parameters are not fixed i.e. they are varied as the iteration moves forward so we can easily switch from a more global approach to a more selective local approach.

The following formulae govern the frequency which the bat emits and the velocity & position of the bat in a fashion similar to PSO.

$$f_i = f_{\min} + \left(f\max - f_{\min}\right) \times k \tag{1}$$

$$v_i(t+1) = v_i(t) + \left(x_i(t) - x_g(t)\right) \times f_i \tag{2}$$

$$x_i(t+1) = x_i(t) + v_i(t) \tag{3}$$

where, f_i is the frequency being emitted by the ith particle, $f_{\max}$ and $f_{\min}$ are the limits, k is a random vector between [0,1], v_i is velocity of the ith particle, x_i is position and x_g is the globally best particle, while the value for Amplitude was taken constant [8].

Steps for bat algorithm:

```
Begin;
    Initialize bat population x_i and v_i;
    Define pulse frequency f_i at x_i;
    Initialize pulse rates r_i and the loudness A_i;
    WHILE (t<Maximum number of Iteration)
        Generate new solutions by adjusting frequency and updating velocity & location;
        IF (rand>r_i)
            Select best solution;
            Generate Local Solution around Best Solution;
        END IF
        Generate new solution by flying randomly;
        IF (rand<A_i & f(x_i)<f(x_g))
            Accept new solutions;
            Increase r_i & Reduce A_i;
        END IF
        Rank bats and find current best x_g;
        t=t+1;
    END WHILE;
END
```

Figure 2 describes the flowchart for BA in accordance with the standard algorithm. The bats are initiated randomly and their pulse rate, loudness (amplitude) are initialized. The frequencies, velocities and positions are then updated using the Eqs. 1–3. After this a random pulse rate is compared with the pulse rate and the algorithm proceeds by generating local solutions around the best solutions. The bats are then ranked and the best is chosen as the optimal solution.

3.3 Cuckoo Search Algorithm

Cuckoo Search Algorithm (CA) employs a methodology that qualifies as a meta-heuristic technique based on the unique behaviour demonstrated by cuckoo birds for reproduction and breeding. This algorithm finds its most fundamental applications in finding the most optimal solution to any mathematical problem with an objective function, in the global domain. It was proposed by Xin-She Yang and Suash Deb in 2009 [4]. Cuckoo birds follow an aggressive and unique strategy for reproduction wherein they strategically place their eggs in nests of other host birds. In addition to that, some cuckoo species raise the relative chances of survival for their own eggs by removing the eggs of other birds from the nests. Generally, the cuckoo chicks hatch marginally before the host eggs and instinctively alienate the host bird's eggs from the nest by pushing them out. They do so to receive more care and food from the host bird, thereby increasing their chances for survival. Host birds can resist this encroachment by the parasitic cuckoos and dispose the alien eggs that they discover to be not of their own or simply alienate them by abandoning the nest. To counter this threat, some cuckoo species cuckoos specialize in mimicry of the eggs of some host

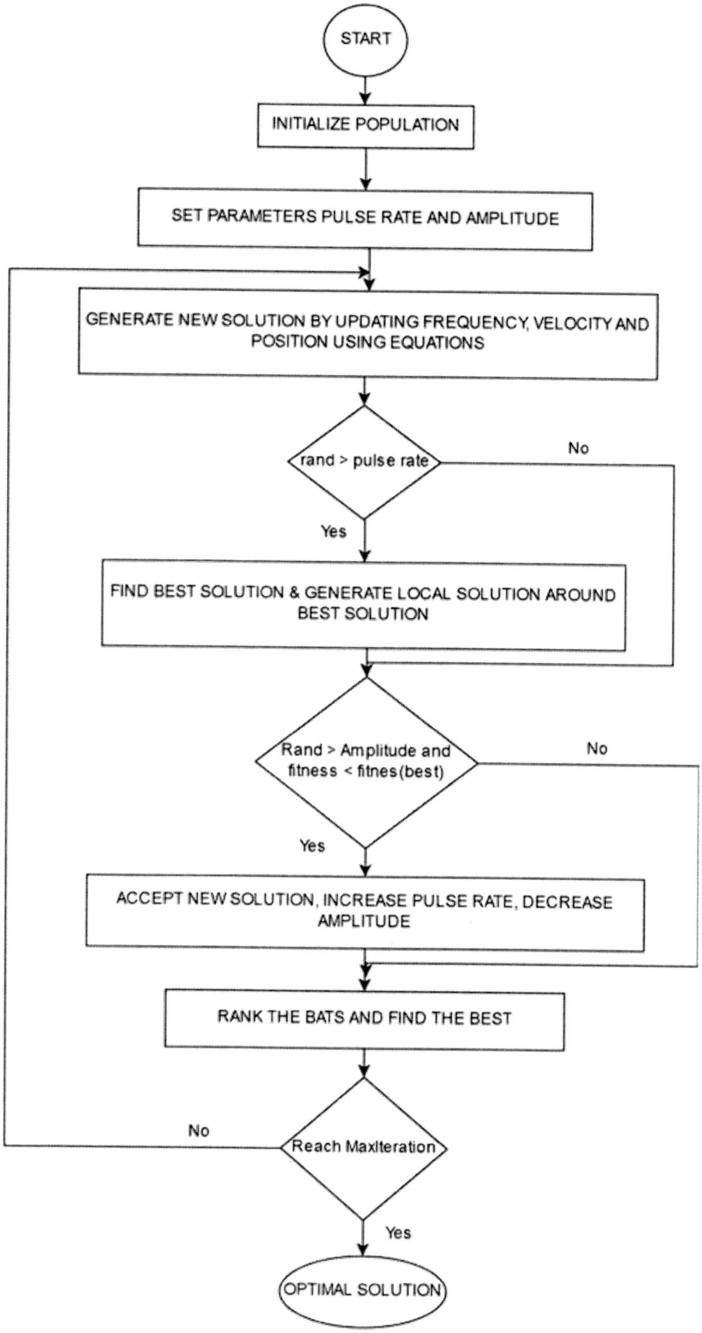

Fig. 2 Flowchart for BA

species by matching their egg color and pattern, thereby increasing reproductively by reducing the probability of abandoning of eggs. Although egg rejection technique is available in previous researches as well [25, 26].

CA represents an optimization technique modelled on the impulsive brood parasitism of the cuckoo bird which nurtures its own eggs by laying them in nests of other birds. CA owes its functionalities on the following idealized, fundamental rules [27]:

Each cuckoo produces one egg at an instance and subsequently places it in a nest chosen in a random manner;
The next generation of cuckoo eggs consists of high quality of eggs which are planted in nests labeled as the best;
There is a fixed quantity of host nests where the eggs can be laid. Additionally, the probability of discovery of a cuckoo egg by the host bird is given by $\rho\alpha\epsilon(0, 1)$. This probability is governed by calculations taking on board the sets of worst nests and other sets of discovered solutions.

Steps for Cuckoo search algorithm:

1. Begin;

2. Objective function: f(x), x={x1, x2 , ... ,xd };

3. Generate an initial population of n host nests;

4. WHILE (t<MaxGeneration) or (stop criterion)

5. Get a cuckoo randomly (say, i) and replace its solution;

6. Choose a nest among n (say, j) randomly;

8. IF (Fi > Fj)

9. Replace j by the new solution;

10. END IF

11. A fraction ($\rho\alpha$) of the worse nests are abandoned and new ones are built;

12. Keep the best solutions/nests;

13. Rank the solutions/nests and find the current best;

14. Pass the current best solutions to the next generation;

15. END WHILE.

Figure 3 depicts the flowchart for the CA in the form of a flowchart. The algorithm begins with the initialization of the population after which the loop begins where we

Fig. 3 Flow of Cuckoo search algorithm

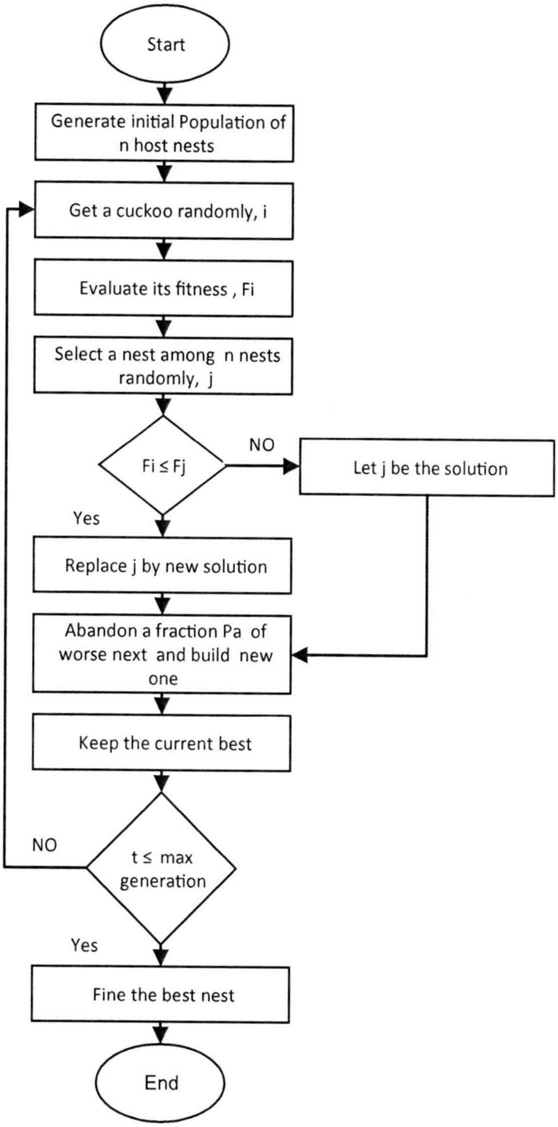

select a cuckoo randomly, evaluate its fitness and then select a nest randomly. If the nest has a higher fitness value, it remains the solution. On the other hand, nests with lower fitness values are replaced with novel sets of solutions and a fraction of $\rho\alpha$ are abandoned for generating new solutions until the stopping/cutoff condition is attained [29].

3.4 *Firefly Algorithm*

Firefly Algorithm (FA) represents a nature-inspired meta-heuristic algorithmic technique based on the principles of swarm intelligence. It was proposed by Yang in late 2007 and idealizes the flashing behaviour of fireflies to perform mathematical optimization. Fireflies typically exhibit rhythmic and transient flashes due to the process of bioluminescence. As ascertained, the fundamental functions of the firefly's flash include attracting both potential prey and mating partners (communication). Additionally, potential predators are reminded of the unpleasant taste of fireflies on being warned by the mechanism of flashing. Firefly–inspired algorithms can be developed by idealizing the fireflies' flashing characteristics to facilitate optimization of problems.

The FA takes inspiration from the natural flashing tendencies of fireflies to operate as a meta-heuristic algorithm. The firefly's flash plays the fundamental role of acting as a signal system for attracting other fireflies. Yang assumed the following parameters for development of the FA [5]:

1. Each firefly is attracted to all other fireflies i.e. fireflies are unisexual in nature;
2. Attractiveness of a firefly is directly related to the brightness of the firefly, and for any two fireflies under considering, the mutually bright firefly attracts the mutually dim firefly. Furthermore, the mutual sense of brightness between fireflies reduces with an increase in distance between them;
3. Absence of relatively brighter fireflies within a firefly's domain cause the firefly to move randomly through the domain.

The brightness is computationally assigned through the objective function.

Steps for Firefly algorithm:

1. *Begin;*
2. *Define an initialize benchmark function f(x),x= (x1 , ...,xd);*
3. *Generate initial population of fireflies xi(i = 1, 2, ...,n);*
4. *Determine light intensity for xi by calculating f(xi);*
5. *Define light absorption coefficient γ*
6. *WHILE(t<MaximumGeneration)*
7. *Make a copy of the generated firefly population for move function;*
8. *FOR i= 1:n all n fireflies*
9. *FOR j= 1 :i all n fireflies*
10. *IF (Ij>Ii),*
11. *Move fireflies i and j according to attractiveness;*
12. *Evaluating new solutions and updating light intensity for next iterations;*
13. *END IF*
14. *END FOR j*
15. *END FORi*
16. *Sorting the fireflies to find the present best;*
17. *END WHILE*
18. *Begin post process on best results obtained;*
19. *END.*

Figure 4 depicts the flow chart representation of the FA. In the steps shown we first define an initialize benchmark function and generate an initial population of fireflies x_i after which we determine light intensity for each and define the light absorption coefficient. Subsequently, the loop iterations commence with generating the copies of the already existent firefly population. Secondly, the fireflies are moved relative to each other based on their mutual attractiveness. Thirdly, the new set of solutions are calculated and the light intensity for the next set of iterations are updated. Finally, the fireflies within the domain are sorted and the best solution in the current iteration is estimated.

3.5 *Particle Swarm Algorithm*

Particle Swarm Optimization (PSO) was developed in 1995 by Eberhart and Kennedy. PSO is a meta-heuristic algorithm that imitates the nature of migrating animals such as a group of birds trying to find their destination. This algorithm itself is simple yet very powerful, it has been applied various fields and many researchers have created variations of it.

Every individual bird among the entire bird population in PSO is a solution and is called a 'particle'. The following are the steps for finding the optimized solution:

- The bird population is first initialized and each bird or particle proceeds to move in a random direction with a particular velocity.

Fig. 4 Flow chart for firefly algorithm

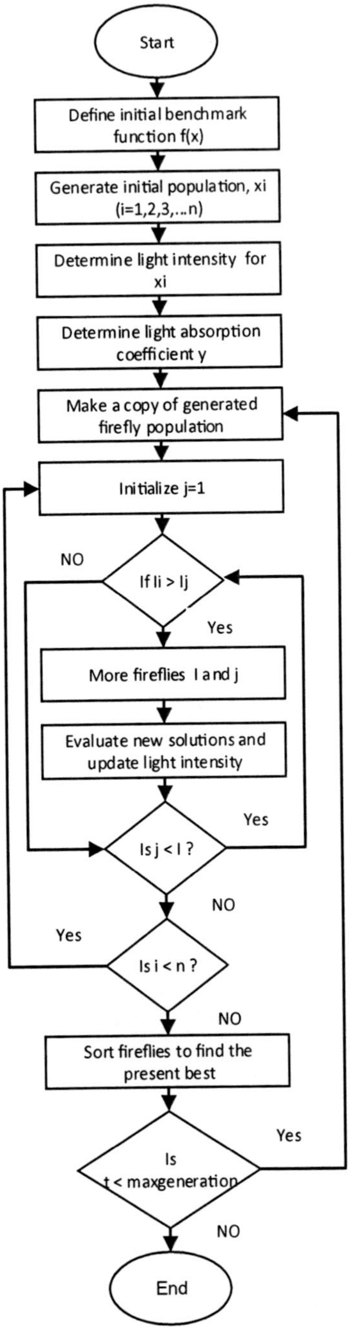

- The fitness function is run on each particle to find the global best bird in the flock.
- The personal best from every particle's individual past and the global best from the entire flock are taken into consideration to find the new velocity of each particle.

The particles are moved to their new position and the steps are repeated till solution is reached.

The Particle Swarm Optimization is based on intelligent behaviour of the swarm. It belongs to a new genre of algorithms where individuals inside the population evolve by competition as well as cooperation contrary to how other algorithms function. PSO's performance has been previously demonstrated to comparable to former algorithms with similar goals.

PSO combines multi agent aspects with parallel searching. In this algorithm the particles represent a theoretical entity that traverse in multiple dimensions for achieving the solution. The values being manipulated in PSO are position and velocity of the particles while the set of positions of the particles represent the solution. The algorithm starts by first initializing the position vectors x_i and velocities v_i with random values. Together, the entire population generated this way is called a "swarm".

The ith particle is represented in N-Dimensional space by a point where N is the number of variables. The position X_i, personal best P_i and velocity V_i are described as follows:

$$X_i(t) = (x_{i1}, x_{i2}, x_{i3} \ldots x_{in})$$

$$P_i(t) = (p_{i1}, p_{i2}, p_{i3} \ldots p_{in})$$

$$V_i(t) = (v_{i1}, v_{i2}, v_{i3} \ldots v_{in})$$

The gth particle is the best particle and its position is given by P_g. Each particle's position and velocity are updated based on the following equations:

$$V_i(t) = \mu \times V_i(t) + k_1 \times r() \times (P_i(t) - X_i(t)) + k_2 \times R() \times \left(P_g(t) - X_i(t)\right) \tag{4}$$

where, $V_{\max} \leq V_i \leq V_{\max}$.

$$X_i(t+1) = X_i(t) + V_i(t+1) \tag{5}$$

The constants k_1 & k_2 are known as learning factors such that $k_1, k_2 > 0$, 'r' and 'R' are two methods that produce random number between [0, 1], V_{max} is the maximum possible velocity [6], μ is called the inertia weight which was proposed as an improvement to thPSO by Yang and Eberhart to limit the effect of the previous velocities on the current velocity. Such that the two terms decrease linearly which balances out global search and local search components. Hence the global search begins with a larger weight and gradually reduces with time to increase the weight of local search. The second term in Eq. (4) denotes the cognitive part i.e. private thinking and the third term in Eq. (4) specifies the social collaboration globally.

The main parameters that affect the optimization are:

- Population Size (N)
- Number of Cycles (t)
- Max change of velocity (v)
- Inertia weight (μ)
- Dimensions (n)

Steps for Particle Swarm Optimization

```
1. Begin;
2. Initialize population of N particles;
3. Evaluate fitness for each particle;
4. Assign weight factor's value;
5.        FOR (each particle);
6.                Assign Pi= personal best of particle;
7.                IF (fitness(Xi)> fitness(Pi))
8.                        Pi = Xi;
10.               END IF
11.       END FOR
12. Assign Pg= best particle's positon;
13.       FOR (each particle)
14.               Update velocity and position;
15.       END FOR;
16. If stop criteria=true END else Go back to step 3;
```

Figure 5 depicts the flow chart for PSO algorithm in which we start by initializing the population after which we evaluate the fitness of each particle. After which the loop begins where in the velocity and distance is calculated for every particle then the particle's personal best is compared with the global best. Loop terminates when the stop criteria is reached and the solution is generated.

3.6 *Artificial Bee Colony Algorithm*

Artificial Bee Colony Algorithm was created by Dervis Karaboga in 2007 as a self-organizing system based on swarm intelligence and division of labour. It is a Bio-Inspired meta-heuristic algorithm that derives its metaphor from the way a hive of bees works together to find relevant sources of nectar in the nearby environment with the help of segregated tasks for individuals conforming to a division of labour [29]. In ABC the three main types of bees are: onlookers, employed bees and scouts. It uses dynamic mechanisms that result in structures at a global level from multiple interactions at the local level. These interactions are based around four main factors namely; multiple interactions, fluctuations and positive & negative feedback. Positive

Fig. 5 Flow chart of PSO

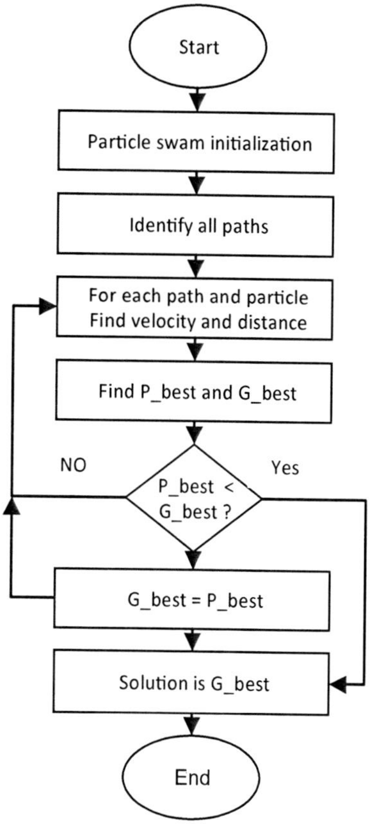

feedback implies the food sources with maximum profitability get accessed more governed by the onlooker bees choosing the best employed bees. Negative feedback allows the system to not get stuck in local optima or an early saturation of the solution. Fluctuations involve the use of random walks with the help of scouts while multiple interactions imply that the swarm works together with all of these interactions that take place simultaneously.

The ABC algorithm is based around three main components that define the core data structures for the implementations which are food source, employed foragers and unemployed foragers. The food sources represent the solution to the problem. Employed foragers are the bees or agents that are focused on one particular food source and go back to the hive to provide information to the unemployed bees. Each discovered food source is assigned one employed forager at the start. The unemployed foragers comprise of two sub categories namely onlookers and scouts. Onlookers are the bees that look at the employed bees and the information they are conveying which is done via a dance in the natural setting and then assign a value of probability to each of the employed bee based on the profitability of that food source. They basically choose which employed bee to follow. So the onlooker bees convert to employed bees and follow the profitable one. After they return there are three possible options for the bee. First option involves it going back to an unemployed bee state, second options involves recruiting more bees to pursue to same path and the third option implies that the bee pursue the same food source without recruiting new bees.

While all these calculations take place there are also random walks taking place to find out new un-identified food sources with the help of scouts. When an employed bees' food source is finished by others it becomes a scout bee which looks in the environment randomly until it finds a solution worth pursuing based on the profitability.

The number of employed bees and number of onlookers is about 50% each at the start of the algorithm but as the program proceeds the mean number of scout's average over conditions at about 5–10%. The experiments done by Karaboga confirmed that in ABC the new bees start searching at rates directly relative to the change in total number of bees and the number of bees currently searching [7].

Steps for artificial bee colony algorithm:

```
Steps for artificial bee colony algorithm
1. BEGIN;
2. Initialize population;
3. Calculate fitness for the population;
4. WHILE (t<MaxIteration or StopCriterion)
5.      FOR (each employed bee);
6.              Produce fresh solutions;
7.              Calculate fitness;
8.              Use selection process (greedy);
10.     END FOR
11. Measure probabilities of the solutions;
12.     FOR (each onlooker bee)
13.             Use Probability to choose solution;
14.             Generate fresh solution;
15.             Calculate fitness;
16.             Apply greedy selection process;
17.     END FOR
18. IF (Abandoned solution for the scout)
19.     Replace with randomly generated solution;
20. END IF
21. Memorize best solution;
22. t=t+1;
21. END WHILE
```

Figure 6 depicts the algorithm in a flowchart for ABC. The entire process is divided into three phases that are scout bee phase, onlooker bee phase and also the employed bee phase. The population is first initialized after which the employed bees look around their assigned food sources based on greedy approach to find new food sources. The onlooker bee evaluates these employed bees and pursues new food sources around the ones assigned to employed bees. Lastly the scout bees come into action when the food source exhausts and find new random food sources.

Fig. 6 Flowchart for ABC

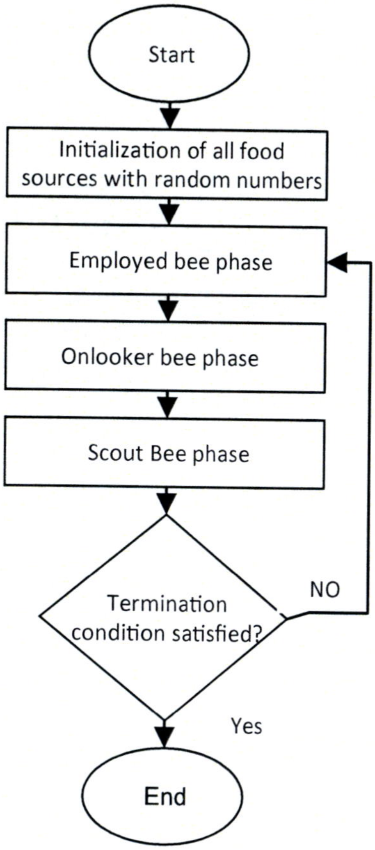

3.7 *Applications of Bio Inspired Algorithms*

Hill climbing is suitable for deployment in any problem where an accurate evaluation function can be associated with the current state of the problem. For example, the eight-queen problem, the circuit design, the Travelling Salesman Problem and a variety of other real-world problems. HCA has been applied to problems based on inductive learning model like the probabilistic hill climbing system which models speed-up and inductive learning. Certain applications of this system have been accommodated into "utility analysis" "explanation-based learning systems" models [30].

Additionally, HCA has been utilised in robotics to handle multiple-robot teams. The Parish algorithm, which allows for efficient and scalable coordination in multi-robot systems provides an instance for the same. The research team devised "a team of robots that must coordinate their actions so as to guarantee location of a skilled evader." [31].

Their algorithm provides robots with a choice of two sets of operation modes: either they work in isolation or they work in teams, by using hill-climbing. Robots implementing the Parish Algorithm are therefore "collectively hill-climbing according to local progress gradients, but stochastically make lateral or downward moves to help the system escape from local maxima".

Cuckoo Search Algorithm is used for a variety of applications involving optimization and computational intelligence. CA shows superior efficiency when compared to other algorithms for design problems involving continuous optimisation of data such as the design of springs and welded beams. Modified CA utilizing effective assignment of the associated algorithms for the algorithm have been used for resolving nonlinear, multimodal problems like mesh generation [32]. Additionally, CA have been used for training NN models [33, 34] designing the optimal design of embedded systems and for obtaining the optimal machine parameters for industrial operations like milling and structural design of vehicles [35]. CA has also been used for scheduling problems and for generating independent paths for test data generation and software testing. CA usually gives more robust solutions as compared to other algorithms like FA and hence has been applied for obtaining efficient solutions of problems like the Knapsack problem in computing, structural optimization [36]. The relatively better performance of the CA merits its usage for evaluation of problems like complex equilibrium based thermodynamic calculations [37], clustering [38] etc.

Firefly Algorithm is a popular algorithm for optimization of parameters and is used for a plethora of applications. FA is used for image processing applications and has been deminstrated to be the methodology that expends the least computational machine time for image compressing problems. Firefly utilises system resources and also produces better optimality in results with regard to image processing applications like feature selection. FA is also used for engineering design problems and has been shown to provide efficient solutions to complex problems(For e.g. NP-Hard problems and Non-linear problems). Thereby, FA has been used to solve issues like antenna design, scheduling and routing problems like the TSP and CMOS LNA parameters [39]. Furthermore, FA outperforms other metaheuristic algoirthms like Artificial Bee Colony and PSO in terms of achieving the most optimised results in a global perspective. FA is also used for applications fir implementing intelligent, automated computing systems by performing classification and clustering. Additionally, FA has been used for training of neural networks and for performing optimisation operations in a dynamic environment.

PSO has a potential for and a history of deployment across different applications. In general, we can say that areas where PSO has generally shown promise in solving problems with no specialized method of solution available, problems that give unsatisfactory results with all specialized methods and multimodal problems.

Some of the applications of PSO include antenna design, pharmaceutical, medical and biological applications. PSO have also been implemented on communication networks as well as clustering, classification and data mining [40, 41]. Other applications include distribution networks and optimization of electronic and electromagnetic designs. Further in recent times they have been implemented over fuzzy and neuro-fuzzy systems and control as well as computer vision, forecasting, artificial NN, financial predictions and many more [42].

Artificial bee colony algorithm has a variety of applications starting from benchmarking optimization to bioinformatics applications. Other than that ABC algorithm has been successfully applied over clustering and mining operations as well as scheduling applications like spanning tree [43]. It has also been implemented in computer vision particularly in the field of image processing [44]. Further it has also been implemented in economic dispatch problems as well as engineering designs sensors and other applications like with levy flight based ABC [45].

3.8 Limitation of Bio Inspired Algorithms

3.8.1 Limitation of Hill Climbing Algorithm

The following are the limitations HCA.

- **Ridges**: Drop offs towards edges. Steps towards southward direction, eastward direction, northward direction and westward direction might go downwards, while step towards the North West direction might go upwards.
- **Plateaus**: In such cases the algorithm has no direction to walk as it only consists of plane regions.
- **Local Maxima**: The algorithm can get stuck at local maxima before achieving global maxima.

3.8.2 Limitation of Bat Algorithm

The following are the limitations of BA.

- The algorithm converges really quickly at the beginning stages itself leading slower rates of convergence at later stages.
- Accuracy is limited if number of function evaluations are not high enough.
- BA has not been as widely implemented as others due to it being a much newer algorithm.

3.8.3 Limitation of Cuckoo Search Algorithm

The following are the limitations of CA.

- It takes more runtime for execution as compared to other meta-heuristic algorithms like Firefly Algorithm.
- It has a relatively low speed of convergence on optimized solutions.
- For achieving the same level of optimization, CA requires more iteration as compared to other meta-heuristic algorithm like ABC optimization and FA.

3.8.4 Limitation of Firefly Algorithm

The following are some of the limitations of FA.

- The implementation is relatively tougher as assigning an apt parameter for optimization (modelled by distance) and its associated characteristic distance for defining the breakpoints of the solution is difficult.
- It can be outperformed on the grounds of accuracy of solutions as compared to other meta-heuristic algorithms like ABC optimizations.
- It is less resource efficient as compared to other algorithms like BA.

3.8.5 Limitation of Particle Swarm Optimization

The following are some of the limitations of PSO.

- The algorithm gets stuck in Local Optima in certain situations as these conditions were not taken into account when the algorithm was formed.
- The algorithm only works good in smaller populations as the best solution from the neighbors is taken into consideration. An alternate version of PSO (1998) works better but merely smaller benchmark functions were used to test it.
- It is difficult to select proper values of inertia weight.

3.8.6 Limitation of Artificial Bee Colony

The following are certain limitations to ABC.

- Lack of use of secondary information
- Requires new fitness tests on the new fitness parameters.
- The possibility of losing relevant information.
- High number of objective function evaluations.
- Slow down when used in sequential processing.
- The population of solutions increases the computational cost.

3.9 Theoretical Comparison

Table 1 depicts the theoretical comparison of the six bio inspired algorithms presented in this study based on the various parameters such as fitness function, termination criteria, population of test suite generation etc. It also lists out which type each algorithm belongs to, whether they are pheromone based and also if they provide global search or just local search [46].

Table 1 Theoretical comparison of bio-inspired algorithms

Parameters	Hill climbing	Bat algorithm	Cuckoo algorithm	Firefly algorithm	Particle swam algorithm	Artificial bee algorithm
Type	Search based	Swarm intelligence search based	Swam intelligence search based	Swam intelligence search based	Swam intelligence search based	Swam intelligence search based
Fitness Function	Based on value of new node chosen	Based on echolocation	Based on cuckoo egg in host nest	Based on intensity of light	Based on velocity and position	Based on nectar of food
Termination Criteria	Maximum node found	Maximum iteration reached	Maximum egg generated by cuckoo	Maximum population of firefly	Best velocity and best Position achieved	All food source exploited
Population of test suite generation	Random	Random	Random	Random	Random	Random
Pheromone	–	–	–	Pheromone	Pheromone	Pheromone
Non pheromone	Non pheromone	Non pheromone	Non pheromone	–	–	–
Local	Local	Local	Local	Local	Local	Local
Global		Global	Global	Global	Global	Global

4 Methodology

This section describes the methodology and framework implemented in detail. The definitions, explanation and the premise is set first followed by the framework and detailed description of test suite generation and their performance analysis.

4.1 Overview

The key difference between heuristics and meta-heuristics is that heuristics involve modification of the base algorithm to suite the problem at hand while meta-heuristics deals with a generalized algorithm that can be applied to various optimization problems without modification []. With this in mind for this study, the six meta-heuristic algorithms were implemented in their standard form as described in Sect. 3. The key changes were made in the optimizing function used for generation of the test data. Each individual population member represented a Test Suite and each potential Test Suite was improved iteratively using that particular multi-objective optimizing function. The best Test Suite was then selected as per the algorithm under question.

The two objectives being optimized are: to maximize path coverage and to maximize branch coverage. While, for a comparative study any optimizing function could have been used (since their relative performances in the same environment are to be studied) in this study we focused on maximizing path coverage and branch coverage which are also two of the major white box testing parameters [47]. The choice of this optimizing function was done in a fashion similar to previous studies [2]. Linear Scalerization was used to create a single objective function that takes account both path coverage and branch coverage. Both were given equal weights for convenience.

A different optimizing function (objective function method) was manually written for each program under test in such a way that given a Test Suite as an input, it returned the scalerization of path coverage and branch coverage. Further the data for performance analysis was stored in parallel. The entirety of the TSG and its analysis was written in Java. A Control Flow Graph (CFG) was also manually generated for each program under test.

4.2 Research Methodology

Figure 7 describes the entire research methodology and process undertaken for this study. The first step was to select six meta-heuristic algorithms. These algorithms were selected on the basis of their popularity and how recently were they developed. Bat Algorithm being the most recent first published in 2010. The next step was to study the past literature and review past concepts and findings as specified in Sect. 2. After this the framework for testing was designed under which the execution and analysis for this research was conducted in its entirety. The next phase was the execution of each algorithm to generate the best test suite for the particular program. This was done a total of hundred times for each algorithm and the best time, worst time, average time etc. were stored. This comprised the Test Suite Generation phase. The performance analysis metrics once stored were then analysed to compare the algorithms relative to each other. This comprised the Comparative Analysis phase. The last phase was the result reporting and deduction of the best algorithm and further analysis and discussion over the matter to conclude the finding for this research.

4.3 Testing Framework

Figure 8 describes the entire Testing Framework as well as the flow for this study. The framework has been divided into three phases namely the Test Suite Design Phase, Test Suite Optimization Phase and Performance Evaluation Phase. Each phase and its sub-steps are described in detail below.

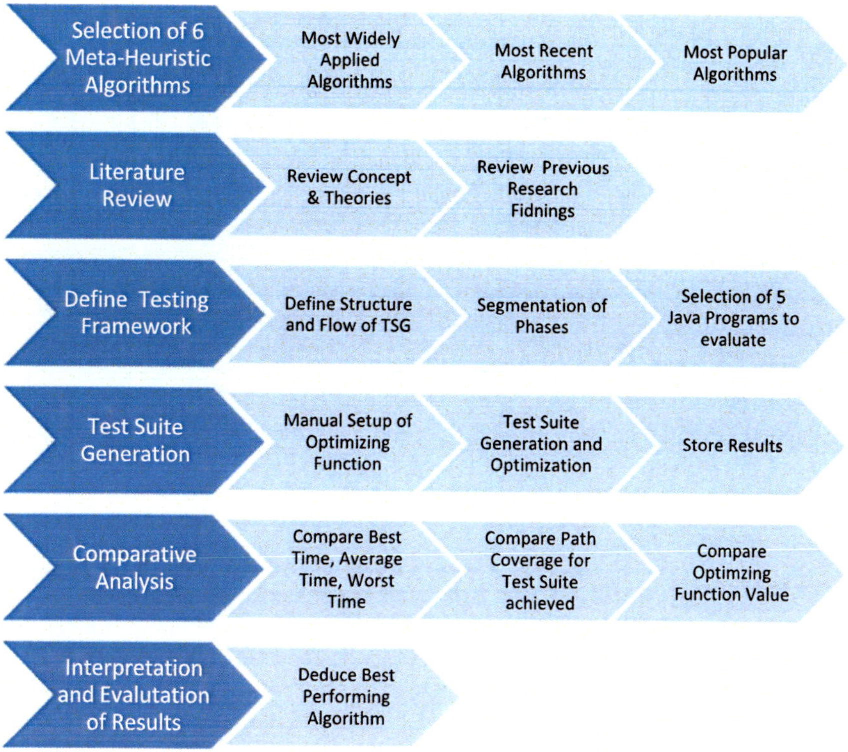

Fig. 7 Research methodology

4.3.1 Test Suite Design Phase

This is the initial phase for our study wherein the program under test is considered to produce the CFG, Constraints and the Objective Function.

Software Under Test

This study focuses on five programs to generate the test suites. These five programs under test are detailed in Table 2. The five programs are of various lengths in terms of lines of code (LOC) ranging from around 20 to around 75 LOCs. The programs are also varied in terms of the type of programming paradigm they use for example P1 uses nested if-else conditionals, P2 uses simple if-else conditionals, P3 uses a do-while loop with nested conditionals, P4 uses a switch case conditional and lastly P5 uses a while loop with nested conditionals. Each of these programs was written in Java.

Table 2 details the Java programs under test and gives their descriptions as well. The programs are labelled from P1 to P5.

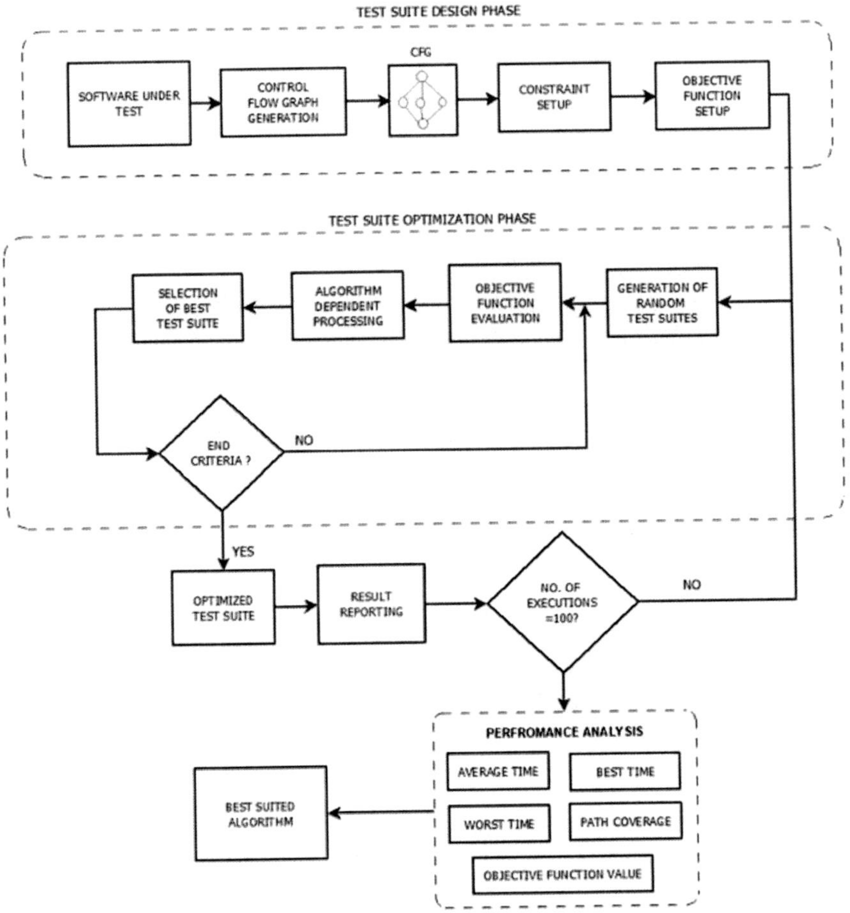

Fig. 8 Software test automation framework

Table 2 Java programs under test

Label	Program Name	Description
P1	Triangle Classification Problem	Identifies the type of triangle based on three input values
P2	Greatest Number Problem	Finds out the largest number from three given input numbers
P3	Prime Number Identification Problem	Identifies whether the given input is prime or not
P4	Days in a Month Problem	Given a Month and Year, the program finds out the number of days in that month
P5	Binary search problem	Searches a sorted array and finds out the location of the specified value

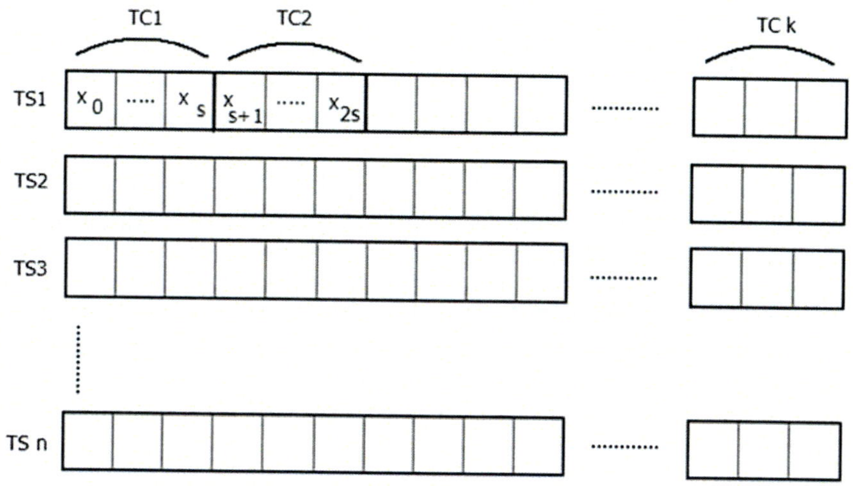

Fig. 9 Population definition of potential test suites

Control Flow Graph

A control flow graph was manually generated for each program. This was done to find out the McCabe's Cyclomatic Complexity of each program. Cyclomatic Complexity was used in this study to account for the minimum number of Test Cases required in a Test Suite for a particular program to conduct Testing [48]. In this study, the number of Test Cases per Test Suite was equal to the Cyclomatic Complexity of the program under test in a fashion similar to Basis Path Testing. In Basis Path Testing each linearly independent path is considered through the program which implies that number of test cases will be equal to the Cyclomatic Complexity of the program [49, 50].

Constraint Setup

The constraints were setup for each program under test to generate the test suites as depicted in Fig. 9. Each element of the population was considered as an Array of Data forming a Test Suite. While each Test Suite comprised of individual collection of Test Data to form a Test Case. These sets of Test Cases were iteratively improved to produce the best Test Suite.

Figure 9 describes the Population Definition of Potential Test Suites for each algorithm. Here k is the McCabe's Cyclomatic Complexity as discussed previously and is the number of Test Cases in a Test Suite. The population size n and was fixed at 1000 for each algorithm. The number of inputs for every program under test is used to determine the size of a Test Case from x_0 to x_s and so on. Here $s + 1$ is the number of inputs in the program under test. TS1 to TSn are the test suites, TC1 to

TCk are the Test Cases in said Test Suites while x_s represents the Test Data. Thus the Dimension Size (D) for each algorithm was equal to k times (s + 1).

$$D = k \times (s + 1) \tag{6}$$

where D is the Dimension Size, k is the Cyclomatic Complexity and s + 1 gives the number of inputs. For example, for the program Triangle Classification Problem, we have 3 inputs adhering to the three sides of a triangle while the Cyclomatic Complexity for that program is 5. Hence in the case of PSO for example the Dimension Size for each bird in the population is 5 times 3 i.e. 15. Each of these Test Suites were subjected the objective function using which the Test Suites were Optimized.

$$V(G) = E - N + 2 \tag{7}$$

where $V(G)$ is the Cyclomatic Complexity, E is the number of edges and N is the number of nodes in the CFG,

Objective Function Setup

This step revolves around manually modifying the program under test to achieve a Java method that takes in as input the Test Suite and returns the Objective Function value to use in each algorithm. This was done by taking into consideration the paths in the program procured with the control flow graph. The Java method for objective function takes in the Test Suite and creates a list of arrays (Array-List) comprising 0 s and 1 s; one array per Test Case within that Test Suite. This list of arrays describes list of the paths taken by the Test Cases within a Test Suite. The size of these arrays equals the total number of paths in the program and a 1 indicates that that path was traversed. The number of unique arrays in the Array-List gives us the count for unique paths traversed in by the entire Test Suite. This gives us the path coverage. Branch Coverage is also calculated in a similar fashion. For example an array such as [1,0,1,0,1] for branches describes the fact that the Test Case in question went through the first branch, didn't go through second, went through the third branch and so on. The path and branch coverage for the Test Suites are calculated in this manner.

The two values undergo Linear Scalerization to produce the value for our fitness function/Objective Function. Here the objective is to maximize both path and branch coverage and to do this we use the Multi-Objective Optimization technique of linear scalerization which is described as follows:

$$f = (w_1 \times f_1 + w_2 \times f_2) + K \tag{8}$$

Equation 8 describes the Linear Scalerization of 2 functions f_1 and f_2 with weights w_1, $w_2 = 1$ along with an offset of K. In this study the weights are considered the same and equal to 1 for simplicity and convenience, while the two functions which are equivalent to the path coverage and branch coverage are given below. For a relative performance evaluation any function can be used and this study uses the multi objective function f which gives the objective function value for this study.

$$f_1 = -\left(\frac{Number\ of\ Paths\ Covered\ by\ Test\ Cases}{Total\ number\ of\ paths} \times 100\right) \tag{9}$$

$$f_2 = -\left(\frac{total\ branches\ \text{covered}}{total\ possible\ branches\ in\ all\ test\ cases} \times 100\right) \tag{10}$$

Equations 9 and 10 describe the formulae which are equivalent for the path and branch coverage. Each algorithm was implemented to minimize the objective function f for the Test Suites. If an function is to be maximized, it is equivalent to minimize its negative. Hence the negative value of the functions f_1 and f_2 are taken so as to maximize the path coverage and branch coverage. In Eq. 8, the value of K is 200 and it is added as an offset to ease the process of comparative analysis of their performances as well ease the process of graphically representing them by making the values non negative. The offset value of K is added to make sure each Objective Function value remains greater than 0 and hence graphing and analyzing was made simpler.

4.3.2 Test Suite Optimization Phase

This is the main phase wherein the best Test Suite is generated using the previous Test Suite Design and Setup phase. This phase is completely dependent on the algorithm under question. The overall idea as shown in Fig. 8 is that a randomized set of Test Suites is first generated which the undergoes the Test Suite Optimization process as described in Sect. 3 using the objective function as described in Sect. 4.3.1.4. Once the end criteria for the algorithm under question are achieved we get the best Test Suite i.e. the Test Suite with the minimum value of Objective Function. This process is repeated 100 times to make sure the optimization results are consistent. The best, worst and average time as well as the path coverage and objective function value are stored for Performance Evaluation in the next phase.

4.3.3 Performance Evaluation

This is the last phase and deals with the Comparative Analysis of the results produced in the previous step. Here the each value is compared to deduce the best suited algorithm when it comes to Test Suite Generation and Optimization Problem. The best suited algorithm can be improved further in future studies to produce an algorithm that works specifically well for the TSG problem.

5 Experimental Results

The following tables and figures show the results for hundred executions of each algorithm to generate the most Optimized Test Suites for the five Java programs under test. The programs under test and their details are as follows.

Table 3 describes the programs under test, their descriptions, LOC as well as their Cyclomatic Complexities. The following sections describe the results for each Program Under Test.

5.1 Triangle Classification Problem

Table 4 describes the results for the Triangle Classification problem. The results indicate that the fastest performing algorithm was BA with an average time of 145.38 ms while the most efficient algorithm was ABC with the second fastest average time of 493.09 ms and the least objective function value of 49.16 as well as the maximum path coverage of 97.8%. BA stood at second best in terms of how optimized the results were with an Objective Function value of 58.4 and 97% path coverage. Hence ABC being the best suited algorithm for this.

Figure 10 describes the CFG for the Triangle Classification Problem. The program has a Cyclomatic Complexity of 5.

Table 3 Java programs under test

Label	Program name	Lines of code (LOC)	Cyclomatic complexity (G)	Description
P1	Triangle Classification Problem	24	5	Identifies the type of triangle based on three input values
P2	Greatest Number Problem	18	4	Finds out the largest number from three given input numbers
P3	Prime Number Identification Problem	25	4	Identifies whether the given input is prime or not
P4	Days in a month problem	75	13	Given a Month and Year, the program finds out the number of days in that month
P5	Binary search problem	46	5	Searches a sorted array and finds out the location of the specified value

Table 4 Results for triangle classification problem

Parameters	Hill climbing	Bat algorithm	Cuckoo Algorithm	Firefly algorithm	Particle Swam algorithm	Artificial bee colony algorithm
Average time (milliseconds)	694.36	145.38	683.84	13,355.18	1589.32	493.09
Best time (milliseconds)	543	100	625	8062	1188	453
Worst time (milliseconds)	1465	350	1000	14,331	3108	828
Average path Coverage (%)	33.2	97	20.0	87.0	96.2	97.8
Average objective function value	66.94	58.4	69.0	60.95	55.44	49.16

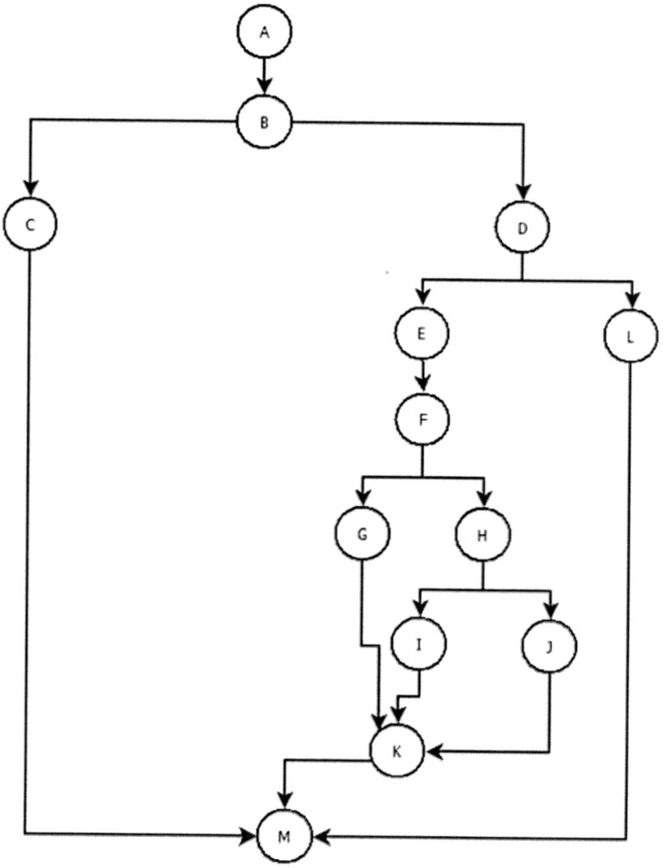

Fig. 10 CFG for triangle classification problem

Table 5 Results for greatest number problem

Parameters	Hill climbing	Bat algorithm	Cuckoo algorithm	Firefly algorithm	Particle swam algorithm	Artificial bee colony algorithm
Average time (milliseconds)	304.25	101.07	587.64	9837.95	1369.68	437.49
Best time (milliseconds)	250	62	546	6158	1116	406
Worst time (milliseconds)	469	188	843	10,672	2388	718
Average path coverage (%)	89.5	95.0	28.0	88.0	95.0	98.0
Average objective function value	26.02	24.9	26.88	26.0	24.9	24.0

5.2 *Greatest Number Problem*

Table 5 describes the results for the Greatest Number problem. The results indicate that the fastest performing algorithm was BA with an average time of 101.07 ms while the most efficient algorithm was ABC with the least objective function value of 24.0 as well as the maximum path coverage of 98%.

Figure 11 describes the CFG for the Greatest Number Problem. The program has a Cyclomatic Complexity of 4.

5.3 *Prime Number Identification Problem*

Table 6 describes the results for the Prime Number Identification problem. The results indicate that the fastest performing algorithm was BA with an average time of 156.77 ms while all algorithms except for CA performed the same.

Figure 12 describes the CFG for the Prime Number Identification Problem. The program has a Cyclomatic Complexity of 4.

5.4 *Days in a Month Problem*

Table 7 describes the results for the Days in a Month problem. The results indicate that the fastest performing algorithm was BA with an average time of 138.24 ms while the most efficient algorithm was ABC with the least objective function value of 88.59 as well as the maximum path coverage of 97.92%. CA again performed the worst efficiently and the slowest algorithm was FA.

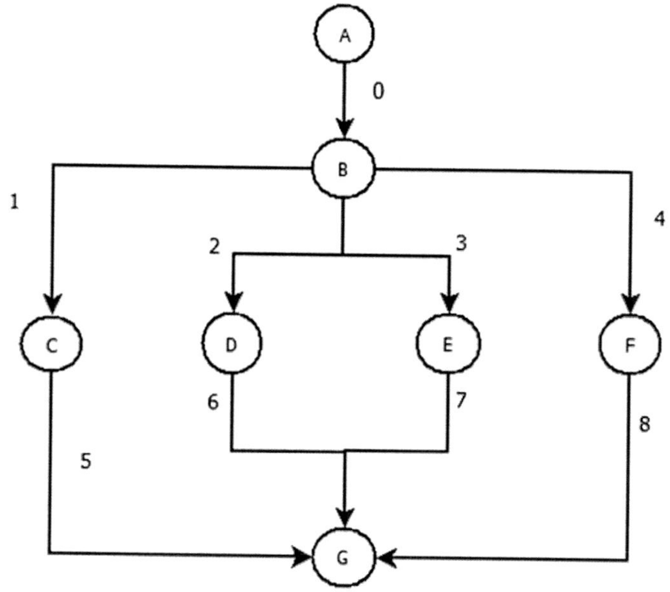

Fig. 11 Path flow diagram for greatest number problem

Table 6 Results for prime number identification problem

Parameters	Hill climbing	Bat algorithm	Cuckoo algorithm	Firefly algorithm	Particle swam algorithm	Artificial bee colony algorithm
Average time (milliseconds)	185.48	156.77	313.04	3124.3	1052.4	530.44
Best time (milliseconds)	140	74	281	2875	630	482
Worst time (milliseconds)	383	441	531	3922	2190	698
Average path coverage (%)	50.0	50.0	25.0	50.0	50.0	50.0
Average objective function value	18.0	18.0	22.98	18.0	18.0	18.0

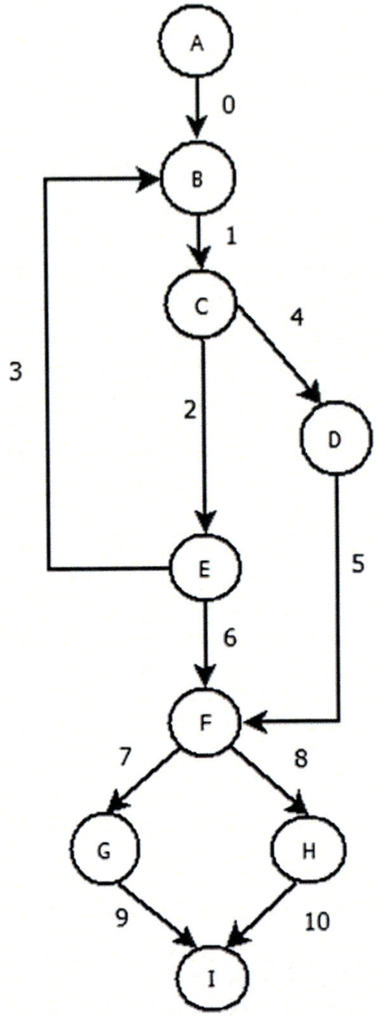

Fig. 12 Path flow diagram for prime number problem

Figure 13 describes the CFG for the Days in a Month Problem. The program has a Cyclomatic Complexity of 13.

5.5 *Binary Search Problem*

Table 8 describes the results for the Days in a Month problem. The results indicate that the fastest performing algorithm was BA with an average time of 143.75 ms while the most efficient algorithm was ABC with the least objective function value of 49.0 as well as the maximum path coverage of 80.0%. FA performed the worst

Table 7 Results for days in a month problem

Parameters	Hill climbing	Bat algorithm	Cuckoo algorithm	Firefly algorithm	Particle swam algorithm	Artificial bee colony algorithm
Average time (milliseconds)	642.87	138.24	2857.84	22,586.08	4313.81	1596.38
Best time (milliseconds)	562	100	2641	17,033	3200	1484
Worst Time (milliseconds)	891	268	3453	31,611	1032	2172
Average Path Coverage (%)	42.15	88.23	54.76	76.66	94.46	97.92
Average objective function value	137.41	100.02	131.21	121.12	93.78	88.59

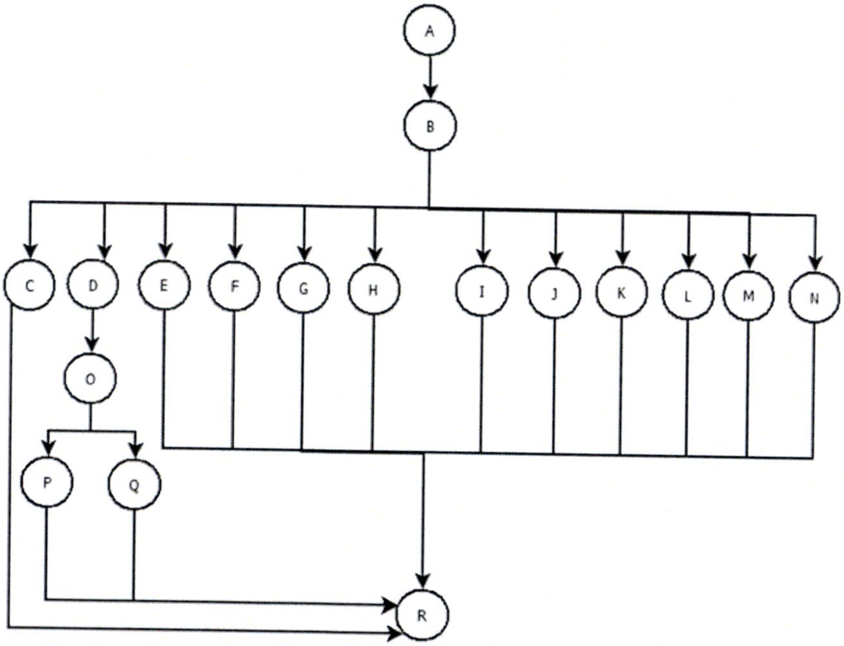

Fig. 13 Path flow diagram for days in a month problem

efficiently and was also slowest algorithm. PSO had the same path coverage as ABC i.e. 80% but the objective value was not the least indicating the results produced were not perfectly optimized. BA performed the fastest but yet again was second best when in terms of efficiency to ABC.

Figure 14 describes the CFG for the Binary Search Problem. The program has a

Table 8 Results for Binary search problem

Parameters	Hill Climbing	Bat Algorithm	Cuckoo Algorithm	Firefly Algorithm	Particle Swam Algorithm	Artificial bee colony Algorithm
Average time (milliseconds)	927.06	143.75	4342.26	20,778.5	4380.36	1381.86
Best time (milliseconds)	782	101	3953	19,860	4006	1224
Worst time (milliseconds)	1603	283	7429	25,642	5019	1469
Average path coverage (%)	56.8	65.0	35.0	75.8	80.0	80.0
Average objective function value	79.62	61.0	131.57	175.045	61.42	49.0

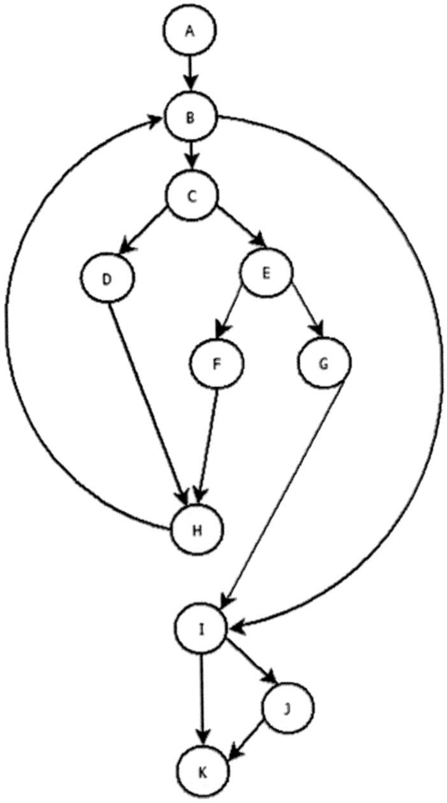

Fig. 14 Path flow diagram for binary search problem

Cyclomatic Complexity of 5.

Figures 15, 16, and 17 describe the average time, best time and worst time respectively. The x-axis represents the five discrete points for the five programs under test while the y-axis represents the time in seconds. Figures 18 and 19 describe the path

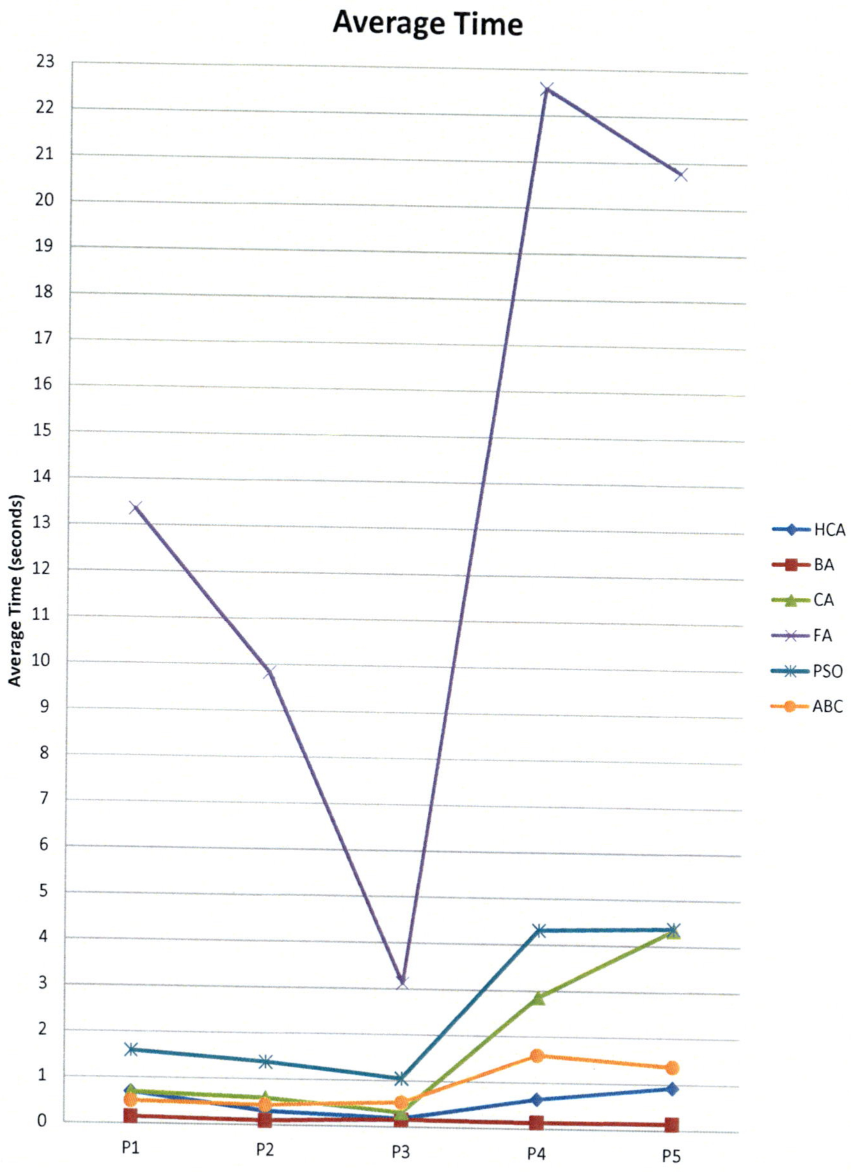

Fig. 15 Graph for average time analysis

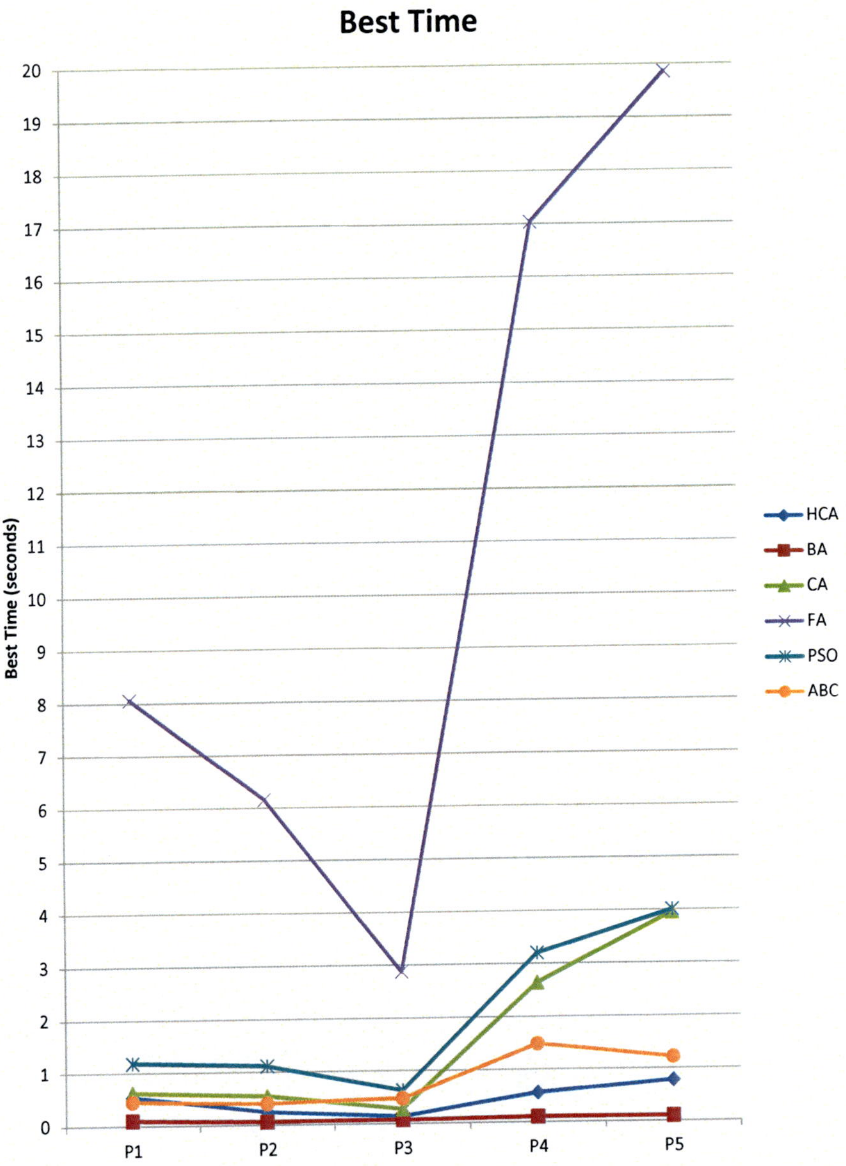

Fig. 16 Graph for best time analysis

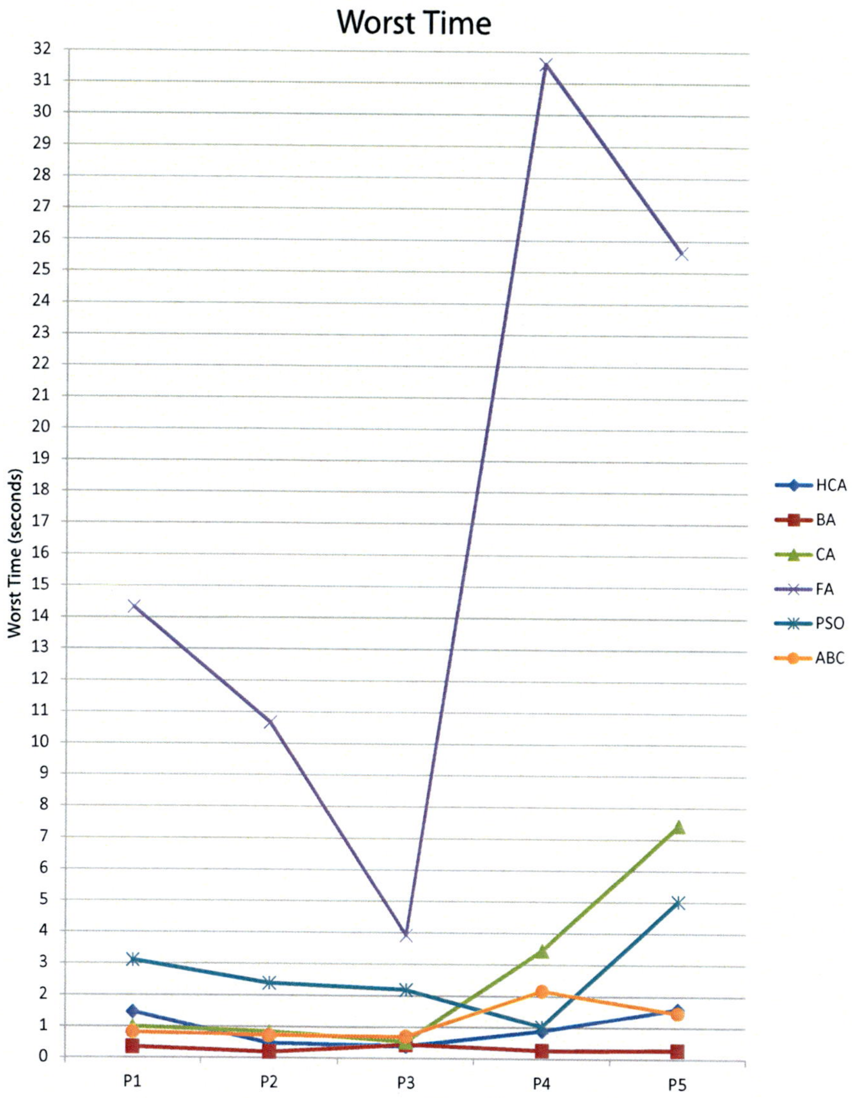

Fig. 17 Graph for worst time

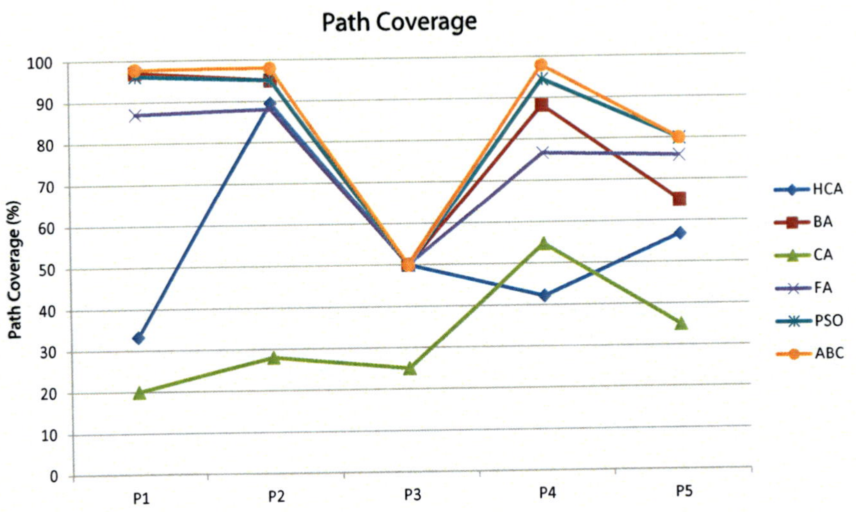

Fig. 18 Graph for path coverage

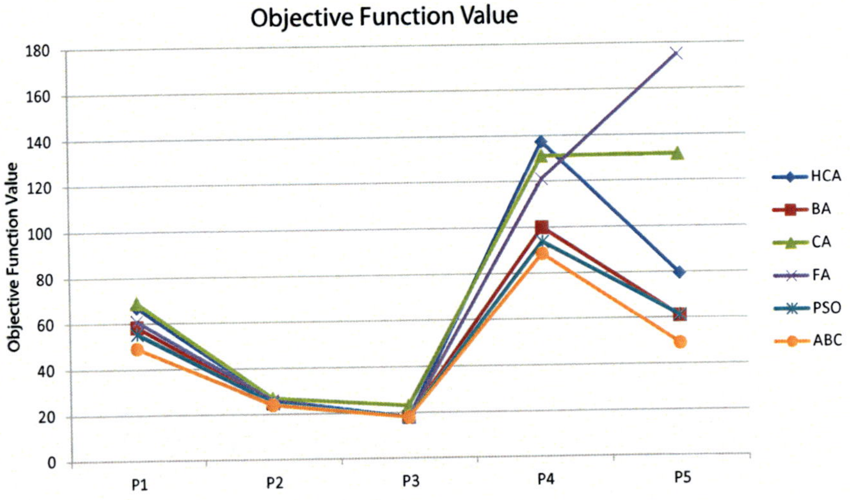

Fig. 19 Graph for objective function value

coverage and objective function values with the x-axis describing the discrete points for each program under test and the y-axis representing their values respectively.

6 Discussion

The purpose of this study is to compare the performance of six major meta-heuristic algorithms to solve the problem of test suite generation and test suite optimization. Each algorithm was implemented over five software programs written in java and the metrics used to evaluate their relative performances were average time, best time, worst time, path coverage and objective function value.

HCA across the five programs under test performed fast but not optimally as indicated by results shown in Tables 4, 5, 6, 7 and 8. Results demonstrated that the average time for HCA were really small with the highest being 927.06 ms for P5 while the smallest being 185.48 ms for P3. This strongly indicates that HCA demonstrates fast results with very slight variation in average time with respect to the LOC of the program. This is evident by the data about time over the two extreme ends of LOCs. In the study the average time for the longest program P4 with an LOC of 75 was 642.87 ms while that for P2 with an LOC of 18 was 304.25 ms. These fast results though were not optimal Test Suites that could be safely facilitated for proper testing. The path coverage for HCA varied from 33.2% for P1 to 89.5% for P2 as shown in Fig. 18. These were the second lowest path coverage only better than CA for P1, P2, P3, and P5 while HA had the lowest path coverage for P4. The objective function values were found to be ranging from 18 to 137.41, the results of which indicate that HCA produced one of the least optimal results for TSG. The study hence demonstrated that faster times are a strong suite for HCA but the results produced aren't ideal when it comes to TSG.

BA was the found to be the fastest algorithm of the six meta-heuristic algorithms undertaken. This characteristic of BA can be attributed to the Automatic Zooming and Parameter Tuning features of BA. The results as shown in Fig. 15 for average time indicate BA as the fastest performing algorithm over all five programs under test. The average time values ranged from 101.07 ms for P2 to 313.04 ms for P3. Overall the best time and worst time for BA were 62 ms for P2 and 441 ms for P3 respectively out of all 100 executions of the algorithm for the programs under test. BA performed very fast even for P4 and P5 that had the highest LOC of the 5 programs. Average time for P4 with 75 LOC was 138.24 ms which was really fast. The Test Suites produced were not as optimal though. The path coverage values ranged from 97% for P1 to 50% for P3. Results indicated that BA provides highly optimal results for smaller LOC while producing slightly worse for higher LOC and more complex programs as evidenced by Cyclomatic Complexity values. The evidence for which is indicated over the results from Tables 3, 4, 5, 6, 7 and 8 that show that BA produced the second most optimal results for P1 and P2 that had smaller LOC, right below ABC, while BA produced third best Test Suites for P4 after ABC and PSO and fourth best Test Suites for P5 after ABC, PSO and FA. Program P3 witnessed the same path coverage of 50% from all six algorithms owing to infeasible paths in the program. The path coverage values for BA over the 5 programs under test were 97%, 95%, 50%, 88.23% and 65% respectively and are relatively optimal across the six algorithms under question though not as good as ABC.

CA surprisingly didn't perform as well when it comes to the generation of Test Suites over the programs under test. It produced less optimal Test Suites while also being among the slower performing algorithms. The average time for CA ranged from 313.04 ms for P3 to 4380.36 ms for P4. It was one of the slower algorithms right after PSO and CA which were much slower as shown in Fig. 15. In addition to the slower time the Test Suites produced were not optimal in fact it produced the least optimal Test Suites among the six algorithms. The path coverage values for CA ranged from the lowest at 20% for P1 to the highest at 54.76% for P4 while the Objective Function values ranged from 22.98 for P3 for to 131.57 for P5 which were not as minimized as other algorithms.

FA was the slowest performing algorithm as indicated by the results shown in Tables 4, 5, 6, 7 and 8 for the programs under test. The performance was slow even in programs with smaller LOC. The average time values ranged from 3124.3 ms for P3 that had a mere 25 LOC to 22,586.08 ms for P4 that had 75 LOC. FA produced mixed results when it comes to optimal Test Suites with path coverage values ranging from 50% for P3 to 87% for P1. It produced more optimal Test Suites than HCA and CA but far less optimal results than ABC, BA and PSO over all programs under test. Results indicate that FA wouldn't be as suited an algorithm for Testing due to its slow and non optimal characteristics relative to other far superior algorithms as shown by the evaluation values.

PSO is capable of producing optimal Test Suites as indicated by the results for path coverage and objective function. It produced better path coverage for programs with higher LOC as described in Tables 7 and 8 for P4 and P5 respectively. The average time for PSO ranged from 1052.4 ms for P3 to 4380.36 ms for P4. The results hence demonstrate that PSO constitutes one of the slower algorithms among the six meta-heuristic algorithms under question. Thus there is a clear trade off for PSO between Time and Optimizing Capabilities. The path coverage values for PSO over the 5 programs under test were 96.2%, 95%, 50%, 94.46% and 80% which constitute a highly optimal set of Test Suites in the realm of TSG. The metrics for the Test Suites are lower compared with ABC over all programs and lower compared with BA over P1, P2 and P3. The trade off must be considered while implementing PSO for TSG and future work can involve improving PSO to hybridize it to produce faster results.

ABC was found to be the most well rounded algorithm for TSG as indicated by the results produced. The path coverage values ranged from the highest 98% for P2 to 50% as the lowest for P3 owing it to the infeasible paths in P3. The algorithm performed well irrespective of LOC and Cyclomatic Complexity of the program under question as described by Table 3 and Figs. 16 and 17. The average time for ABC represented the fastest time of 437.49 ms for P2 which had an LOC of 18 i.e. the smallest LOC among the programs under test as described in Table 5. ABC was the third fastest algorithm not much behind BA and HCA that performed faster but produced non optimal results. BA and HCA had average times at 101.07 and 304.25 ms respectively for P2 which is similar to that of ABC and concurrently the Test Suites generated by BA & HCA were not as optimal when compared to ABC as given in Table 5. Similarly the slowest average time for ABC was found to be 1596.38 ms for P4 which had an LOC of 75 i.e. the largest LOC among the programs under test which was again comparable with other algorithms as described in Table 7. This is evidence for the fact that ABC performs well irrespective of the program's LOC and complexity. Thus as evident by the results ABC offers the best trade off between time and optimality when it comes to TSG as ABC is reasonably fast while also producing the best and most optimal Test Suite results.

The results hence indicated that ABC performed in the most balanced way. ABC produced the most optimal results in relatively faster time when compared to the other algorithms. The path coverage values for the five programs for ABC were [97.8%, 98%, 50%, 97.92%, 80%]. When compared with other algorithms these were the best path coverage attained. Even though ABC wasn't the fastest algorithm it gave the best results. ABC was found to be the second and third fastest algorithm depending on the program under test right after BA which was the fastest as depicted in Fig. 15. BA was the fastest but the path coverage produced was lower than that of ABC. The path coverage values for BA were [97%, 95%, 50%, 88.23%, 65%].

FA was found to be the slowest of the algorithms and the results produced were not comparable to ABC and BA's test suites. HCA while performing really fast produced non optimal results while PSO performed in the middle.

In general the results can be summed up to say that ABC, BA and PSO were the better optimal Test Suite Generators (in that order) while CA, HCA and FA produced non optimal Test Suites. This stands for the Product Metrics: Path Coverage and Objective Function Value of the Test Suites generated. On the other hand BA, HCA and ABC were the faster algorithms with similar processing times for the Process Metrics: Average Time, Best Time & Worst Time. FA, PSO and CA were among the slower performing algorithms. Hence ABC and BA being the choice of algorithms for TSG while PSO can be improved in the future for the special case of TSG.

6.1 Theoretical Contribution

The study presents extensive results for analysis between the relative performances of six major meta-heuristic algorithms over the same problems. The algorithms chosen were the most popular and most widely implemented algorithms of the many algorithms. With so many algorithms to choose from the process of creating a tool that exploits the best features of each algorithm is a daunting task. And the first step in the process of automating the Test Suite Optimization phase is to analyse and understand which algorithms work the best for the problem of Test Suite Generation and Optimization. This would really narrow down the research to the best performing algorithms.

6.2 Implication for Practice

The final tools to be built for Test Automation require a set of algorithms that work the best together in producing comprehensive Test Suites and hence with this study the research can be narrowed to two or three algorithms. To be specific ABC performed the most optimally and BA performed the fastest and hence the Automatic Zooming and Parameter Control features of BA can be combined with the division of labour in ABC to produce an even better algorithm that's a hybrid of the two and produces the best results.

Further researchers and developers can choose algorithms based on their particular requirements. If time is a more relevant factor then the algorithm of choice would be BA. On the contrary if quality of test suites being more important would imply ABC to be the choice of algorithm.

7 Conclusion and Future Work

The purpose of the study was to find the best suited algorithms for the task of test suite optimization and compare their relative performance under similar conditions. This research concludes that BA and ABC would be the best alternatives to further improve for Test Suite Generation. Other algorithms couldn't produce qualitatively comparable Test Suites as compared to the corresponding path coverage metrics for BA [97%, 95%, 50%, 88.23%, 65%] and ABC [97.8%, 98%, 50%, 97.92%, 80%]. FA was found to be slowest and HCA & CA produced highly non optimal Test Suites out of the six algorithms namely, PSO, ABC, BA, HCA, FA and CA. In the correct order: ABC, BA & PSO were the most optimal algorithms (ABC being the most optimal) while CA, HCA & FA were less optimal. On the other hand BA, HCA & ABC were the faster algorithms while FA, PSO & CA were among the slower performing algorithms. The results presented in this study indicate the merits and demerits of each of the six individual algorithms which can aid in choosing algorithms based on the relevant requirement.

Even though the study details an extensive list of results for six algorithms there can be further improvement. The algorithms were tested over just five programs under test and hence a more extensive set of problems need to be solved in the future to see how they perform in even larger sets with larger LOC and a diverse set of Cyclomatic Complexities. Further the algorithms were tested using the linear scalerization of path coverage and branch coverage to form a multi objective function. In the future the optimization function can be broadened to include more parameters. The study focuses on comparing their relative performance in the same environment.

Key challenges in future studies include testing with a wide variety of Multi Objective functions, and testing the hypothesis that hybridized versions of ABC and BA can overcome their individual trade-offs of time and optimality.

The key aspects and lessons learned involve especially the trade-offs of BA and PSO. Another important learning through the study includes the slow processing of FA as well as the lack of optimality in CA which seemed surprising. These learning along with other aspects of the six algorithms can be used by future researches in the field of Test Suite Generation to improve the overall Testing Process and make Software Test Automation more accessible.

Lastly the future researchers can combine the characteristics of the studied algorithms to produce hybrid algorithms of desired capability. The results presented in this study paint a broad picture based on the results of individual merits of various algorithms.

References

1. Banker, R.D., Datar, S.M., Kemerer, C.F., Zweig, D.: Software complexity and maintenance costs. Commun. ACM **36**(11), 81–95 (1993)
2. Li, A., Zhang, Y.: Automatic generating all-path test data of a program based on pso. In: Software Engineering, 2009. WCSE'09. WRI World Congress on, vol. 4, pp. 189–193. IEEE (2009, May)
3. Skalak, D.B.: Prototype and feature selection by sampling and random mutation hill climbing algorithms. In: Machine Learning Proceedings 1994, pp. 293–301
4. Yang, X.-S., Deb, S.: Cuckoo search via Lévy flights. In: Nature & Biologically Inspired Computing, 2009. NaBIC 2009. World Congress on, pp. 210–214. IEEE (2009)
5. Yang, X.-S.: Firefly algorithm. In: Engineering Optimization, pp. 221–230 (2010)
6. Kennedy, J., Eberhart, R.: Particle swarm optimization. In: Proceedings of the IEEE International Conference on Neural Networks, Perth, Australia, 1942–1948. IEEE Service Center, Piscataway, NJ (1995)
7. Karaboga, D., Basturk, B.: A powerful and efficient algorithm for numerical function optimization: artificial bee colony (ABC) algorithm. J. Global Optim. **39**(3), 459–471 (2007)
8. Yang, X.-S.: A new metaheuristic bat-inspired algorithm. In: Nature inspired cooperative strategies for optimization (NICSO 2010), pp. 65–74. Springer, Berlin Heidelberg (2010)
9. Nguyen, D.C., Perini, A., Tonella, P.: A goal-oriented software testing methodology. Lect. Notes Comput. Sci. **4951**, 58–72 (2008)
10. Ngo, M.N., Tan, H.B.K.: Heuristics-based infeasible path detection for dynamic test data generation. Inf. Softw. Technol. **50**(7), 641–655 (2008)
11. Kanmani, S., Maragathavalli, P.: Search-based software test data generation using evolutionary testing techniques. Int. J. Softw. Eng. (IJSE) **1**(5), 10–22 (2010)
12. Yang, X.S.: A new metaheuristic bat-inspired algorithm. In: Nature Inspired Cooperative Strategies for Optimization (NICSO 2010), pp. 65–74 (2010)
13. Mansouri, P., Asady, B., Gupta, N.: The combination of bisection method and artificial bee colony algorithm for solving hard fix point problems. Soft Comput. Models Ind. Environ. Appl., 33–41 (2013)
14. Sharma, C., Sabharwal, S., Sibal, R.: A survey on software testing techniques using genetic algorithm (2014). arXiv preprint arXiv:1411.1154
15. Varshney, S., Mehrotra, M.: (2016). Search-based test data generator for data-flow dependencies using dominance concepts, branch distance and elitism. Arab. J. Sci. Eng. (Springer Science & Business Media BV), **41**(3)
16. Panichella, A., Kifetew, F., Tonella, P.: Automated test case generation as a many-objective optimisation problem with dynamic selection of the targets. IEEE Trans. Softw. Eng. (2017)

17. Kumar, S., Yadav, D.K., Khan, D.A.: A novel approach to automate test data generation for data flow testing based on hybrid adaptive PSO-GA algorithm. Int. J. Adv. Intell. Paradigms **9**(2–3), 278–312 (2017)
18. Mann, M., Tomar, P., Sangwan, O.P.: Bio-inspired metaheuristics: evolving and prioritizing software test data. Appl. Intell. 1–16 (2017)
19. Khari, M., Kumar, P., Burgos, D., Crespo, R.G.: Optimized test suites for automated testing using different optimization techniques. Soft Comput., 1–12 (2017)
20. Malhotra, R., Khari, M.: Test suite optimization using mutated artificial bee colony. In: Proceedings of the International Conference on Advances in Communication, Network, and Computing, CNC, pp. 45–54. Elsevier (2014)
21. Khari, M., Kumar, P.: An extensive evaluation of search-based software testing: a review. Soft Comput., 1–14 (2017)
22. Lourenço, H.R., Martin, O.C., Stützle, T.: Iterated local search. In: Handbook of Metaheuristics, pp. 320–353. Springer, Boston, MA (2003)
23. Wang, H., Wang, D., Yang, S.: A memetic algorithm with adaptive hill climbing strategy for dynamic optimization problems. Soft. Comput. **13**(8–9), 763–780 (2009)
24. Tsamardinos, I., Brown, L.E., Aliferis, C.F.: The max-min hill-climbing Bayesian network structure learning algorithm. Mach. Learn. **65**(1), 31–78 (2006)
25. Lotem, A., Nakamura, H., Zahavi, A.: Rejection of cuckoo eggs in relation to host age: a possible evolutionary equilibrium. Behav. Ecol. **3**(2), 128–132 (1992)
26. Rajabioun, R.: Cuckoo optimization algorithm. Applied soft computing **11**(8), 5508–5518 (2011)
27. Basu, M., Chowdhury, A.: Cuckoo search algorithm for economic dispatch. Energy **60**, 99–108 (2013)
28. Civicioglu, P., Besdok, E.: A conceptual comparison of the Cuckoo-search, particle swarm optimization, differential evolution and artificial bee colony algorithms. Artif. Intell. Rev. **39**(4), 315–346 (2013)
29. Karaboga, D., Gorkemli, B., Ozturk, C., Karaboga, N.: A comprehensive survey: artificial bee colony (ABC) algorithm and applications. Artif. Intell. Rev. **42**(1), 21–57 (2014)
30. Cheriguene, S., Azizi, N., Zemmal, N., Dey, N., Djellali, H., Farah, N.: Optimized tumor breast cancer classification using combining random subspace and static classifiers selection paradigms. In Applications of Intelligent Optimization in Biology and Medicine, pp. 289–307. Springer International Publishing (2016)
31. Yıldız, A.R.: An effective hybrid immune-hill climbing optimization approach for solving design and manufacturing optimization problems in industry. J. Mater. Process. Technol. **209**(6), 2773–2780 (2009)
32. Coelho, L.S., Guerra, F., Batistela, N.J., Leite, J.V.: Multiobjective cuckoo search algorithm based on Duffing's oscillator applied to Jiles-Atherton vector hysteresis parameters estimation. IEEE Trans. Magn. **49**(5), 1745–1748 (2013)
33. Moravej, Z., Akhlaghi, A.: A novel approach based on cuckoo search for DG allocation in distribution network. Int. J. Electr. Power Energy Syst. **44**(1), 672–679 (2013)
34. Sur, C., Shukla, A.: Discrete cuckoo search optimization algorithm for combinatorial optimization of vehicle route in graph based road network. In: Proceedings of the Third International Conference on Soft Computing for Problem Solving, pp. 307–320. Springer, New Delhi (2014)
35. Durgun, İ., Yildiz, A.R.: Structural design optimization of vehicle components using cuckoo search algorithm. Materials Testing **54**(3), 185–188 (2012)
36. Gandomi, A.H., Yang, X.S., Alavi, A.H.: Cuckoo search algorithm: a metaheuristic approach to solve structural optimization problems. Eng. Comput. **29**(1), 17–35 (2013)
37. Bulatović, R.R., Đorđević, S.R., Đorđević, V.S.: Cuckoo search algorithm: a metaheuristic approach to solving the problem of optimum synthesis of a six-bar double dwell linkage. Mech. Mach. Theory **61**, 1–13 (2013)
38. Senthilnath, J., Das, V., Omkar, S. N., Mani, V.: Clustering using levy flight cuckoo search. In: Proceedings of Seventh International Conference on Bio-Inspired Computing: Theories and Applications (BIC-TA 2012), pp. 65–75. Springer, India (2013)

39. Kumar, R., Rajan, A., Talukdar, F.A., Dey, N., Santhi, V., Balas, V.E. Optimization of 5.5-GHz CMOS LNA parameters using firefly algorithm. Neural Comput. Appl. **28**(12), 3765–3779 (2017)
40. Vimalarani, C., Subramanian, R., Sivanandam, S.N.: An enhanced PSO-based clustering energy optimization algorithm for wireless sensor network. Sci. World J. (2016)
41. Sousa, T., Silva, A., Neves, A.: Particle swarm based data mining algorithms for classification tasks. Parallel Comput. **30**(5–6), 767–783 (2004)
42. Satapathy, S.C., Naik, A., Parvathi, K. A teaching learning based optimization based on orthogonal design for solvin (2013)
43. Singh, A.: An artificial bee colony algorithm for the leaf-constrained minimum spanning tree problem. Appl. Soft Comput. **9**(2), 625–631 (2009)
44. Karaboga, N.: A new design method based on artificial bee colony algorithm for digital IIR filters. J. Franklin Inst. **346**(4), 328–348 (2009)
45. Sharma, H., Bansal, J.C., Arya, K.V., Yang, X.S.: Lévy flight artificial bee colony algorithm. Int. J. Syst. Sci. **47**(11), 2652–2670 (2016)
46. Elbeltagi, E., Hegazy, T., Grierson, D.: Comparison among five evolutionary-based optimization algorithms. Adv. Eng. Inform. **19**(1), 43–53 (2005)
47. Nidhra, S., Dondeti, J.: Black box and white box testing techniques-a literature review. Int. J. Embed. Syst. Appl. (IJESA) **2**(2), 29–50 (2012)
48. Watson, A.H., Wallace, D.R., McCabe, T.J.: Structured Testing: A Testing Methodology Using the Cyclomatic Complexity Metric (Vol. 500, No. 235). US Department of Commerce, Technology Administration, National Institute of Standards and Technology
49. Yan, J., Zhang, J.: An efficient method to generate feasible paths for basis path testing. Inf. Proces. Letters **107**(3–4), 87–92 (2008)
50. Zhonglin, Z., Lingxia, M. (2010, August). An improved method of acquiring basis path for software testing. In: Computer Science and Education (ICCSE), 2010 5th International Conference on, pp. 1891–1894. IEEE

Humanoid Robotics and Fuzzy Control

Comparative Study of Conventional and Interval Type-2 Fuzzy Logic Controllers for Velocity Regulation in Lego Mindstorms Ev3 Humanoids

Fevrier Valdez, Oscar Castillo, Camilo Caraveo, and Cinthia Peraza

Abstract Recently, fuzzy logic has allowed complex systems to be designed and controlled more effectively than traditional approaches, since it provides a simple way to handle noisy or imprecise information. Nowadays, one of the great problems in the speed control cases is to make the decision to use a fuzzy logic control (FLC) system instead of a conventional controller system as a Proportional Integral (PI) and Integral-Derivative (PID). This chapter presents a comparative study using three types of controllers, FLC, PI and PID, applied to the speed control of a Robot built using the ev3 Lego Mindstorms kit. Matlab and Simulink were used to validate the performance of the speed control obtained with the controller proposed in this paper. This particular type of robot has some similarities to Humanoid robots as it needs to move autonomously avoiding obstacles, so we plan to later implement the fuzzy controllers on larger and more complex Humanoid robots.

Keywords Fuzzy logic control · Proportional integral · Proportional integral derivative · ev3 Lego Mindstorms · Closed loop control

1 Introduction

There are several methods to control system control the classic ones are the proportional control (P) that determines the relation of the current error [1], proportional integral control (PI) this generates a proportional correction to complete the of the errors as can be observed in [2–5] and the proportional Integral Derivative (PID) determines the reaction of the time in which the occurs error, these controllers are used to solve a problem when there is no knowledge of the process where its objective is to obtain greater precision in the control of the system, some related works in the literature can be observed in [6–8].

A system of fuzzy logic control (FLC) is widely used to solve any problem knowledge-based fuzzy inference proposed by Zadeh in 1965 [9], unlike classical

F. Valdez · O. Castillo (✉) · C. Caraveo · C. Peraza
Tijuana Institute of Technology, 22379 Tijuana, Mexico
e-mail: ocastillo@tectijuana.mx

C. Kahraman and E. Bolturk (eds.), *Toward Humanoid Robots: The Role of Fuzzy Sets*, Studies in Systems, Decision and Control 344,
https://doi.org/10.1007/978-3-030-67163-1_7

logic where you have only true or false, an inference system fuzzy handles simple linguistic variables that adapt better to the real world, for example: Minimum, Maximum, etc. [10], fuzzy logic has been widely used as shown in [11–17].

The era of robots has been increasing in recent years [18], there are various types of robots in this case a Ev3 Lego Mindstorms which is the third generation of sets of robotics belonging to the company LEGO is used, these have been used to solve certain problems as shown in [19–21]. This particular robot has been considered as a Humanoid in some studies, although a relatively simple one [22–24].

Today is complex make the decision to use a PID controller or a fuzzy logic system (FLC), in this paper the 3 types of control PI, PID and FLC applied to the speed control of an Ev3 Lego Mindstorms with the main objective to achieve speed control robot.

There are different methods and techniques to find the best parameters for PI, PID controllers such as the Ziegler Nichols method [25–27], for the case of FLC there are evolutionary computation techniques to find the best parameters by means of some algorithm [28–32]. In this case the values of the parameters of these drivers were found to trial and error.

The metrics used are the Integral of Squared error (ISE), Integral of the Absolute value of the Error (IAE), Integral of Time-weighted Squared Error (ITSE), Integral of the Time multiplied by the Absolute value of the Error (ITAE) and the root mean square error (RMSE).

This chapter is comprised of the following sections: the conventional control system is presented in Sect. 3, the theory of fuzzy control system is shown in Sect. 4, the theory of robotics kit Ev3 Lego Mindstorms is observed in Sect. 5, Sect. 6 shown the methodology, Sect. 7 shown the simulations and results, conclusions are presented in Sect. 8.

2 Humanoid Robots

Humanoid robotics is an emerging and challenging research field, which has received a lot of attention during the past years and will continue to play a central role in robotics research and in many applications of the twenty-first century. Regardless of the application area, one of the common problems tackled in humanoid robotics is the understanding of human-like information processing and the underlying mechanisms of the human brain in dealing with the real world [22].

Humanoid robots are now used as research tools in several scientific areas. On one side, researchers study the human body structure and behavior (biomechanics) to build humanoid robots. On another side, the attempt to simulate the human body leads to a better understanding of it. Human cognition is a field of study which is focused on how humans learn from sensory information in order to acquire perceptual and motor skills. This knowledge is used to develop computational models of human behavior and it has been improving over time.

A humanoid robot can be defined as a robot with its body shape built to resemble the human body. The design may be for functional purposes, such as interacting with human tools and environments, for experimental purposes, such as the study of bipedal locomotion, or for other purposes [23].

In planning and control, the essential difference between humanoids and other kinds of robots (like industrial ones) is that the movement of the robot must be human-like, using legged locomotion, especially biped gait. The ideal planning for humanoid movements during normal walking should result in minimum energy consumption, as it does in the human body. For this reason, studies on dynamics and control of these kinds of structures has become increasingly important.

The question of walking biped robots stabilization on the surface is of great importance. Maintenance of the robot's gravity center over the center of bearing area for providing a stable position can be chosen as a goal of control [33]

To maintain dynamic balance during the walk, a robot needs information about contact force and its current and desired motion. The solution to this problem relies on a major concept, the Zero Moment Point (ZMP).

Another characteristic of humanoid robots is that they move, gather information (using sensors) on the "real world" and interact with it. They do not stay still like industrial manipulators and other robots that work in highly structured environments. To allow humanoids to move in complex environments, planning and control must focus on self-collision detection, path planning and obstacle avoidance [24].

3 Control Models

This section describes the principle and analysis of a closed loop control, proportional-integral (PI) plus a derivative (D) controller.

3.1 Closed-Loop Control

There are two types of forms in a control algorithm to decide on an action: the open loop system the parameters of the algorithm are preset and do not change while the system is running, the closed loop system sensors measure the error between the desired state of the system and its real state, and this error is used to decide what action to take. To achieve speed control in a Ev3 Lego Mindstorms, closed circuit

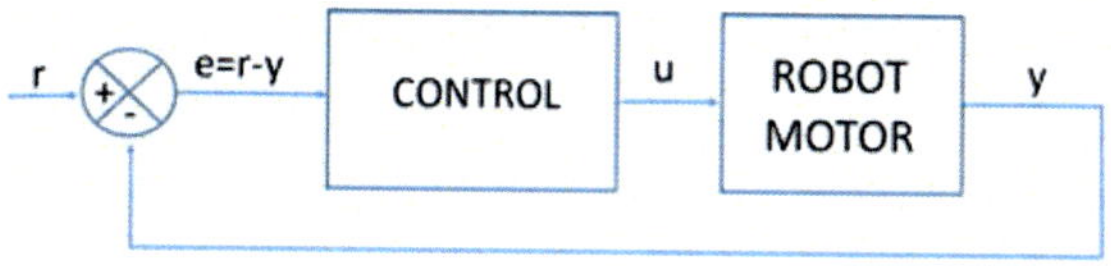

Fig. 1 Closed loop control system

control systems were used. Figure 1 shows the general scheme of a closed loop control system [34–37].

The parameter r represents the reference value, the variable transformed into a control value is represents by u, the parameter y represents the output which is calculated with the equation $e = r - y$.

3.2 Proportional-Integral Controller

A proportional-integral controller can reach the reference distance even in the presence of friction or a moving object by taking into account the accumulated error over time, the PI controller adds the integral of the error from the time when the algorithm starts to run until the present time, this controller is given by Eq. (1).

$$r(t) = k_p e(t) + k_i \int_0^t e(\tau) d\tau \tag{1}$$

where r is the controller output, t is the time, k_p is the proportional gain, e is the error between the reference value and the system output and $d\tau$ is the derivative time.

3.3 Proportional Integral-Derivative

A Proportional Integral Derivative controller (PID), is a feedback control mechanism that calculates the deviation or error between a measured value and the value to be obtained, to apply a corrective action that adjusts the process. The calculation algorithm of the PID control is given in three different parameters: the proportional (k_p), the integral (k_i) and the derivative (k_d), as shown in the Eq. (2).

$$u(t) = k_p e(t) + k_i \int_{\tau=0}^{t} e(\tau) d\tau + k_d \frac{de(t)}{dt} \tag{2}$$

4 Fuzzy Logic Controller

The control algorithms PI and PID used exact mathematical computations to determine the signals used to control the behavior of a system. An alternate method is to use fuzzy logic control (FLC) algorithms based upon rules, these systems have been

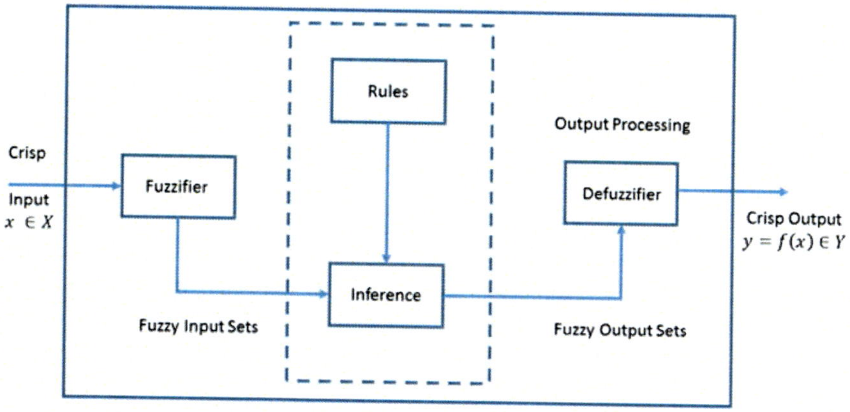

Fig. 2 Diagram of a fuzzy logic system

Fig. 3 Diagram of an FLC

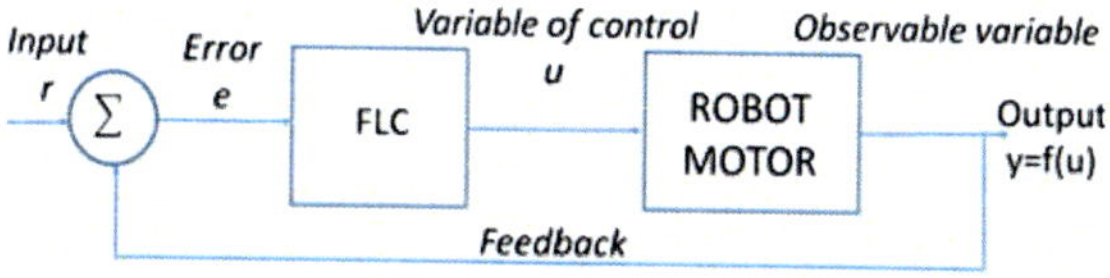

widely used as shown in [38–43]. A control system contains rules; these rules are expressed in terms of linguistic variables. FLC was created by Lotfi Zadeh, based on terms of fuzzy set infefigrence. A fuzzy logic controller consists of three phases that are: fuzzification, apply rules and defuzzification, these elements are represented in the Fig. 2.

Based on the theory of fuzzy sets the FLC control systems fuzzy logic is created, the general diagram of an FLC is shown in the Fig. 3.

5 Robotics Kits

The LEGO Mindstorms robotics kits were introduced in 1998. A kit consists of standard LEGO bricks and other building components, together with motors and sensors, and a programmable brick which contains the computer that controls the components of the robot. LEGO has three generations of robotics sets, the most current is the LEGO Mindstorms EV3, this is the third generation of robotic sets, the acronym Ev indicates the evolution and the number three indicates the generation. Currently these robots have been widely used for their ease of use and programmable blocks, although their main disadvantage is that they are expensive and the ability to implement a robust system, these robots have been widely used as shown in [44–47]. In this chapter a LEGO MINDSTORMS EV3 robot has been used for

Fig. 4 LEGO Mindstorms EV3 Robot

experimentation, this robot has several characteristics therefore can be assembled in different ways, in this case the robot used is shown in Fig. 4.

The mathematical model for modeling the motor robot uses two fundamental laws of physics, the first Kirchhoff's law is based on the energy conservation and load on the electrical circuits shown in Eq. (3) and the second is Newton's law based on the mechanical equations shown in Eq. (4).

$$L\frac{di(t)}{dt} = v_{in}(t) - v_{emf}(t) - Ri(t) \tag{3}$$

$$J\frac{dw(t)}{dt} = T_m(t) - bw(t) \tag{4}$$

where:
R, L = Motor equivalent circuit resistance and inductance respectively.
J = Moment of inertia of the rotor.
b = Damping coefficient of the rotor.
V_{in} = Input voltage.
V_{emf} = Back emf.
Tm = Motor torque.
$T_m(t) = K_m i(t)$.
$v_{em}f(t) = K_{emf}\omega(t)$.

6 Methodology

This article is composed of 3 different control methods to achieve the speed control of a LEGO MINDSTORMS EV3 robot that is shown in Figs. 4 and 5 a general illustration of the proposal is presented.

The main objective is to obtain the speed control of this robot by means of a PI, PID and FLC controller. The design of the controllers in this case was done manually. In this article were used Matlab 2015 and Simulink to perform the experimentation. The speed control was tested with two different speed signal references, a step and a signal generator for each method PI, PID and FLC. The step reference signal *r(t)*, is given by Eq. (5) where *t* is a sample time, the signal generator is given by Eq. (6) where amplitude is 200, frequency is 0.8, waveform is square and *t* is a simulation time.

$$r(t) = \begin{cases} 2 * \frac{180}{pi} & t > 0 \\ 0 & t \leq 0 \end{cases} \tag{5}$$

$$r(t) = Amp * Waveform(Freq, t) \tag{6}$$

The objective function for FLC is the root mean square (RMSE) that is represented in the Eq. (7) and the objective function for PI and PID controller are the Settling Time, although there are other control metrics that were also used which are the following: ISE (Integral of Squared error), IAE (Integral of the Absolute value of the Error), ITSE (Integral of Time-weighted Squared Error), ITAE (Integral of the Time multiplied by the Absolute value of the Error), respectively presented in Eqs. (8)–(11).

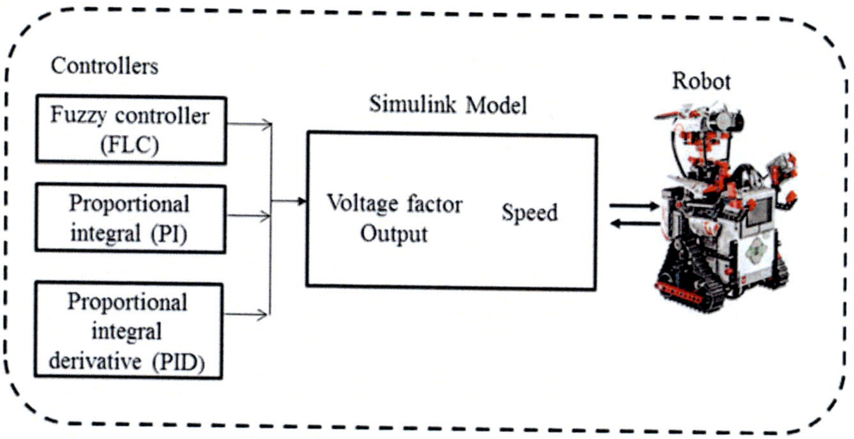

Fig. 5 General diagram of the proposed approach

$$RMSE = \sqrt{\frac{1}{N}\sum_{t=1}^{N}(x_t - \hat{x}_t)^2} \tag{7}$$

$$ISE = \sum_{t=1}^{N}\left|(x(t) - \hat{x}(t))^2\right| \tag{8}$$

$$IAE = \sum_{t=1}^{N}\left|x(t) - \hat{x}(t)\right| \tag{9}$$

$$ITSE = \sum_{t=1}^{N} t\left((x(t) - \hat{x}(t))^2\right) \tag{10}$$

$$ITAE = \sum_{t=1}^{N} t\left(x(t) - \hat{x}(t)\right) \tag{11}$$

6.1 *Proportional Integral Controller for Lego Mindstorms ev3*

The two aforementioned speed reference signals, is applied to a PI controller model in order to achieve the speed control in the Lego Mindstorms ev3 robot, Fig. 6 shows the diagram of this controller.

Where *r(t)* is the reference signal (step and signal generator), *e(t)* is the difference between the reference signal and the actual output signal (*error*), *u(t)* is the controller output and *y(t)* is the output, in this case the speed of the robot. The parameters used for speed control with the step reference signal are shown in Table 1 and the parameters used for speed control with the signal generator reference signal are shown in Table 2.

Fig. 6 Schema for PI controller

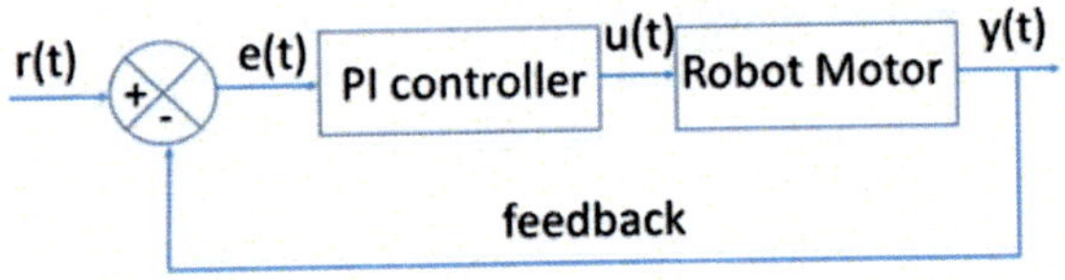

Table 1 Parameters used for step reference

Parameters	P	I
Value	0.0000009	0.5

Table 2 Parameters used for signal generator reference

Parameters	P	I
Value	0.0000009	0.3

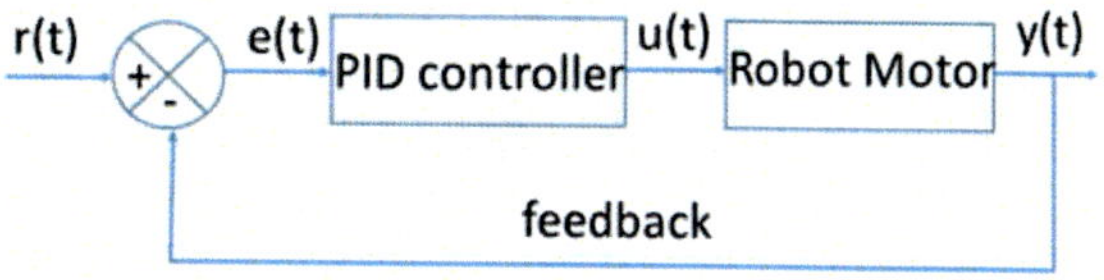

Fig. 7 Diagram of PID controller

Table 3 Parameters used for step reference

Parameters	P	I	D
Value	0	0.5	0

Table 4 Parameters used for signal generator reference

Parameters	P	I	D
Value	0.05	0.2	0.001

6.2 *Proportional Integral Derivative Controller for Lego Mindstorms ev3*

The PID controller model used to achieve the speed control of the Lego Mindstorms ev3 robot, can be seen in Fig. 7.

In this case, the same speed reference signals mentioned above were used. The parameters used for speed control with the step reference signal are shown in Table 3 and the parameters used for speed control with the signal generator reference signal are shown in Table 4.

6.3 *Fuzzy Logic Controller for the Lego Mindstorms Ev3*

A fuzzy logic controller is used in order to control the speed of the robot, also the two speed reference signals were used in the above Eqs. (3) and (4). The FLC control model it can be seen in Fig. 8.

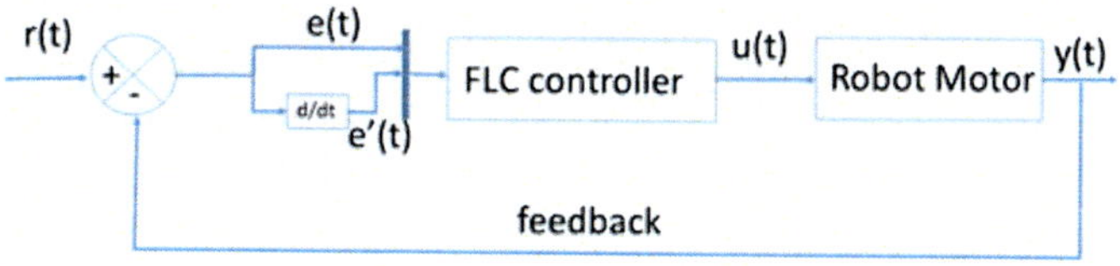

Fig. 8 Schema of fuzzy logic controller

The fuzzy system controller contains two inputs, the *error e(t)* and the *error change e′(t)* and one output the control signal *(Voltage) y(t)*, the error is calculated by means of the Eq. (12) and the error change is calculated using Eq. (13)

$$e(t) = r(t) - y(t) \tag{12}$$

$$e'(t) = e(t) - e(t-1) \tag{13}$$

The inputs and output are composed of trapezoidal and triangular membership functions, which were created with Eqs. (14) and (15), respectively.

$$f(x; a, b, c, d) = \begin{cases} 0, & x \le a \\ \frac{x-a}{b-a}, & a \le x \le b \\ 1, & b \le x \le c \\ \frac{d-x}{d-c}, & c \le x \le d \\ 0, & d \le x \end{cases} \tag{14}$$

$$f(x; a, b, c) = \begin{cases} 0, & x \le a \\ \frac{x-a}{b-a}, & a \le x \le b \\ \frac{c-x}{c-b}, & b \le x \le c \\ 0, & c \le x \end{cases} \tag{15}$$

The parameters of the membership functions for speed control with the step reference signal are shown in Table 5. The parameters of the membership functions for speed control with the signal generator reference signal are shown in Table 6. The FLC contains 15 rules which are show in Fig. 9.

Table 5 shows the parameters of the triangular and trapezoidal membership functions used to achieve speed control with the step reference.

Table 6 shows the parameters of the triangular and trapezoidal membership functions used to achieve speed control with the signal generator reference.

These are the rules have been used to control the motor speed of the Lego robot EV3.

The speed step reference signal is illustrated in Fig. 10 and the speed signal generator reference is illustrated in Fig. 11, where the objective is to approach the speed desired reference.

7 Simulations and Results

Different tests were carried out in order to find the results closest to the reference speed, the robot physically maintains a good speed without overlap or sudden movements. In the following sections the best results obtained by each speed reference

Table 5 FLC Membership functions for step reference

Input *Error*				
MF	a	b	c	d
NegV	−915.7	−625.2	284.4	6.609
CeroV	−287.4	0	292.6	–
PosV	−1.391	308.2	590.4	870.40
Input *Error Change*				
ErrNeg	−400.1	−98.01	−40	−10.01
ErrNegM	−40	−20	0	–
SinErr	−17.69	0	15.6	–
ErrMaxM	−7.677	20	40	–
ErrMax	4.497	40	98.01	400.1
Output *Voltage*				
MDis	0	0	5	11.34
MDism	6.576	9.937	13.53	–
Man	10.41	12.5	14.78	–
Aumm	11.41	15.06	17.56	–
Aum	14.25	20.53	25.52	25.52

Table 6 FLC Membership functions for signal generator reference

Input *Error*				
MF	a	b	c	d
NegV	−549	−375	−224.6	3.975
CeroV	−239	0	224	
PosV	−0.795	226.2	354	522
Input *Error Change*				
ErrNeg	−200	−49	−39	−16.27
ErrNegM	−34.25	−17.33	−7.8	–
SinErr	−19.45	−0.264	11.24	–
ErrMaxM	−0.398	14.42	30	–
ErrMax	7.804	32.95	49	200
Output *Voltage*				
MDis	−71	−71	−52.03	−6.59
MDism	−43.08	−23.99	−3.572	–
Man	−11.1	0.75	13.7	–
Aumm	6.574	27.33	41.53	–
Aum	9.98	48.27	74	74

No.	Inputs		Output
	Error	*Error change*	*Voltage*
1	NegV	ErrNeg	Dis
2	NegV	SinErr	Dis
3	NegV	ErrMax	Dis_m
4	CeroV	ErrNeg	Aum_m
5	CeroV	ErrMax	Dis_m
6	PosV	ErrNeg	Aum_m
7	PosV	SinErr	Aum
8	PosV	ErrMax	Aum
9	CeroV	SinErr	Man
10	NegV	ErrNeg_M	Dis
11	CeroV	ErrNeg_M	Aum_m
12	PosV	ErrNeg_M	Aum
13	PosV	ErrMax_M	Aum
14	CeroV	ErrMax_M	Dis_m
15	NegV	ErrMax_M	Dis

Fig. 9 Fuzzy rules

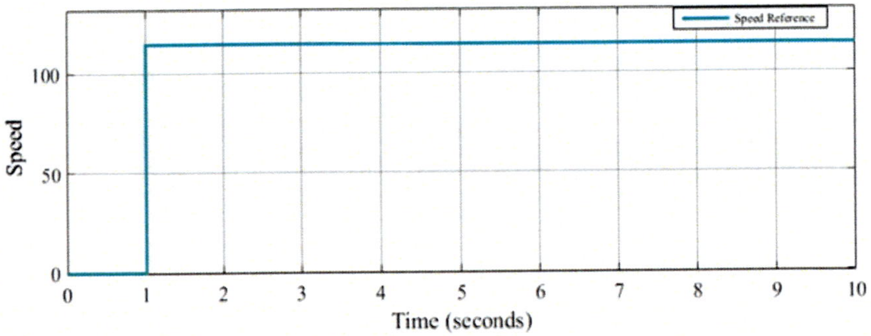

Fig. 10 Step reference signal

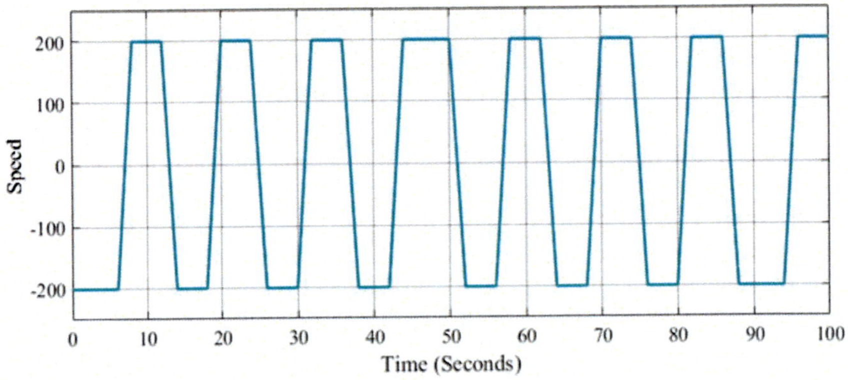

Fig. 11 Signal generator reference

is shown, it is important to note that adjusting these parameters have been found manually by trial and error.

7.1 Simulations for Step Reference

Tables 7 and 8 show the results obtained when adjusting the parameters of the PI and PID controller respectively for the speed reference of the step, in these tables we can see the results of the PID controller and the settling time which is the objective function in the case of these two controllers.

Figures 12 and 13 show the simulation obtained from the PI and PID controller respectively for the speed reference of the step, the blue line represents the reference speed and the orange line represents the actual speed.

Table 9 shows the results obtained by manually adjusting the parameters of the fuzzy logic controller for the speed reference of the step, RMSE error obtained is shown. Figure 14 shows the best simulation obtained.

Table 7 Results for PI controller

Metrics	ITAE	ITSE	IAE	ISE	Settling time
Value	1.13E+04	2.59E+06	2034	4.63E+05	1.98

Table 8 Results for PID controller

Metrics	ITAE	ITSE	IAE	ISE	Settling time
Value	1.131e+04	2.589e+06	2033	4.628e+05	1.98

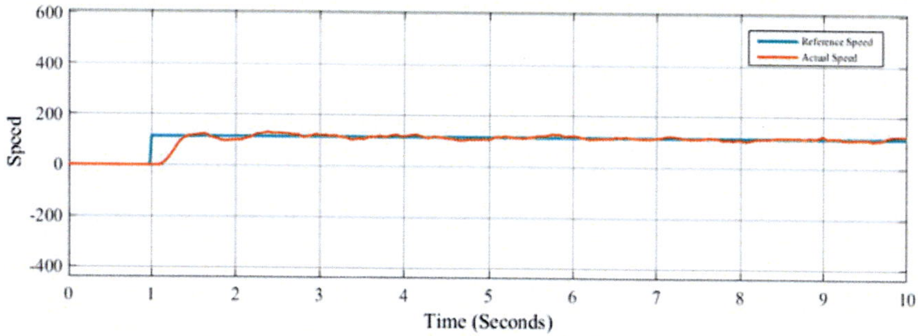

Fig. 12 Simulation for PI controller

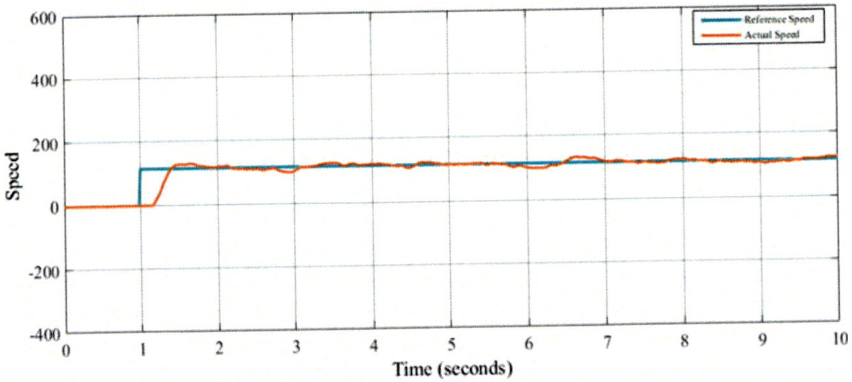

Fig. 13 Simulation for PID controller

Table 9 Results for FLC controller

Metrics	ITAE	ITSE	IAE	ISE	RMSE
Value	1.045e+04	2.199e+06	1927	3.98e+05	25.6765

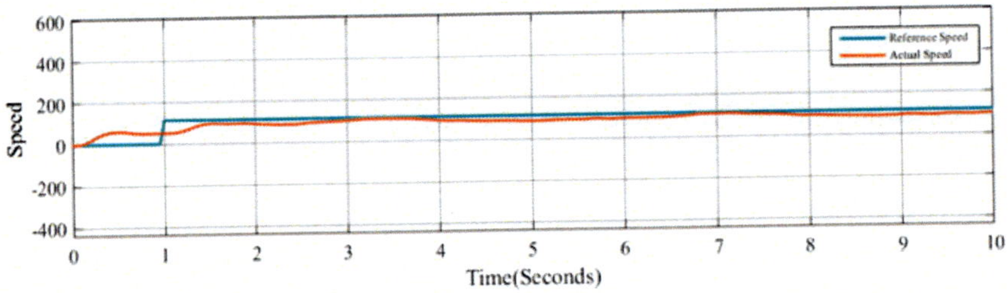

Fig. 14 Simulation for FLC controller

7.2 *Simulations for Signal Generator Reference*

Tables 10 and 11 show the results obtained when adjusting the parameters of the PI and PID controller respectively for the speed reference of the signal generator, in these tables we can see the results of the PID controller and the settling time which is the objective function in the case of these two controllers.

Table 10 Results for PI controller

Metrics	ITAE	ITSE	IAE	ISE	Settling time
Value	1.795e+04	6.932e+06	3592	1.37e+06	1.98

Table 11 Results for PID controller

Metrics	ITAE	ITSE	IAE	ISE	Settling time
Value	1.68e+04	6.062e+06	3411	1.22e+06	1.98

Figures 15 and 16 show the simulation obtained from the PI and PID controller respectively for the speed reference of the signal generator, the blue line represents the reference speed and the orange line represents the actual speed.

Table 12 shows the results obtained by manually adjusting the parameters of the fuzzy logic controller for the speed reference of the signal generator, RMSE error obtained is shown. Figure 17 shows the best simulation obtained.

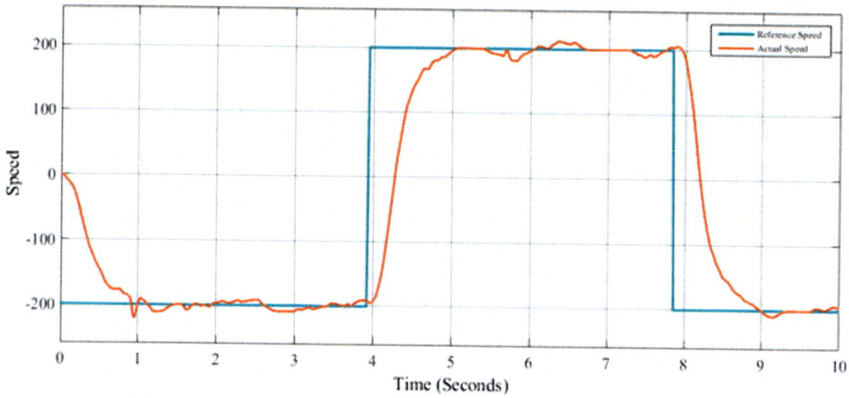

Fig. 15 Simulation for PI controller

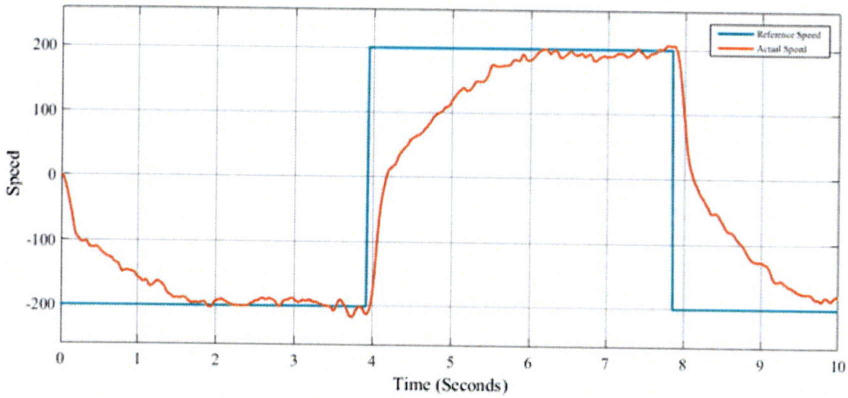

Fig. 16 Simulation for PID

Table 12 Results for FLC controller

Metrics	ITAE	ITSE	IAE	ISE	RMSE
Value	1.245e+04	3.336e+06	2511	6.659e+05	112.49

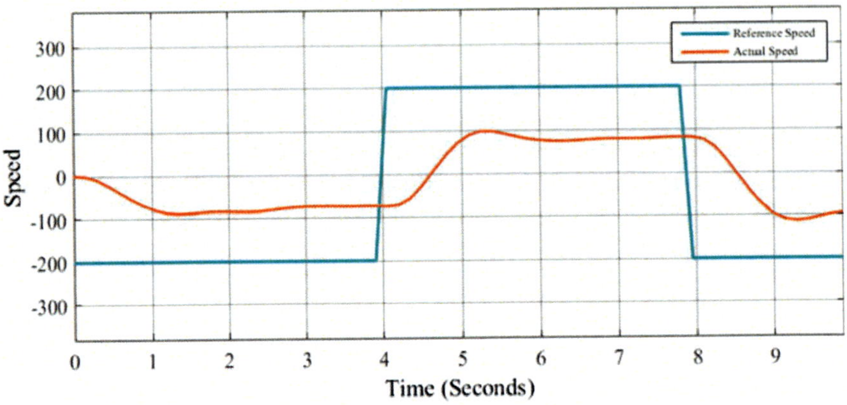

Fig. 17 Simulation for FLC

8 Conclusions

The main objective of this chapter was the design and integration of the following controllers (PI, PID, FLC) applied and modeled in a robot built using the Mindstorms Lego kit version ev3. The authors of this work consider important to mention to the reader that our approach and contribution was not the minimization of the obtained errors, rather the application and integration of different types of controls systems. The design of these controllers was done manually with the help of an expert in the area.

The speed control was performed with two different reference signals that are step and signal generator. The results obtained for the step reference speed are favorable but for the case of the signal generator reference speed they are not favorable, but in real time the robot manages to maintain a stable speed without frights which is the objective of this work.

As future work is considered consider it necessary to use some optimization meta-heuristics or other intelligent search techniques to find the best values for the controllers in order to improve the results shown. In addition, we plan to later implement the fuzzy controllers on larger and more complex Humanoid robots.

References

1. Ang, K.H., Chong, G., Li, Y.: PID control system analysis, design, and technology. IEEE Trans. Control Syst. Technol. **13**(4), 559–576 (2005)
2. Raviraj, V.S.C., Sen, P.C.: Comparative study of proportional-integral, sliding mode, and fuzzy logic controllers for power converters. IEEE Trans. Ind. Appl. **33**(2), 518–524 (1997)
3. Cheng, Y., Nan, Q., Wang, R., Dong, T., Tian, Z.: Fuzzy proportional integral derivative control of a radiofrequency ablation temperature control system, pp. 1–5 (2017)
4. Colonna, P., et al.: Organic rankine cycle power systems: from the concept to current technology, applications, and an outlook to the future. J. Eng. Gas Turbines Power **137**(10), 100801 (2015)
5. Ibrahim, Z., Levi, E.: A comparative analysis of fuzzy logic and PI speed control in high-performance AC drives using experimental approach. IEEE Trans. Ind. Appl. **38**(5), 1210–1218 (2002)
6. Bingol, M.C., Akpolat, Z.H., Koca, G.O.: Robust control of a robot arm using an optimized pid controller. In: Březina, T., Jabłoński, R. (eds.) Mechatronics 2017, vol. 644, pp. 484–492. Springer International Publishing, Cham (2018)
7. Bhandari, A.S., Chaudhuri, A., Roy, S., Negi, S., Sharad, M.: Single chip self-tunable N-input N-output PID control system with integrated analog front-end for miniature robotics, pp. 109–114 (2017)
8. Sa-ngiamvibool, W.: Optimal fuzzy logic proportional integral derivative controller design by bee algorithm for hydro-thermal system. IEEE Trans. Ind. Inform. pp. 1–1 (2017)
9. Zadeh, L.A.: Fuzzy sets. Inf. Control **8**(3), 338–353 (1965)
10. Zadeh, L.A.: The concept of a linguistic variable and its application to approximate reasoning—I. Inf. Sci. **8**(3), 199–249 (1975)
11. Caraveo, C., Valdez, F., Castillo, O.: Optimization mathematical functions for multiple variables using the algorithm of self-defense of the plants. In: Melin, P., Castillo, O., Kacprzyk, J. (eds.) Nature-Inspired Design of Hybrid Intelligent Systems, vol. 667, pp. 631–640. Springer International Publishing, Cham (2017)
12. Amador-Angulo, L., Castillo, O.: A new fuzzy bee colony optimization with dynamic adaptation of parameters using interval type-2 fuzzy logic for tuning fuzzy controllers. Soft Comput. **22**(2), 571–594 (2018)
13. . Olivas, H., Valdez, F., Castillo, O., Melin, P.: Theory and background. In: Dynamic parameter adaptation for meta-heuristic optimization algorithms through type-2 fuzzy logic. Springer International Publishing, Cham, pp 3–10 (2018)
14. Amador, L., Castillo, O.: Theory and background. In: Optimization of type-2 fuzzy controllers using the Bee Colony Algorithm. Springer International Publishing, Cham, pp. 7–11 (2017)
15. Bernal, E., Castillo, O., Soria, J., Valdez, F.: Imperialist competitive algorithm with dynamic parameter adaptation using fuzzy logic applied to the optimization of mathematical functions. Algorithms **10**(1), 18 (2017)
16. Barraza, J., Melin, P., Valdez, F., Gonzalez, C.: Fuzzy fireworks algorithm based on a sparks dispersion measure. Algorithms **10**(4), 83 (2017)
17. Ochoa, P., Castillo, O., Soria, J.: Interval type-2 fuzzy logic dynamic mutation and crossover parameter adaptation in a fuzzy differential evolution method. In: Hadjiski, M., Atanassov, K.T. (eds.) Intuitionistic Fuzziness and Other Intelligent Theories and Their Applications, vol. 757, pp. 81–94. Springer International Publishing, Cham (2019)
18. Poikselka, K., Vallivaara, I., Roning, J.: Evolutionary robotics on Lego NXT Platform, pp. 1137–1144 (2015)
19. Azlan, N.Z., Zainudin, F., Yusuf, H.M., Toha, S.F., Yusoff, S.Z.S., Osman, N.H.: Fuzzy logic controlled miniature LEGO robot for undergraduate training system, pp. 2184–2188 (2007)
20. Akmal, M.A., Jamin, N.F., Ghani, N.M.A.: Fuzzy logic controller for two wheeled EV3 LEGO robot, pp. 134–139 (2017)
21. Nair, S., Coronado, E., Frye, M., Goldaracena, T., Arguello, C.: Particle swarm optimization for the control of a swarm of biological robots, pp. 1–4 (2015)

22. Carpenter, J., Davis, J., Erwin-Stewart, N., Lee. T., Bransford, J., Vye, N.: Gender representation in humanoid robots for domestic use. Int. J. Soc. Robot. (special issue) **1**(3), 261–265 (2009)
23. Carpenter, J., Davis, J., Erwin-Stewart, N., Lee. T., Bransford, J., Vye, N.: Invisible machinery in function, not form: user expectations of a domestic humanoid robot. In: Proceedings of 6th Conference on Design and Emotion, Hong Kong, China (2008)
24. Pereira, M.S., Nijs, Y., Shahid, S., Swerts, M.: Children's lying behaviour in interactions with personified robots. In: Proceedings of British HCI 2016—Fusion, Bournemouth, UK (2016)
25. Åström, K.J., Hägglund, T.: Revisiting the Ziegler-Nichols step response method for PID control. J. Process Control **14**(6), 635–650 (2004)
26. Hang, C.C., Åström, K.J., Ho, W.K.: Refinements of the Ziegler-Nichols tuning formula. IEE Proc. Control Theory Appl. **138**(2), 111 (1991)
27. Valério, D., da Costa, J.S.: Tuning of fractional PID controllers with Ziegler–Nichols-type rules. Signal Process. **86**(10), 2771–2784 (2006)
28. Sivanandam, S.N., Deepa, S.N.: Genetic algorithm optimization problems. In: Introduction to genetic algorithms, pp. 165–209. Springer, Berlin (2008).
29. Ünal, M., Ak, A., Topuz, V., Erdal, H.: Ant colony optimization (ACO). In: Optimization of PID Controllers Using Ant Colony and Genetic Algorithms, vol. 449, pp. 31–35. Springer, Berlin (2013)
30. Poli, R., Kennedy, J., Blackwell, T.: Particle swarm optimization: an overview. Swarm Intell. **1**(1), 33–57 (2007)
31. Geem, Z.W., Kim, J.H., Loganathan, G.V.: A new heuristic optimization algorithm: Harmony search. Simulation **76**(2):60–68 (2001)
32. Brest, J., Zumer, V., Maucec, M.S.: Self-adaptive differential evolution algorithm in constrained real-parameter optimization. In: 2006 IEEE International Conference on Evolutionary Computation, Vancouver, BC, Canada, pp. 215–222 (2006)
33. Bazylev, D.N., et al.: Approaches for stabilizing of biped robots in a standing position on movable support. Sci. Tech. J. Inf. Technol. Mech. Opt. **15**(3), 418 (2015)
34. Ben-Ari, M., Mondada, F.: Robots and their applications. In: Elements of Robotics. Springer International Publishing, Cham, pp. 1–20 (2018)
35. Kast, C., et al.: Development of a modular bionic prototype arm prosthesis integrating a closed-loop control system. In: Lhotska, L., Sukupova, L., Lacković, I., Ibbott, G.S. (eds.) World Congress on Medical Physics and Biomedical Engineering 2018, vol. 68/2. Springer, Singapore, pp. 751–753 (2019)
36. Liu, Y.-J., Gong, M., Tong, S., Chen, C.L.P., Li, D.-J.: Adaptive fuzzy output feedback control for a class of nonlinear systems with full state constraints. IEEE Trans. Fuzzy Syst., pp. 1–1 (2018)
37. Rinehart, J., Ma, M., Calderon, M.-D., Cannesson, M.: Feasibility of automated titration of vasopressor infusions using a novel closed-loop controller. J. Clin. Monit. Comput. **32**(1), 5–11 (2018)
38. Melendez, A., Castillo, O., Alanis, A., Soria, J.: Reactive and tracking control of a mobile robot in a distributed environment using fuzzy logic, pp. 1–5 (2010)
39. Navabi, H., Sadeghnejad, S., Ramezani, S., Baltes, J.: Position control of the single spherical wheel mobile robot by using the fuzzy sliding mode controller. Adv. Fuzzy Syst. **2017**, 1–10 (2017)
40. Peraza, C., Valdez, F., Melin, P.: Optimization of intelligent controllers using a type-1 and interval type-2 fuzzy harmony search algorithm. Algorithms **10**(3), 82 (2017)
41. Caraveo, C., Valdez, F., Castillo, O.: A new optimization meta-heuristic algorithm based on self-defense mechanism of the plants with three reproduction operators. Soft Comput. (2018)
42. Peraza, C., Valdez, F., Garcia, M., Melin, P., Castillo, O.: A new fuzzy harmony search algorithm using fuzzy logic for dynamic parameter adaptation. Algorithms **9**(4), 69 (2016)
43. Peraza, C., Valdez, F., Castro, J.R., Castillo, O.: Fuzzy dynamic parameter adaptation in the harmony search algorithm for the optimization of the ball and beam controller. Adv. Oper. Res. **2018**, 1–16 (2018)

44. Lins, A.A., de Oliveira, J.M., Rodrigues, J.J.P.C., de Albuquerque, V.H.C.: Robot-assisted therapy for rehabilitation of children with cerebral palsy—A complementary and alternative approach. Comput. Hum. Behav. (2018)
45. Rengifo Sanchez, A.F., Segura-Quijano, F.E., Quijano, N.: An affordable set of control system laboratories using a low-cost robotic platform. IEEE ASME Trans. Mechatron. 1–1 (2018)
46. Rodriguez-Garavito, C.H., Arevalo-Castiblanco, M.F., Patiño-Forero, A.A.: Implementation of a non-linear fuzzy Takagi-Sugeno controller applied to a mobile inverted pendulum. In: Graña, M., López-Guede, J.M., Etxaniz, O., Herrero, Á., Sáez, J.A., Quintián, H., Corchado, E. (eds.) International Joint Conference SOCO'18-CISIS'18-ICEUTE'18, vol. 771, pp. 344–353. Springer International Publishing, Cham (2019)
47. Bobtsov, A.A., et al.: Using of LEGO Mindstorms NXT technology for teaching of basics of adaptive control theory. IFAC Proc. **44**(1), 9818–9823 (2011)

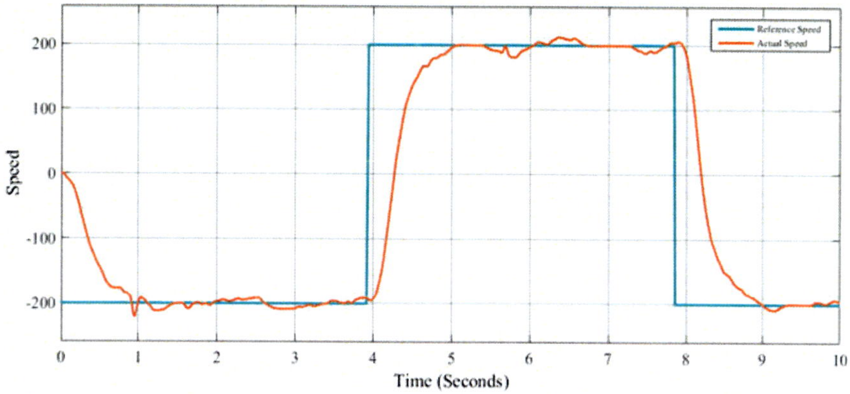

Control Strategies Based on Interval Type-2 Fuzzy Logic for Autonomous Mobile and Humanoid Robots

Felizardo Cuevas, Oscar Castillo, and Prometeo Cortes

Abstract The main purpose considered in this chapter is to maintain a specific location and behavior for a robot that uses type-2 fuzzy logic for controlling its behavior. In this chapter, we propose a combination of behaviors by following a trajectory without leaving or losing it and avoiding obstacles in an omnidirectional mobile platform. The results of the simulation show the advantages of the proposed approach. We describe the previous knowledge concerning type-2 fuzzy logic, the virtualization of the mobile robot and its modeling according to real situations. The proposed control system is developed in Matlab-Simulink, the system can model and guide a mobile robot, successfully in simulated and real environments. We also discuss how the proposed techniques could be extended to humanoid robots.

Keywords Autonomous mobile robots · Type-2 fuzzy systems · Type-2 fuzzy logic controller · Omnidirectional mobile robot · Humanoid robots

1 Introduction

One of the basic problems that a mobile robot faces when navigating from one place to another using sensors in real environments is the uncertainty, which is due to the large number of inaccuracies in the readings obtained in a real environment [1].

The imprecision that results from the information provided by the different sensors in the mobile robot, will result in inefficient behavior.

A mobile robot must be capable of performing correctly in a real environment, which means navigating autonomously and this requires a control strategy that allows it to handle the uncertainty that exists in the environment where it performs, especially in real time, with low computational load [2, 3].

Therefore the navigation of a mobile robot must take into consideration:

- The location of the robot in initial position.

F. Cuevas · O. Castillo (✉) · P. Cortes
Tijuana Institute of Technology, Tijuana, BC, México
e-mail: ocastillo@tectijuana.mx

C. Kahraman and E. Bolturk (eds.), *Toward Humanoid Robots: The Role of Fuzzy Sets*, Studies in Systems, Decision and Control 344,
https://doi.org/10.1007/978-3-030-67163-1_8

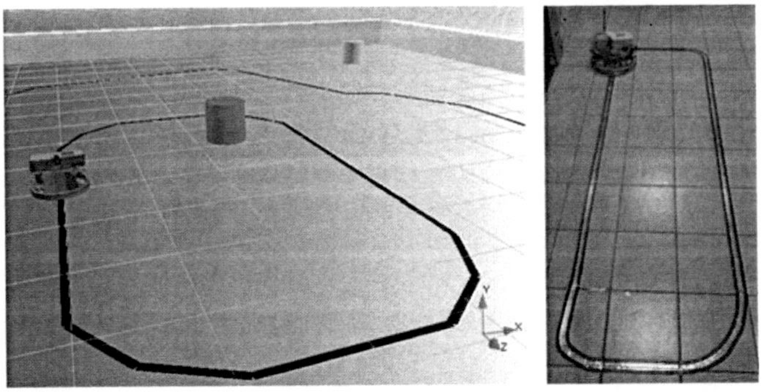

Fig. 1 Navigation scenario where mobile platform moves. **a** Simulation, **b** real life

- The perception of the information provided by the sensors is almost entirely unreliable. The sensors with their low range that in combination with the characteristics of the environment and other conditions induce noise, errors in the responses of the sensors leading us to inaccurate data.

Currently, one of the main challenges of mobile robotics is the design of models that reliably perform complex tasks in face of the uncertainty of the environment [4]. For this reason, it is interesting to carry out research works in this area. Of course, this makes the realization of works in this field interesting.

Figure 1 shows a navigation scenario where the mobile platform moves and follows the path shown by the line to be pursued.

Today in the design of controllers for mobile robotics, is one of the areas where the use of fuzzy logic is playing a very important role for the design of controllers that are based on behavior and expert knowledge. Allowing robustness due to the vagueness of the information and disturbances that affect the correct performance of the sensors [5].

Behavior is understood as a small part of the control designed to achieve an objective in a set of situations with restrictions. It is here where the need arises in mobile robots when performing a specific task, the development and cooperation of diverse behaviors.

Each behavior is implemented for the control of a task such as, the following of a line, evasion of obstacles and search for an objective.

The main contribution of this work is the implementation of a Type-2 Fuzzy Logic Controller for path tracking and obstacle avoidance present in the development of an autonomous mobile robot (AMR) at low speeds. Starting from the study of the kinematic model and adapting the technique of linear and angular speed control, with the objective of minimizing the orientation and position error within the line that serves as a path to the vehicle. The mathematical models and the control routines were simulated and validated in Matlab/Simulink, and later implemented in an Omnidirectional Mobil Robot with the same trajectory in the simulation and analyzed the results.

This chapter is structured as follows: Sect. 2 presents the reference of the three-wheel omnidirectional robot kinematics. Section 3 presents the design, implementation and simulation of the interval type-2 fuzzy logic controller of the mobile robot. Section 4 presents results and discussions of the achievements. Section 5 concludes with some final clarifications, conclusions and future work.

2 Kinematic Model

The most practical way to get the information of a movement and its characteristics, is by its geometric representation, which has its application in the analysis of movements in mobile robots. Among the applications of kinematics is the possibility of its application as an initial mathematical model for modeling the controller, to raise the equations for calculating its odometry, or its simulation that generates its kinetic behavior of the mobile robot. The following statements show as the limitations in the construction of the kinematic model [6]:

- The displacement of the robot is assumed to be on a flat surface.
- The structure of the robot does not have flexible elements.
- The steering axes of the wheels are almost always present on the wheels, making them perpendicular to the surface where they are applied.
- Friction is ignored.

Table 1 contains the orientation, position and distribution of the omnidirectional wheels that take the point of symmetry that generally coincides with the center of the robotic platform [7].

2.1 *Kinematic Behavior*

The kinematics of the wheels can be viewed as a principle where the wheels when making contact with the surface of the ground acts as a point of union in a plane with three degrees of freedom [8].

Table 1 Initial configuration parameters of kinematics omnidirectional platform

	Wheel 1	Wheel 2	Wheel 3
α_i	180°	60°	−60°
β_i	0°	0°	0°
γ_i	0°	0°	0°
δ_i	(0, 0, 0)	(0, 0, 0)	(0, 0, 0)
λ_i	(−L, 0, 0)	$\left(\frac{L}{2}, \frac{L\sqrt{3}}{2}, 0\right)$	$\left(\frac{L}{2}, \frac{-L\sqrt{3}}{2}, 0\right)$

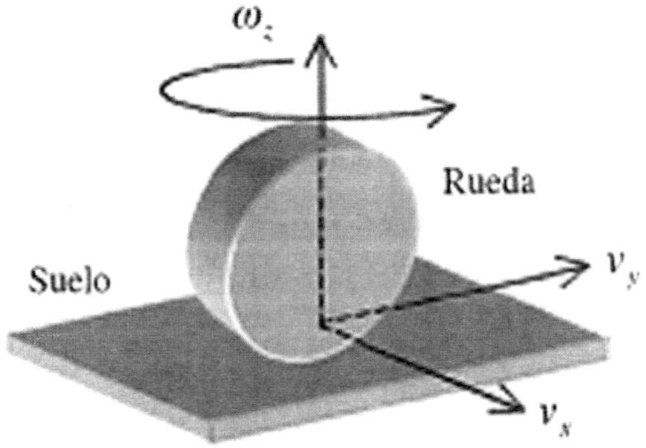

Fig. 2 Representation of wheel contact with the ground

Fig. 3 Omnidirectional wheels types

The wheels act as a flat union and this is described below. The wheel is considered to be a rigid element, and it is always in contact with the ground in a unique position that serves as the origin of the reference system explained in Fig. 2. The directions v_x and v_y determine the direction or direction of the wheel, v_x and v_y represent the linear velocities x–y, and w_z is the angular velocity when the robot turns. The conventional wheel in question, has an element v_x is null, although there are other wheels that provide a different behavior, as shown in Fig. 3.

An omnidirectional wheel is constituted as a standard wheel, one that is equipped with a ring with rollers, 120° one from the other, with active actuator, passive direction and the rotation axes are perpendicular to the forward direction [9, 10].

2.2 Kinematic Model

Location and position of the omnidirectional robot is represented by a vector $(x, y, \emptyset)^T$. The global speed of OMR is represented with the vector $(\dot{x}, \dot{y}, \dot{\emptyset})^T$ and the angular speed for each wheel, is presented with the vector $(\dot{\Theta}_1, \dot{\Theta}_2, \dot{\Theta}_3)^T$, as shown in Fig. 4 based on the model detailed in [5], and the kinematic model of the robotic platform is presented as follows:

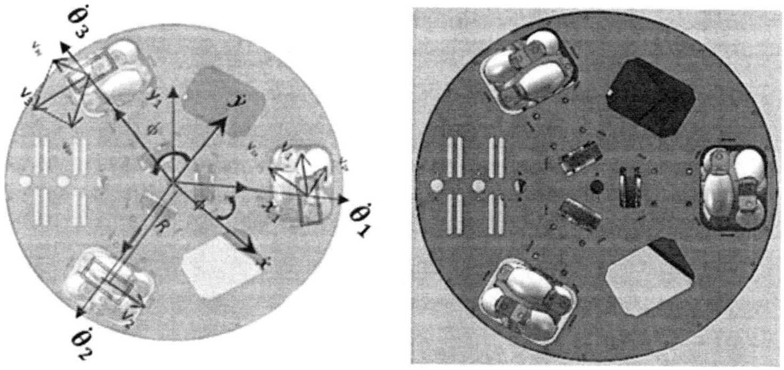

Fig. 4 Kinematic scheme of the robotic platform

$$P(\phi) = \begin{bmatrix} -\sin(\phi + \varphi_1)\cos(\phi + \varphi_1)R \\ -\sin(\phi + \varphi_1)\cos(\phi + \varphi_1)R \\ -\sin(\phi + \varphi_1)\cos(\phi + \varphi_1)R \end{bmatrix} \tag{1}$$

$$\begin{bmatrix} \dot{\theta}_1 \\ \dot{\theta}_2 \\ \dot{\theta}_3 \end{bmatrix} = \frac{1}{r}\begin{bmatrix} V_1 \\ V_2 \\ V_3 \end{bmatrix} = \frac{1}{r}P = \frac{1}{r}P\begin{bmatrix} \dot{x} \\ \dot{y} \\ \# \end{bmatrix} \tag{2}$$

With:
$\dot{\theta}_n$: Angular speed of the wheel n.
V_i: Linear speed of the wheel i.
r: Radial distance of the wheel.
R: Difference between the wheel and the center of the platform.
ϕ: Angular speed.
φ_n: Angular delimitation of the wheel n.
$P(\phi)$: Change matrix calculation through the angular speed of the wheels and the global speed array $(\dot{x}, \dot{y}, \dot{\phi})^T$.

3 Interval Type-2 Fuzzy Systems

A fundamental requirement of autonomous mobile robots is a good behavior to avoid obstacles. This action helps the mobile robot to move without colliding in an unstructured environment [11]. In this work, the behaviors of avoiding obstacles and path tracking, are basic behaviors that use three infrared sensors and an inductive sensor, in the front and side of the mobile robot. The response signals of the sensors located on the front and side of the mobile robot are taken into account in this work and these imprecise values that are the response of the system, which uncertainty must be taken into account in order to create a model capable of tolerating high levels of imprecision in its environment.

In the interest of considering the development of an efficient mobile robot controller, a T2FLS is used [12]. It is expected that the T2FLS control algorithm will produce an efficient controller, where we have the ability to overcome uncertainty and having this ability the robot can plan its movements running efficiently in an unknown environment [13].

For the present work, it is proposed to use the architecture of a type-2 fuzzy Inference System, as illustrated in Fig. 2.

The upper and lower MFs for interval type-2 fuzzy set can be written in Eqs. (4) and (5). Assuming that we have N rules in a fuzzy rule base of Type 2, they take the following form [14] of Eq. (6):

$$F_{\mathrm{A}}^{-}(x) = \begin{cases} 0, & x < l_1 \\ (x - l_1)/(p_1 - l_1), & l_1 < x < 0 \\ 1, & x \geq 0 \\ (r_2 - x)/((r_2 - p_2), & x \geq 0 \\ 0, & x \geq r_2 \end{cases} \quad (3)$$

$$F\_{\mathrm{A}}(x) = \begin{cases} 0, & x < l_2 \\ (x - l_1)/(p_1 - l_2), & x \leq \frac{r_1(p_2 - l_2)}{(p_2 - l_2)} + \frac{l_2(r_1 - p_1)}{(r_1 - p_1)} \\ (r_2 - x)/((r_2 - p_2), & x > \frac{r_1(p_2 - l_2)}{(p_2 - l_2)} + \frac{l_2(r_1 - p_1)}{(r_1 - p_1)} \\ 0, & x > r_2 \end{cases} \quad (4)$$

$$\begin{gathered} R^i\colon If\ x_1\ is\ X_{\mathrm{i}}^l\ and \ldots and\ \ldots, x_p\ is\ X_p^l \\ then\left(y\ is\ Y^l\right) \end{gathered} \quad (5)$$

where $X_{\mathrm{i}}^l (i = 1, \ldots p)$ and Y^l are type-2 fuzzy sets, and $x = \left(x^1, \ldots., x^l\right)$ and y are linguistic variables.

The firing set is defined by interval:

$$F^i(x) = \left[f^i\_(x), f^{-i}(x) \equiv \left[f^i\_, f^{-i}\right]\right. \quad (6)$$

- This IT2FLS handles the uncertainty in the system, while the T1FLS simply does not. Although it is described similarly as a T1FLS, an IT2FLS has some differences when it is described in a block diagram, presented in Fig. 2, the change being shown as it can be seen in the output zone or area, where a Type-Reducer reduces IT2FS to a T1FS, which is then introduced to a Defuzzifier to obtain a crisp result [15] (Fig. 5).

There are many ways to implement an IT2FLS type reducer and at the moment [14], one of these techniques is the most used, the center of gravity (cos), where Y_{cos} is an interval, described by the points $[Y_l^i, Y_r^i]$, which are calculated with Eqs. (8) and (9).

Fuzzification

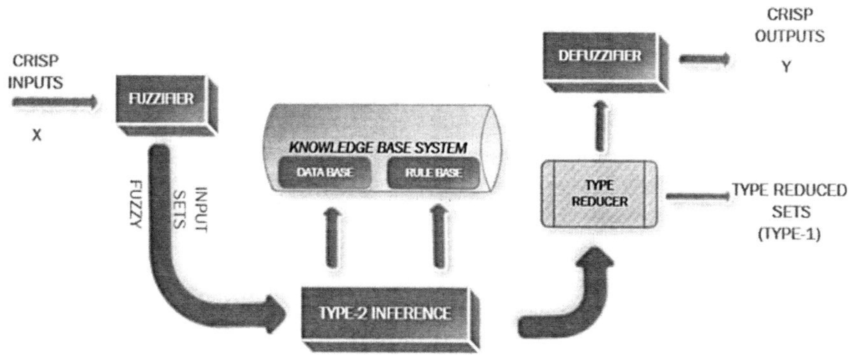

Fig. 5 IT2FLS architecture

$$y_l = \sum_{i=1}^{M} f_l^i y_l^i / f_l^i \tag{7}$$

$$y_l = \sum_{i=1}^{M} f_r^i y_r^i / f_r^i \tag{8}$$

The defuzzification technique is performed by Eq. (10), which achieves a crisp output value.

$$y_{(x)} = \frac{y_l + y_r}{2} \tag{9}$$

The variability in each of the actions is modeled in interval type-2 fuzzy sets (T2FS). The T2FS linguistic inputs variables and their ranges are used for path tracking and orientation obstacle avoidance, as shown in Fig. 6a, b, with two outputs which are the linear and angular speed, are shown in Fig. 6c, d.

We know that the T2FS is located in a region built by a main type-1 membership function (T1MF). T2FS is obtained by using fuzzy sets to partition the input domains of the base line T1FS with footprint of uncertainty (FOU) as shown in Fig. 6.

Consequently, the T1MF extends to T2MF by adding the FOU in the antecedent and consequent parts of each rule.

Therefore, membership functions have values distributed with uncertainty. As well as those that belong to the antecedents also in the consequent parts. From the design, five MFs are for the Inductive sensor input, 3 MFs for the Infrared, 5 MFs for the Linear speed output and 7 MFs for the Angular Speed output, are used.

Fixed values are assigned in a range of 0–10 V for Inductive sensor, 0–2.5 V for Infrared sensors to measure obstacle distance, 30–120 mm/s for Linear velocity and −30 to 30° for Angular velocity depending on the displacement of the mobile robot.

In this work, seventy-four rules were developed by T2FLS in path tracking behavior and to avoid obstacles. Rules are used in the control of angular and linear speed, shown in Table 2.

So if an obstacle comes too close to the robot, it must be able to avoid it and change its behavior by slowing down, stopping if necessary and turning.

In this work, the number of rules are Sixty in total. These are obtained from the combination of four inputs with five and three membership functions as shown in Table 2.

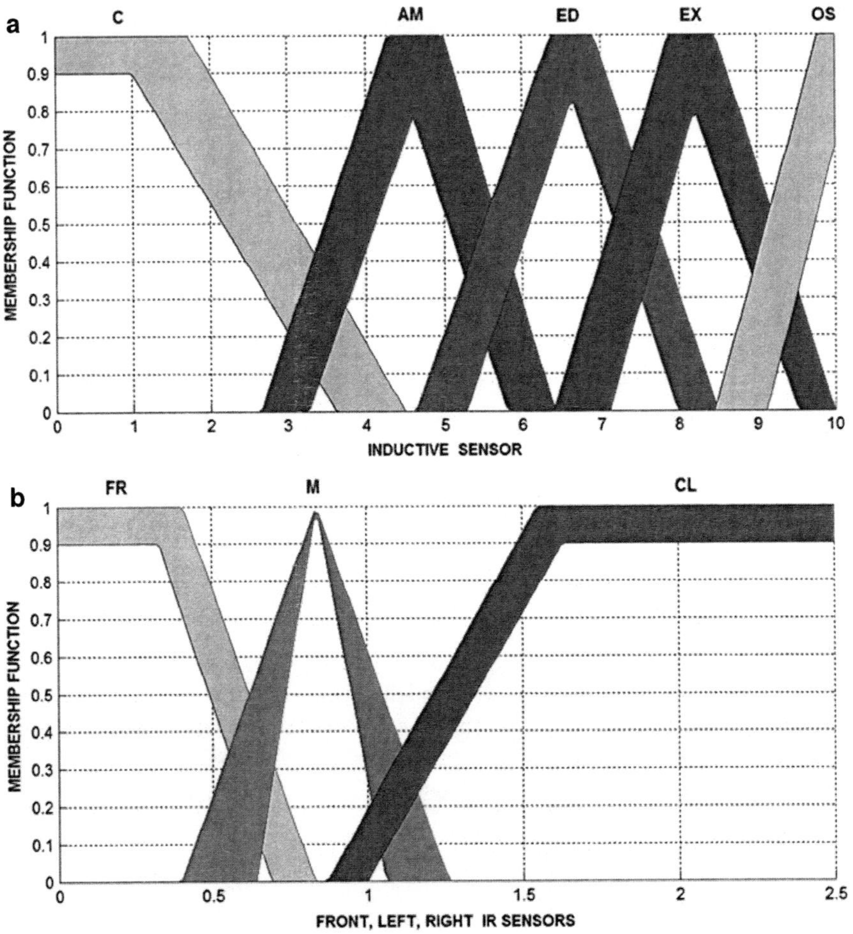

Fig. 6 **a** Inductive sensor as input MFs, **b** infrared left, front and right sensors input MFs, **c** linear Speed output MFs, **d** angular speed as output MFs

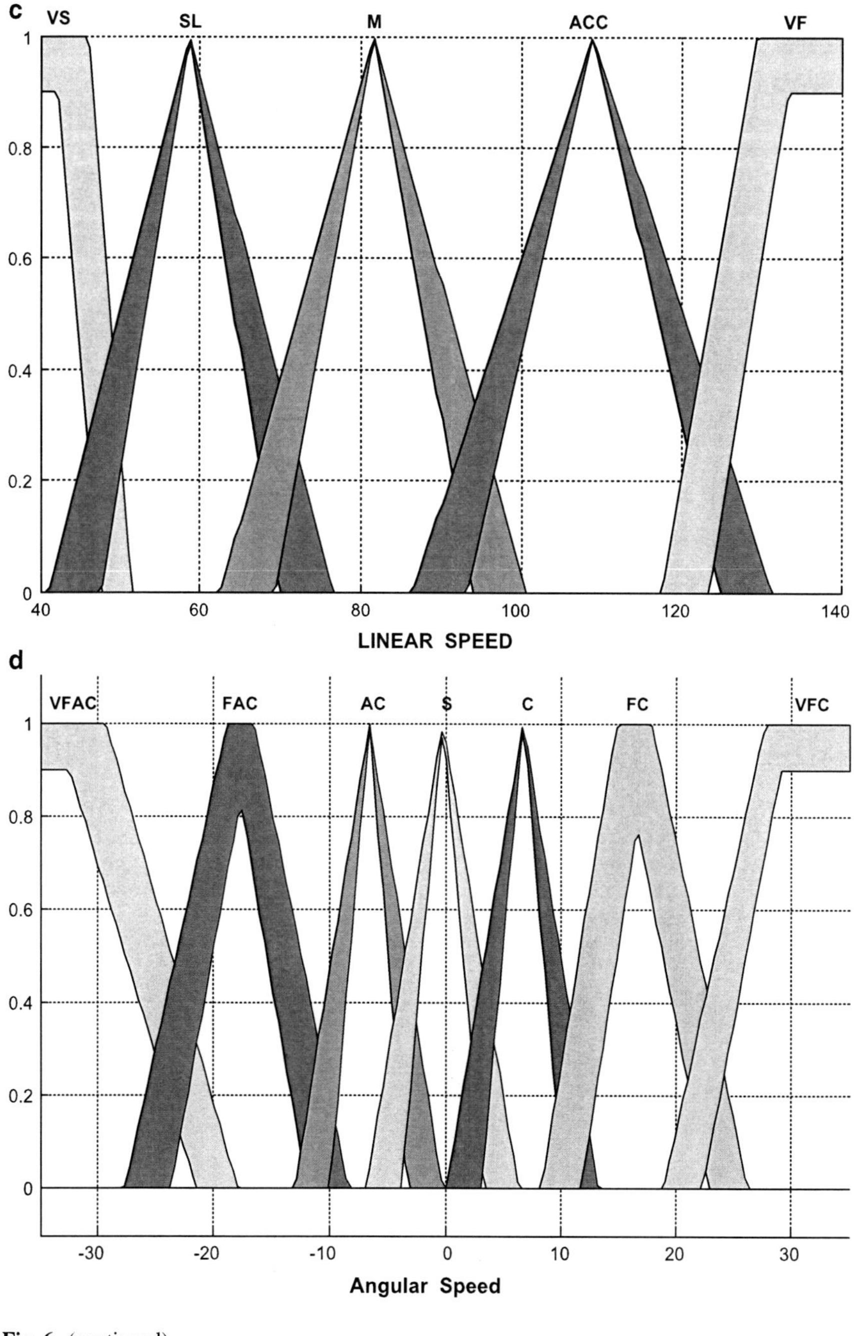

Fig. 6 (continued)

Table 2 Path following-obstacle avoidance behavior rule base

Inductive	Proximity	Proximity	Proximity	Linear	Angular
Sensor	Sensor	Sensor	Sensor	Speed	Speed
C	CL	CL	CL	VS	VFAC
C	CL	CL	M	VS	VFC
C	CL	M	FR	SL	C
C	CL	M	CL	SL	VFC
C	M	FR	M	M	S
C	M	FR	FR	M	S
C	M	CL	CL	ACC	VAC
C	M	CL	M	ACC	VFC
C	FR	M	FR	VF	FC
C	FR	M	CL	VF	FAC
C	FR	FR	M	ACC	S
C	FR	FR	FR	VF	S
CC	CL	CL	CL	VS	VFAC
CC	CL	CL	M	VS	VFC
CC	CL	M	FR	SL	C
CC	CL	M	CL	SL	VFC
CC	M	FR	M	M	S
CC	M	FR	FR	M	S
CC	M	CL	CL	ACC	VAC
CC	M	CL	M	ACC	VFC
CC	FR	M	FR	VF	FC
CC	FR	M	CL	VF	FAC
CC	FR	FR	M	ACC	S
CC	FR	FR	FR	VF	S
FL	CL	CL	CL	VS	VFAC
FL	CL	CL	M	VS	VFC
FL	CL	M	FR	SL	C
FL	CL	M	CL	SL	VFC
FL	M	FR	M	M	S
FL	M	FR	FR	M	S
FL	M	CL	CL	ACC	VAC
FL	M	CL	M	ACC	VFC
FL	FR	M	FR	VF	FC
FL	FR	M	CL	VF	FAC
FL	FR	FR	M	ACC	S

(continued)

Table 2 (continued)

Inductive	Proximity	Proximity	Proximity	Linear	Angular
FL	FR	FR	FR	VF	S
E	CL	CL	CL	VS	VFAC
E	CL	CL	M	VS	VFC
E	CL	M	FR	SL	C
E	CL	M	CL	SL	VFC
E	M	FR	M	M	S
E	M	FR	FR	M	S
E	M	CL	CL	ACC	VAC
E	M	CL	M	ACC	VFC
E	FR	M	FR	VF	FC
E	FR	M	CL	VF	FAC
E	FR	FR	M	ACC	S
E	FR	FR	FR	VF	S
FU	CL	CL	CL	VS	VFAC
FU	CL	CL	M	VS	VFC
FU	CL	M	FR	SL	C
FU	CL	M	CL	SL	VFC
FU	M	FR	M	M	S
FU	M	FR	FR	M	S
FU	M	CL	CL	ACC	VAC
FU	M	CL	M	ACC	VFC
FU	FR	M	FR	VF	FC
FU	FR	M	CL	VF	FAC
FU	FR	FR	M	ACC	S
FU	FR	FR	FR	VF	S

4 Results and Discussion

In this work an evaluation is made to analyze the performance of the mobile robot based on T2FLS shown in Fig. 5, when compared to T1FLS with similar number of rules [13]. The responsiveness of the movement of the mobile robot to follow the path and avoid obstacle is done through the use of simulation. The evaluation is carried out in the environment of the simulator of the omnidirectional FESTO platform [14]. The experiments in each environment will have their own parameters whose specification is be explained later.

The data that should be stored in this simulation is the voltage of the input sensors, the linear and angular velocities. All the data is obtained from the real experiment

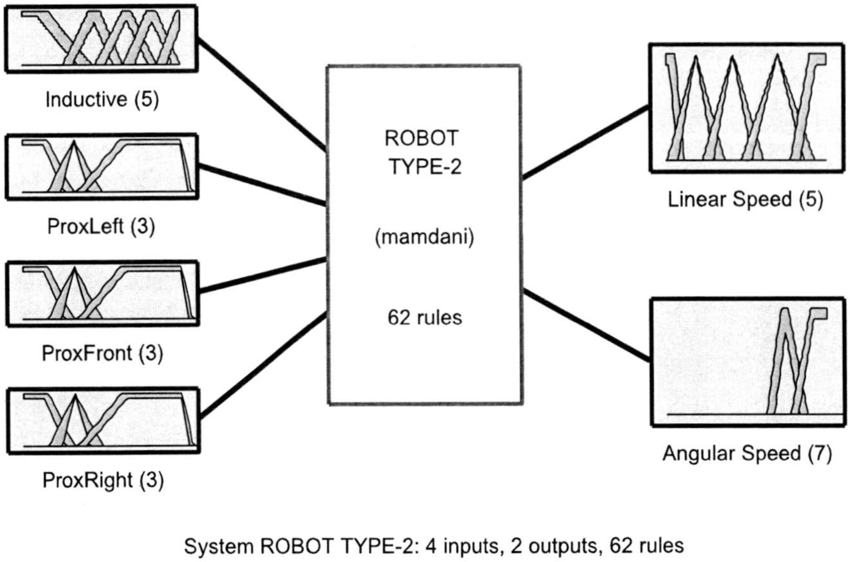

Fig. 7 Interval type-2 fuzzy system with 4 inputs and 2 outputs

and the simulation with the help of Simulink, each output data is recorded per unit of time, which in this case is 1 mm/s (Fig. 7).

The control surface of T2FLS with 60 rules is included, whose inputs are an inductive and infrared sensors and the output is the linear speed. In the development of T2FLS a smooth surface is obtained, which is characterized by the slopes on its surface and each slope has a gradual change represented by the linear velocity, which is shown in Fig. 8.

5 Humanoid Robots

Humanoid robotics is an emerging and challenging research field, which has received significant attention during the past years and will continue to play a central role in robotics research and in many applications of the twenty-first century. Regardless of the application area, one of the common problems tackled in humanoid robotics is the understanding of human-like information processing and the underlying mechanisms of the human brain in dealing with the real world.

Humanoid robots are now used as research tools in several scientific areas. Researchers study the human body structure and behavior (biomechanics) to build humanoid robots. On the other side, the attempt to simulate the human body leads to a better understanding of it. Human cognition is a field of study which is focused on how humans learn from sensory information in order to acquire perceptual and

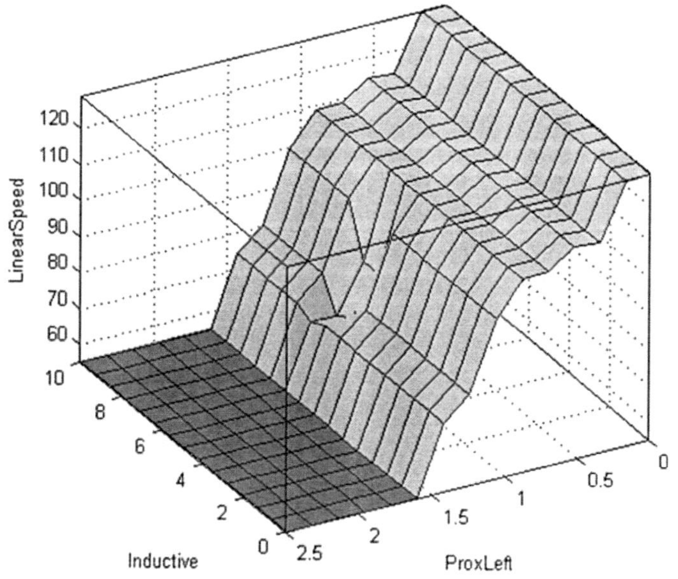

Fig. 8 The control surface of T2FLS with 74 rules

motor skills. This knowledge is used to develop computational models of human behavior and it has been improving over time.

A humanoid robot is a robot with its body shape built to resemble the human body. The design may be for functional purposes, such as interacting with human tools and environments, for experimental purposes, such as the study of bipedal locomotion, or for other purposes.

In planning and control, the essential difference between humanoids and other kinds of robots (like industrial ones) is that the movement of the robot must be human-like, using legged locomotion, especially biped gait. The ideal planning for humanoid movements during normal walking should result in minimum energy consumption, as it does in the human body. For this reason, studies on dynamics and control of these kinds of structures has become increasingly important.

The question of walking biped robots stabilization on the surface is of great importance. Maintenance of the robot's gravity center over the center of bearing area for providing a stable position can be chosen as a goal of control [16].

To maintain dynamic balance during the walk, a robot needs information about contact force and its current and desired motion [17]. The solution to this problem relies on a major concept, the Zero Moment Point (ZMP).

Another characteristic of humanoid robots is that they move, gather information (using sensors) on the "real world" and interact with it. They do not stay still like industrial manipulators and other robots that work in highly structured environments. To allow humanoids to move in complex environments, planning and control must focus on self-collision detection, path planning and obstacle avoidance [18].

The proposed techniques in previous sections, although designed for autonomous mobile robots can be extended to humanoid robots as both type of robots require moving autonomously and avoiding obstacles in Dynamic environments. This is precisely our future work, but we anticípate that the proposed approach with type-2 fuzzy logic could be easily modified to deal with humanoid robots.

6 Conclusions

The proposed approach implements a type-2 fuzzy controller that produces the linear and angular velocity values in path tracking, with the ability to avoid obstacles that arise in its development using a self-contained omnidirectional mobile robot, so that the robot can travel the road without losing it and with the least possible error. In the initial tests, the effectiveness of the T2FLC is demonstrated when implementing it in an omnidirectional mobile robot, and its use provides an advantage in the efficiency of the route line, the smoothness in its follow-up and the decision on the correct acceleration and braking according to the type of direction change of the line being pursued. In the path that the robot follows with the T2FLC it is possible to observe how the values of linear and angular velocity are changing, as well as conditioned of the inductive sensor signal. In future work we will use Genetic Algorithm and PSO [19–21] to optimize the membership functions of the fuzzy systems [22–25] and we will work with neural networks and perform more experiments [26–28]. Also, type-2 fuzzy logic could be added as in [29–32]. In addition, we plan to later implement the fuzzy controllers on larger and more complex Humanoid robots.

References

1. Oltean, S., Dulau: Position control of Robotino mobile robot using fuzzy logic. In: 2010 IEEE International Conference on Automation, Quality and Testing, Robotics (AQTR), pp. 1–6 (2010)
2. Mac, T.T., Copot, C., Keyser, R.D.E., Tran, T.D., Vu, T.: MIMO fuzzy control for autonomous mobile robot. J. Autom. Control Eng. **4**, 65–70 (2016)
3. Cuevas, F., Castillo, O.: Design and implementation of a fuzzy path optimization system for omnidirectional autonomous mobile robot control in real-time. In: Studies in Computational Intelligence, vol. 749 (2018)
4. Prakash Moon, A., Jajulwar, K.K.: Design of adaptive fuzzy tracking controller for Autonomous navigation system. ijrter.com. Int. J. Recent Trends Eng. Res. (IJRTER) **2**, 268–275 (2016)
5. Li, X., Zell, A.: Motion control of an omnidirectional mobile robot. In: Informatics in Control, Automation and Robotics, pp. 181–193. Springer, Berlin
6. Parhi, D.R., Deepak, B.B.V.L.: Kinematic model of three wheeled mobile robot. Mech. Eng. Res. **3**, 307–318 (2011)
7. Zhao, Y., BeMent, S.L., Kinematics, dynamics and control of wheeled mobile robots. In: Proceedings 1992 IEEE International Conference on Robotics and Automation, pp. 91–96 (1992)

8. Garcia Sillas, D., Gorrostieta Hurtado, E., Vargas Soto, E., Rodríguez Reséndiz, J., Tovar Arriaga, S.: Kinematics modeling and simulation of an autonomous omni-directional mobile robot. Ing. E Investig. **35**, 74–79 (2015)
9. Song, J.B., Byun, K.S.: Design and control of an omnidirectional mobile robot with steerable omnidirectional wheels. In: Mobile Robots, Moving Intelligence, pp. 224–240 (2006)
10. Kanjanawanishkul, K.: Omnidirectional wheeled mobile robots : wheel types and practical applications. Int. J. Adv. Mechatron. Syst. **6**, 289 (2015)
11. Njah, M., Jallouli, M.: Wheelchair obstacle avoidance based on fuzzy controller and ultrasonic sensors. In: International Conference on Computer Applications Technology (ICCAT), vol. 1–5 (2013)
12. Castillo, O., Melin, P.: A review on the design and optimization of interval type-2 fuzzy controllers. Appl. Soft Comput. J. **12**, 1267–1278 (2012)
13. Castillo, O., Melin, P., Pedrycz, W.: Design of interval type-2 fuzzy models through optimal granularity allocation. Appl. Soft Comput. J. **11**, 5590–5601 (2011)
14. Festo Didactic GmbH & Co. KG, Festo Didactic GmbH & Co. KG: Robotino Manual, pp.7–25, Denkendorf (2007)
15. Wu, D.: On the fundamental differences between interval Type-2 and Type-1 Fuzzy Logic Controllers. IEEE Trans. Fuzzy Syst. **20**(5), 832–848 (2012)
16. Bazylev, D.N., et al.: Approaches for stabilizing of biped robots in a standing position on movable support. Sci. Tech. J. Inf. Technol. Mech. Opt. **15**(3), 418 (2015)
17. Carpenter, J., Davis, J., Erwin-Stewart, N., Lee. T., Bransford, J., Vye, N.: Gender representation in humanoid robots for domestic use. Int. J. Soc. Robot. (special issue) **1**(3), 261–265. Springer, The Netherlands (2009)
18. Carpenter, J., Davis, J., Erwin-Stewart, N., Lee. T., Bransford, J., Vye, N.: Invisible machinery in function, not form: user expectations of a domestic humanoid robot. In: Proceedings of 6th Conference on Design and Emotion. Hong Kong, China (2008)
19. Ochoa, P., Castillo, O., Soria, J.: Presented at the differential evolution with dynamic adaptation of parameters for the optimization of fuzzy controllers (2014)
20. Rodríguez, L., Castillo, O., Soria, J., Melin, P., Valdez, F., Gonzalez, C.I., Martinez, G.E., Soto, J.: A fuzzy hierarchical operator in the grey wolf optimizer algorithm. Appl. Soft Comput. J. **57** (2017)
21. Martínez-Soto, R., Castillo, O., Aguilar, L.T., Rodriguez, A.: A hybrid optimization method with PSO and GA to automatically design Type-1 and Type-2 fuzzy logic controllers. Int. J. Mach. Learn, Cybern (2015)
22. Amador-Angulo, L., Castillo, O.: Optimization of the Type-1 and Type-2 fuzzy controller design for the water tank using the Bee Colony Optimization. In: 2014 IEEE Conference on Norbert Wiener in the 21st Century (21CW). pp. 1–8. IEEE (2014)
23. Bhoyar, D.V., Chilke, P.B.J., Kemekar, S.S.: Design and analysis of fuzzy PID controllers using genetic algorithm, pp. 135–138 (2016)
24. Caraveo, C., Valdez, F., Castillo, O.: Optimization of fuzzy controller design using a new bee colony algorithm with fuzzy dynamic parameter adaptation. Appl. Soft Comput. **43**, 131–142 (2016)
25. Pérez, J., Valdez, F., Castillo, O.: Modification of the bat algorithm using type-2 fuzzy logic for dynamical parameter adaptation. In: Studies in Computational Intelligence (2017)
26. Castillo, O., Martínez-Marroquín, R., Melin, P., Valdez, F., Soria, J.: Comparative study of bio-inspired algorithms applied to the optimization of type-1 and type-2 fuzzy controllers for an autonomous mobile robot. Inf. Sci. (Ny) **192**, 19–38 (2012)
27. Amador-Angulo, L., Mendoza, O., Castro, J., Rodríguez-Díaz, A., Melin, P., Castillo, O.: Fuzzy sets in dynamic adaptation of parameters of a bee colony optimization for controlling the trajectory of an autonomous mobile robot. Sensors **16**, 1458 (2016)
28. Melin, P., Astudillo, L., Castillo, O., Valdez, F., Garcia, M.: Optimal design of type-2 and type-1 fuzzy tracking controllers for autonomous mobile robots under perturbed torques using a new chemical optimization paradigm. Expert Syst. Appl. **40**, 3185–3195 (2013)

29. Melin, P., González, C.I., Castro, J.R., Mendoza, O., Castillo, O.: Edge-detection method for image processing based on generalized type-2 fuzzy logic. IEEE Trans. Fuzzy Systems **22**(6), 1515–1525 (2014)
30. González, C.I., Melin, P., Castro, J.R., Castillo, O., Mendoza, O.: Optimization of interval type-2 fuzzy systems for image edge detection. Appl. Soft Comput. **47**, 631–643 (2016)
31. González, C.I., Melin, P., Castro, J.R., Mendoza, O., Castillo, O.: An improved sobel edge detection method based on generalized type-2 fuzzy logic. Soft Comput. **20**(2), 773–784 (2016)
32. Ontiveros, E., Melin, P., Castillo, O.: High order α-planes integration: a new approach to computational cost reduction of general type-2 fuzzy systems. Eng. Appl. AI **74**, 186–197 (2018)

Present Applications of Humanoid Robots and Fuzzy Control

Omer Cetin

Abstract In this chapter, after given short information about humanoid robots and fuzzy control, recent applications and research papers of humanoid robots related with fuzzy control are reviewed. Studies about the humanoid robots and fuzzy logic-based control have been grouped under four major topic; the first one is stability and reliability control, the second one is walking pattern detection and the third one is navigation, and the final one is obstacle avoidance. In fact, these four topics are basic components of autonomous humanoid robots as similar as the other type robots. But the implementation of the control approach for each problem is special for humanoid robots in this chapter. As a result, this chapter reveals that various adaptations of the fuzzy logic-based control approach are used effectively to solve the different control problems of humanoid robots. Fuzzy logic-based control approaches are one of the basic control approaches that are widely used in humanoid robots, just like other robotic systems. Studies on this subject in the literature are gathered, grouped and presented in a structure that is easily accessible to the reader within this chapter.

Keywords Humanoid robots · Fuzzy control · Stability · Walking pattern detection · Navigation · Obstacle avoidance

1 Introduction

With their unique features, robots are classified usually like aerospace, military, security, research, consumer, drones, disaster response, entertainment, education, exoskeletons, humanoids, industrial, medical, underwater, etc. This classification depends on the usage field, physical appearance, shape and movement style. In its most general definition, a humanoid robot is an electro-mechanical device that possesses the general robot characteristics and additionally in appearance like the human body. The major difference between common robots and humanoids is the shape of the robot. But this little difference makes humanoids hard to implement

O. Cetin (✉)
Department of Computer Engineering, Hezârfen Aeronautics Space Technologies Institute, National Defense University, Istanbul, Turkey
e-mail: o.cetin@hho.edu.tr

C. Kahraman and E. Bolturk (eds.), *Toward Humanoid Robots: The Role of Fuzzy Sets*, Studies in Systems, Decision and Control 344,
https://doi.org/10.1007/978-3-030-67163-1_9

when compared with other type robots. Humanoid robots use two legs to stabilize themselves and move around, while other robots make use of structures such as wheels, pallets, multiple legs, wings, which are relatively much easier to implement for the same purpose. Beyond the similarity of the human legs to the structures that provide their movements, humanoid robots aim to behave like human beings in every subject. An autonomous humanoid robot is a kind of robot that can adapt itself to the environmental changes while it tries to reach its goal [1].

Every robot, regardless of its physical appearance, needs control structures to display autonomous capabilities. However, the physical appearance, in other words, the basic mechanisms that enable it to perform the movement, is one of the main reasons that affect the control structure of the robot. Beyond the similarity of structures to human legs, humanoid robots aim to act as human beings in all matters. Humanoid robots are being developed to emulate physical behaviors and mental tasks of humans perform [2]. Scientists from different research fields combine their researches to create a humanoid robot as human-like as possible [3, 4]. In fact, this approach will provide two-way benefits; both accurate identification and analysis of human behavior, and ensuring that the robot can act like humanoid behaviors as much as possible [2]. If humanoid robots can do that, they can finally work together with humans. For this purpose, they use their arms, hands and even fingers to perform their duties. They also use these limbs to stabilize, to stand up again when they fall, and to hold, lift, move or even shape the task. They have human-like sensors. They can hear sounds, perceive odors and temperatures, taste and see their surroundings. They can speak and touch with human-like sounds in order to communicate with the outside world.

After a detailed definition of humanoid robot term, now control structures of a humanoid robot can be discussed. An example of humanoid robot's close loop control can be seen in Fig. 1. It is an overview of the hierarchical control system. Different control loops are hierarchically applied to a humanoid robot as seen from the figure. The inner loop is joint control that is used for determining the actuators' speeds and move direction for each joint accordingly to obtain *stability* and *reliability control* of a humanoid robot. One outer loop is called local motion planning and it is used for stabilization and *walking pattern detection.* The outermost loop is named global motion planning as seen from the figure and it is used for *navigation control* and *obstacle avoidance*. Closed-loop control in the solution of problems shown by the above Fig. 1 has a plurality of parameters and some constraints like internal forces and moments for a smooth transition, geometric constraints for walking on a rough surface, etc. due to some requirements like human walking dynamics especially for *stability* and *reliability control* of humanoid robots. The *path-planning control* level deals with global motion and obstacle crossing. Path planning and gait selection require high-level sensor information. The gait synthesizer takes the steps of the gait, each step length, step speed and foot lift size without route planning, and then generates consecutive common reference commands in the form of a control loop during the gait.

Some methods have been proposed in the literature for all these defined control problems such as the *adaptive neuro-fuzzy system* [5, 6], *genetic algorithm* [7],

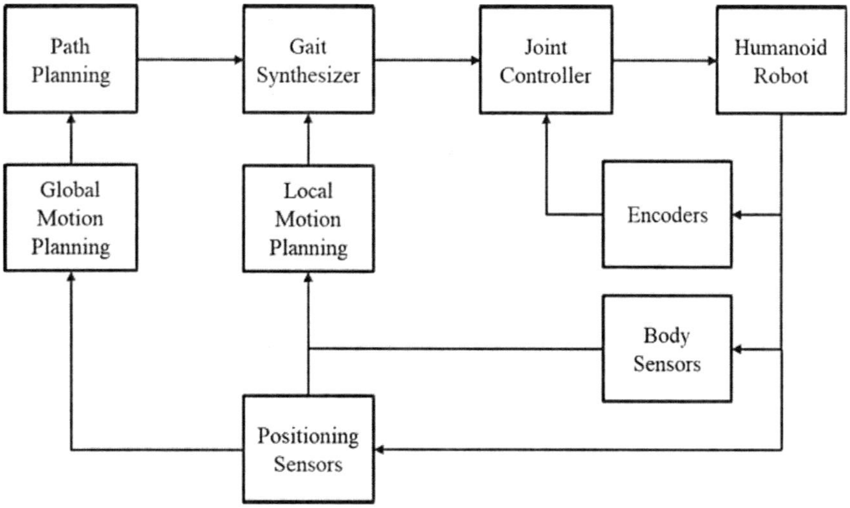

Fig. 1 Humanoid robot close loop control diagram

optimal preview control [8], and *incremental fuzzy control*, etc. An interesting area of applications for fuzzy controllers is the robotics control, and a relatively new application is its use to control biped or humanoid robots. In this chapter, detailed information has been given about the recent research activities about stability and reliability control, walking pattern detection, navigation control and obstacle avoidance of a humanoid robot by using *fuzzy based control* likewise organized brought together under the following titles.

2 Stability and Reliability Control of Humanoid Robot with Fuzzy Sets

The motion of the humanoid robot should be smooth and stable to ensure that the humanoid robot walks successfully. Besides, the robot must be able to sense its environment and to control itself according to conditions like unexpected external disturbances. A humanoid robot may encounter different types of the floor during its march, as shown in Fig. 2. These floor shapes may be protruding floors, suppressed floors, etc.

While the humanoid robot is willing to walk in the orbit generated by path planning, it is to decide an action model to ensure stability for various environments, as shown in Fig. 2. During the walking, sole of humanoid robot's feet is always parallel to the surface and monitoring the ground reaction force. If reaction forces detected during a step, it means that feet have been reached to the surface. If the surface is not flat or it is not at the same level as the fixed feet surface, it means that the surface

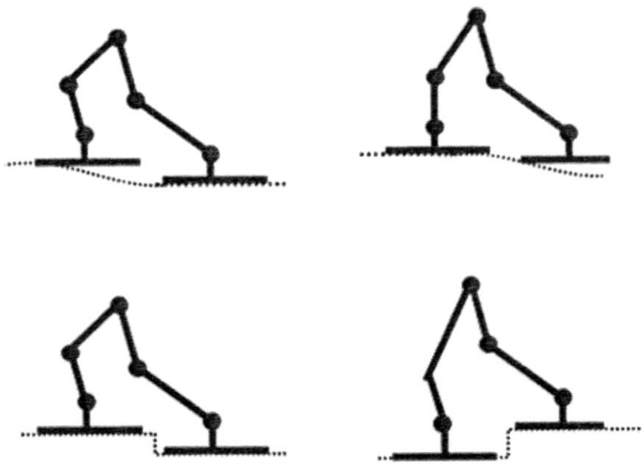

Fig. 2 Various irregular ground types

is irregular as seen from Fig. 2., and stability has become a more difficult problem than a flat surface. In this case, a humanoid robot can decide to change its planned walking trajectory or try to balance itself.

There has been an active study of various methods of stability and reliability of humanoid robots in recent years. As one of the methods, some researchers have used fuzzy logic [9, 10]. Fuzzy logic is mainly used as a part of executive control level control systems for generation and adjustment of PID gains, fuzzy control supervision, supervised and increased error signals, and direct fuzzy control. In the article [9], the logic control PID is implemented at the local control level to adjust the gains of the local PID controller, while the full control structure also includes forward feed control. It has shown that in cases where local subsystems are stabilized by fuzzy regulators, the aggregation separation method applies to the stability analysis of the complete dual system. Many types of research about the stabilization problem of humanoid robots have searched ZMP (Zero Moment Point) [9, 11, 12]. Researchers have proposed different controllers or compensators approaches from the difference between desired and obtained ZMP values [11]. Vukobratovic has proposed ZMP, and the ZMP stability criteria indicate that the robot will not fall if the ZMP's foot support remains in the convex shell [9].

Fuzzy set assessment is discussed for the two-legged humanoid gait synthesis problem using feedback and reinforcement learning in [10]. The fuzzy control rules for the first gait are generated using a human heuristic balancing scheme, and simulation studies have shown that the fuzzy gait synthesizer can roughly follow the desired structure. There is no numerical feedback teaching signal, only when the humanoid robot falls or almost falls, there is only a feedback signal, such as evaluative failure or success, which is a disadvantage of the proposed method, the lack of practical training data. Therefore, dynamic balance knowledge accumulates with reinforcement learning that constantly improves gait during gait. Exactly is learning fuzzy reinforcement using the fuzzy critical signal. Furthermore, for humanoid gait, "almost lacking", "almost success", "slower", "faster" and so on and they all such as linguistic signals. In this case, the reinforcement is based on the learned walk synthesizer modified Generalized approximate reasoning for intelligent control architecture that consists of three components as seen from Fig. 3: the action selection network (ASN), the action evaluation network (AEN), and the stochastic action modifier (SAM). ASN and AEN map a state vector to the proposed action using fuzzy inference and a fault signal to a scalar score indicating state goodness orderly. After them, to produce an ideal walk with desired features for humanoid, SAM uses both the recommended action and the internal reinforcement. The boost signal is generated based on the difference between the desired ZMP and the actual ZMP in the x and y plane. The proposed architecture was confirmed with simulations, and even terrain for humanoid walking is considered, so it must be verified for irregular and sloping terrain. In Fig. 3, X_{zmp}, Y_{zmp}, ZMP are the coordinates; $\theta^{d}{}_{zmp}$, $\theta^{d}{}_{zm}$ are the desired joint angles of biped gait.

The center of gravity (COG) can be defined as usual kinematics and it represents dynamic properties such as mass and center of mass. Also, a COG Jacobian can be used as an index for stability [14]. Zhong and Chen [15] presented a control system for humanoid robots walking on rough terrain, utilizing a particular swarm optimized algorithm for the development of artificial neural networks and fuzzy logic controllers. Samant et al. [16] proposed a method for the movement of the humanoid robot in an uneven environment and used the fuzzy logic-based experimental setup

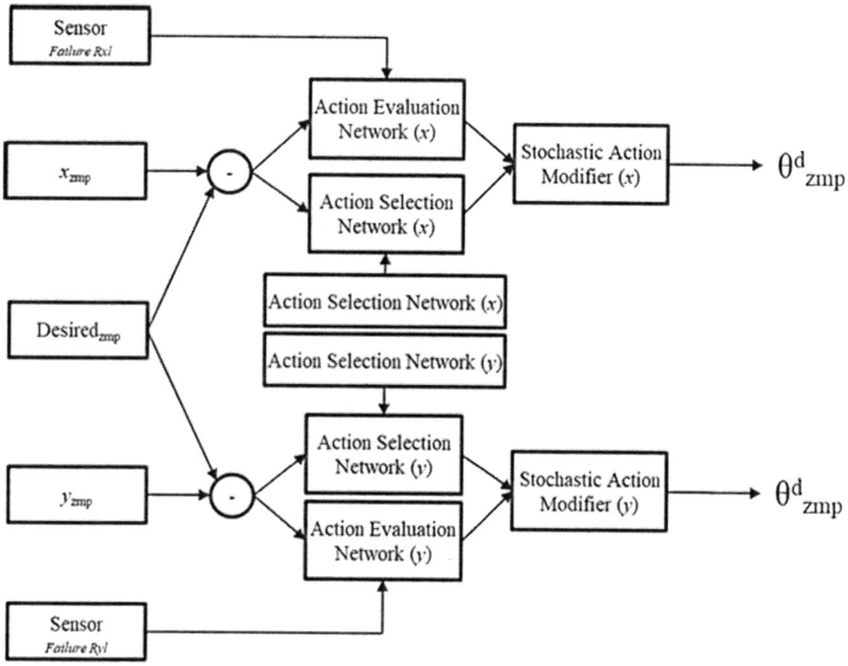

Fig. 3 The gait synthesizer architecture based on reinforcement learning [13]

to verify its success. Wang et al. [17] proposed fuzzy logic and a recursive gait control method to solve high energy consumption in humanoid robot applications. Besides some fuzzy-based methods have been proposed for control of the stabilities of humanoid in literature like in [18]. The trajectory of the ZMP is an important criterion for the balance of a walking robot but the complex dynamics involved make humanoid robot control difficult. An *adaptive neuro-fuzzy system (ANFS)* modeling at the ZMP trajectory of a humanoid robot has been presented by Kim et al. in [5]. They tried to establish relationships between process parameters and explain laws by including them in a walking humanoid robot.

During the entire gait phase, actual ZMP data was obtained from humanoid while walking on flat ground and on slopes. The success of ANFS depends on the membership functions used and the relevant part of the fuzzy rule. The ZMP orbit created using ANFS closely matches the ZMP orbit measured in [5].

There are some heavy works for humanoid robots that require the whole-body cooperative motion like typical examples of pushing a wall and twisting a valve. Hwang and friends presented two main issues in their work [6]. The first one is the simple *genetic algorithm* for the determination of optimal configuration for the pushing task of the humanoid robot and the second one is the generalization of the parameters of the simple genetic algorithm by the *multi-layer neural network* based on the *backpropagation algorithm*. Cuevas et al. applied the *fuzzy PD incremental control algorithm* to implement a biped balance control successfully [18].

The fuzzy PD incremental control algorithm with only four control rules can be seen from Fig. 4. Computationally economic fuzzy PD incremental controller algorithm has been applied for balance control of humanoid robot successfully as seen from Fig. 4. The algorithm proposed in [19], can be used in other robots with a similar dynamic and with other degrees of freedom. For this, will be necessary to only adjust the controller's gain parameters by using a genetic algorithm to find the controller's gain parameters. Park at all. designed a fuzzy based balance control for humanoid robot using 3D images as sensor information in [20].

As seen from this title of chapter, one of the popular usage field of fuzzy based control sets is stability and reliability control of humanoid robots. Examples of the effective use of fuzzy logic-based approaches in order to ensure the balance and stability of the humanoid robots are presented under this heading with different studies.

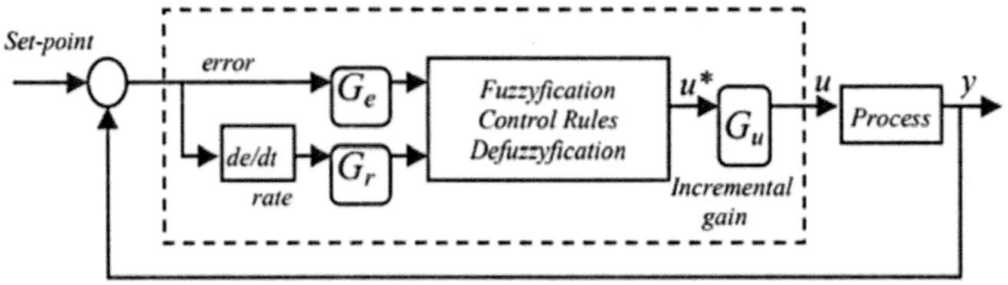

Fig. 4 Fuzzy PD incremental controller algorithm for balance control of humanoid robot [18]

3 Walking Pattern Detection of Humanoid Robot with Fuzzy Set

A humanoid robot walks with the trajectory that was previously produced, assuming even the terrain. From this point of view, priority can be defined as follows; if the gait course is removed first and then different values are measured from the expected sensor during gait, the robot must be repositioned using the stabilization algorithm. Figure 5 shows the walking algorithm.

If the stabilization and stability are not achieved at the desired level during the application of the determined gait route, the gait pattern is recalculated to be determined again. As the humanoid robot walks, it measures surface reaction forces as input signal of the controller. When the control of humanoid robot is interrupted by an unexpected state, or when one step ends unexpectedly, the new trajectory must be created according to the changing state. The re-created trajectory undergoes

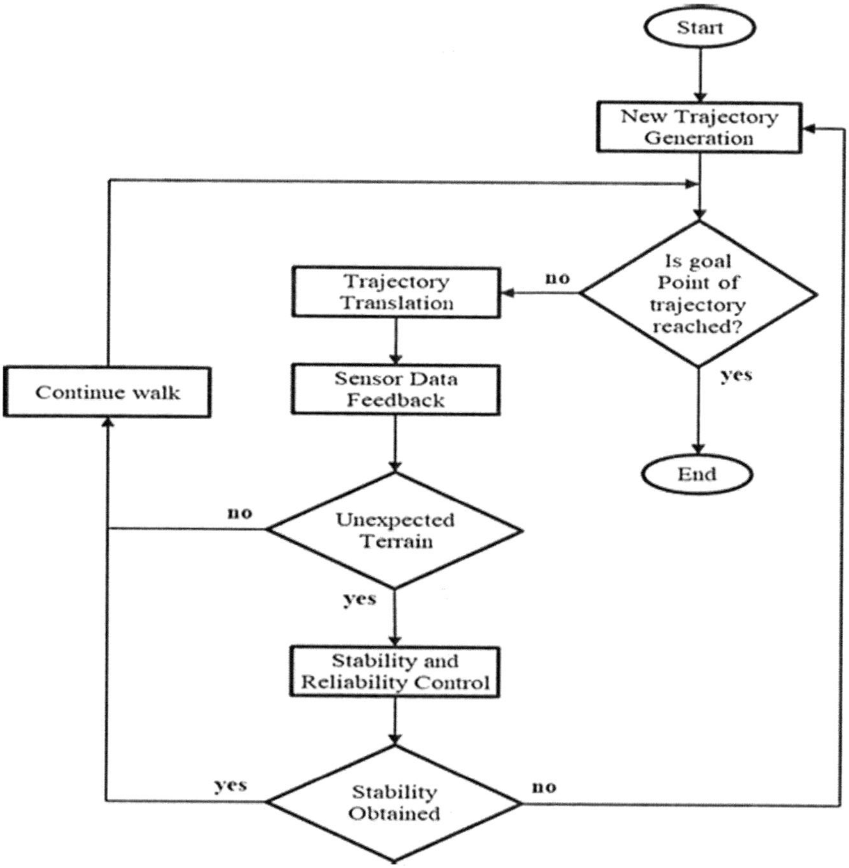

Fig. 5 Walking pattern determination of humanoid robot

stability verification, and once the stability of the trajectory is guaranteed, the robot can continue to walk. A method has been provided by Huang and friends [21] that provides for producing humanoid gait patterns based on human gait characteristics to produce more natural gait patterns and to determine parameters.

In [22], according to the 3-axis accelerometer sensor information, a fuzzy control method was designed and implemented successfully for a humanoid robot that can balance itself while walking even in an inclined surface.

The second usage area of fuzzy logic-based control approaches in humanoid robots is the determination of the walking pattern as seen from this part of chapter. Stability can be obtained during walking of humanoid robot by using fuzzy based control techniques.

4 Navigation of Humanoid Robot with Fuzzy Set

Navigation and road planning are of great importance in humanoid robot research as the other kind of robot architectures. Humanoid robots are started to use with other mobile robots, to benefit the advantage of improving spontaneous sustainability, especially in a dispersed environment. Furthermore, they have found a position to work with human cooperatively today. It means that incorporating the biomechanics of the human movement, so the importance of navigation and path planning is in the top level in robotics research.

Many researchers are investigating the navigation problem in an unknown or known environment for different mobile robotic studies. Fuzzy based control approach for the robot navigation used by many researchers like by Parhi et al. [23, 24]. Several intelligent algorithms inspired by nature for the navigation control of mobile robots have been developed by Mohanty and Parhi [25, 26]. When the road planning approaches in the literature are examined, it is seen that there are two types of road planning approaches for mobile robots, whose features are known and unknown.

Fakoor et al. [27] proposed a dynamic path planning approach for the humanoid robot based on sensor information in an unknown environment. As another example of unknown environment path planning of humanoid robots, Boukezzoula and friends have proposed a real-time decision system based on the approach of the fuzzy system and sensor fusion with the optical data method [28]. Although it was not a direct navigation problem, Lei and Qiang [29] used fuzzy logic to improve the accuracy of detection of the speed of the ball in soccer for a playing football humanoid robot. Flaherty et al. [30] determined the forward and inverse kinematics of a humanoid robot. Pierezan et al. [31] developed a modified self-adaptive different evolution (MSaDE) approach for the humanoid robot and set up a series of experiments to confirm its effectiveness.

Humanoid robots are programmed to be simulated to work in a similar way to human beings. Various researchers [32–35] tried to design the control architecture of mobile robot path planning and confirmed efficiency with appropriate simulation and experimental approaches. Lei et al. [36] analyzed the pose imitation of humanoid robot with a human. They evaluated the imitation analysis using metric-based analysis of pose similarity. A navigation controller for a humanoid robot has been developed using fuzzy based avoid obstacles in the environment and safely reach the goal position [37, 38]. The sensor data about distances of obstacles and bearing angle towards the target position are inputs of the controller and velocities to avoid from the obstacles are controller outputs.

Although some of the fuzzy logic-based approaches listed above for humanoid robots have been reported in the literature, it is difficult to come across special approaches to solve humanoid robots that distinguish humanoid robots from other types of robots. Even if it is assumed that humanoid robots need a special navigation approach, approaches developed for robots with limited maneuverability can often be studies that can respond to the navigation needs of humanoid robots.

5 Obstacle Avoidance of Humanoid Robot with Fuzzy Set

A real-time obstacle avoidance problem for humanoid robots in unknown or known environment based on the obtained sensor information is another popular search filed in the robotics. The humanoid robot can walk across a designated zone without colliding with obstacles like humans do. Therefore, effective and convenient methods must be developed to keep the distance between obstacles and the robot. One of the activities considered to play an important role in accelerating the research on humanoid robots is undoubtedly robot football leagues such as RoboCup and HuroCup. The most important problems that need to be solved among the activities such as soccer league in which such unmanned robots in a collective manner must perform the tasks as a team is that the robot avoids obstacles (other robots), searches, traces and moves the ball to the goal point. Just like typical human behavior, a humanoid robot has multiple options, such as walking around, jumping or pushing around the obstacle, and re-planning its route to avoid obstacles. The researches on these alternatives of obstacle avoidance for the humanoid robot can be listed as stepping over obstacles [39–41], avoidance control of robot arm [42, 43], passing through a structural element like door or corridor [44], and passing under obstacle [45].

Obstacle avoidance is one of the important problems in the humanoid robot design and as seen literature review, fuzzy control is one of the options to provide obstacle avoidance and track the targets. The fuzzy logic approach is a successful, effective and popular solution for the mobile robots' obstacle avoidance problem as seen from the literature [46–48]. As an example, work in [49], a fuzzy logic controller was suggested for the small-sized robot that planner generated a path to the destination while avoiding obstacles in soccer field. In this research, a fuzzy system working with infrared (IR) sensors and electronic compass sensor values as inputs is designed and implemented to avoid obstacles for a humanoid robot. But it is not a real-time obstacle avoidance algorithm based on the visual sensor information in unknown environment. Besides, in [22] as related another example, the decision-making based on a fuzzy set was designed and implemented to decide the behaviors of robot by using 3D stereo vision sensor data. The fuzzy-based control approach makes the robot enable to apply of obstacle avoidance successfully.

In [50], a humanoid robot with a vision-based sensor and fuzzy based obstacle avoidance algorithm is proposed to be implemented to illustrate the effectiveness of the real-time approach. In this study, it is mentioned that only the image sensors are mounted in a position equivalent to the humanoid robot since it is not allowed to place active sensors (emitting light or sound or electromagnetic waves in the environment to measure reflections) to ensure that the humanoid robot behavior is similar to the real human. Based on the obtained visual obstacle information from the scanning lines as shown in Fig. 6, the minimum obstacle distance d_i in the moving direction for the i-th motion is chosen as the first input variable of the fuzzy system as seen in Fig. 7. The angle φ_i from the electronic compass is the second input variable of the fuzzy system. In the proposed method, f_i is the output of the fuzzy system. The linguistic values of close, normal, and far are chosen membership sets for the input variable d_i, small, normal, and big are determined membership sets for input variable φ_i, bad, normal, and good are chosen membership sets for the output variable f_i. The triangular membership functions and fuzzy singleton-type membership function are respectively used for the input and output variables for fuzzy sets.

Furthermore, various a fuzzy-based obstacle avoidance algorithm was proposed especially for humanoid robots. Dadios et al. [51] proposed a humanoid robot using artificial intelligence and demonstrated that the robot was capable of walking on obstacles, balancing itself, and avoiding obstacles. Pandey et al. [52, 53] used fuzzy logic as a navigation algorithm to avoid obstacles for mobile robots in complex environments.

By using different type sensors with fuzzy based obstacle avoidance algorithm, effective solution can be implemented for humanoid robots as seen the examples in this part of chapter.

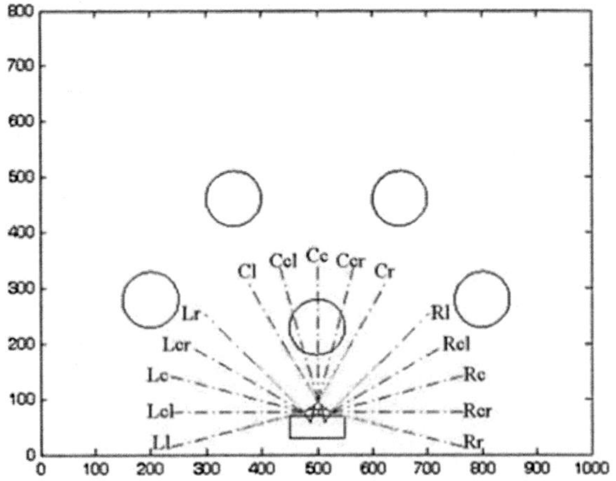

Fig. 6 Description of vision based input signals for fuzzy based obstacle avoidance system [50]

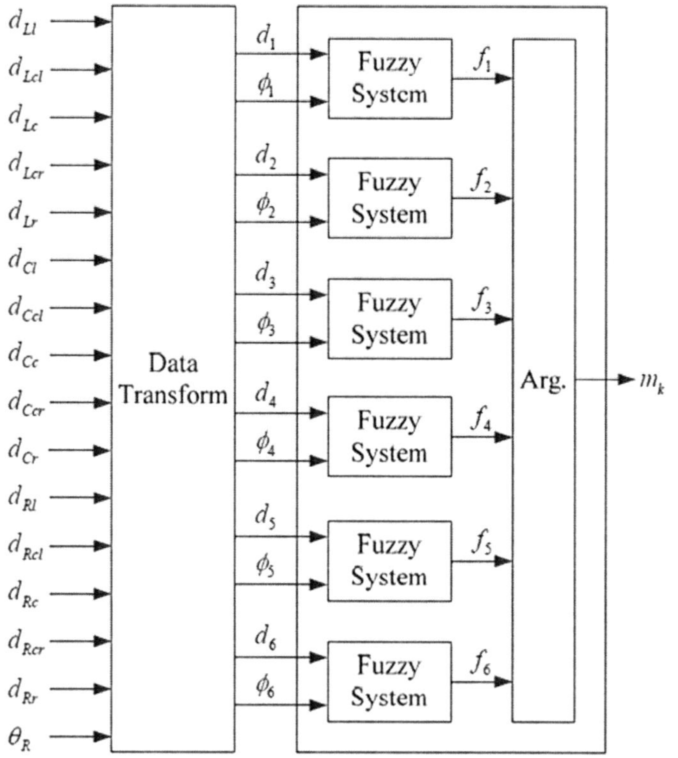

Fig. 7 Fuzzy based system block diagram for obstacle avoidance of humanoid robot [50]

6 Conclusions

This chapter focuses on the recent applications and researches related with different level intelligent fuzzy-based control techniques (neural networks, fuzzy logic, and genetic algorithms) and their hybrid methods (neuro-fuzzy networks, neuro-genetic and fuzzy-genetic algorithms) in the field of humanoid robots. This chapter represents an attempt to report on a literature survey of the basic principles and concepts based on fuzzy control especially related with humanoid robotics. Recent research related to fuzzy-based control approaches for humanoids are categorized under the titles as stability and reliability control, walking pattern determination, navigation control, and obstacle avoidance.

Most common fuzzy-based research were applied to humanoid robots related to *stability and reliability control*. There are numerous examples of different fuzzy techniques to solve the stability and reliability problem of humanoid robots. The shared and main purpose of suggested approaches is balancing the humanoid robot while walking over various defective surface types. In some research fuzzy used for adjustment of PID gains or as feedback and reinforcement learning tool, but it is possible to see fuzzy logic as the only algorithm to provide stability of a humanoid. It is not easy to fully distinguish between *walking pattern determination* algorithm and balance - stability control algorithm as seen from the general control approaches mentioned above. Nevertheless, it is possible to come across studies in the literature that deal with these two problems differently, as can be seen from the examples above. Although fuzzy logic controllers are limited to the number of studies on gait pattern alone, the problem of gait pattern is also widely studied in balance and stability control studies. As seen from the extensive survey of the literature in the previous titles, it can be noticed that various researchers have work popularly about the fuzzy-based stability—reliability control and *obstacle avoidance* for the humanoids. However, a few numbers of works have been mentioned about the fuzzy-based *navigation* of humanoid robots. Also given examples of fuzzy-based navigation algorithms are limited to specific environment conditions only. It is clear that hard to implement a robust control technique that can navigate the humanoid robots in complex terrains irrespective of the environmental conditions. As explained under the title, usually navigation problem of humanoids does not handle a special case of robot type and most of the researcher uses common algorithms that is designed for limited maneuverability robots.

Academic studies in the literature mentioned within the scope of this chapter show that fuzzy logic-based control techniques are used effectively in solving different control problems of humanoid robots. Compared with other control techniques in the literature, their ease of modeling, ability to produce fast reactions and ease of application are the reasons for preference of fuzzy logic-based controller approach. In addition, these advantages are undoubtedly the biggest indication that fuzzy logic-based control approaches will continue to be among the widely used techniques in the future to solve different control problems of humanoid robots.

As a result, fuzzy logic has become a widely used technique in the different control structures required by humanoid robots with increasing number of studies in recent years. Within the scope of this book chapter, humanoid robot control approaches based on different levels of fuzzy logic are gathered and presented to the reader. The number of studies carried out on the development of humanoid robots will undoubtedly increase with an even greater momentum. With this acceleration, it is a fact that there will be an increase in studies on fuzzy logic-based control approaches of humanoid robots. In this way, it is aimed that readers interested in the humanoid robot control approach based on fuzzy logic will reach the literature faster and easier.

References

1. Kagami, S., Kanehiro, F., Tamiya, Y., Inaba, M., Inoue, H.: AutoBalancer: an online dynamic balance compensation scheme for humanoid robots. In: International Workshop on the Algorithmic Foundations of Robotics (WAFR) (2000)
2. Bruce, A., Nourbakhsh, I., Simmons, R.: The role of expressiveness and attention in human-robot interaction. In: Proceedings, AAAI Fall Symposium (2001)

3. Breazeal, C., Velasquez, J.: "Toward teaching a robot infant" using emotive communication acts. In: Proceedings of Simulation of Adaptive Behavior, Workshop on Socially Situated Intelligence, vol. 98, pp. 25–40, Zurich, Switzerland (1998)
4. Breazeal, C., Velasquez, J.: Robot in society: friend or appliance? In: Proceedings of Agents 99 Workshop on Emotion-Based Agent Architectures, pp. 18–26, Seattle, WA (1999)
5. Kim, D., Seo, S., Park, G.: Zero-moment point trajectory modeling of a biped walking robot using an adaptive neuro-fuzzy system. Proc. IEE Control Theory **152**, 411–426 (2005)
6. Hwang, Y.-K., Choi, K.-J., Hong, D.-S.: Self-learning control of cooperative motion for humanoid robots. Int. J. Control Autom. Syst. **4**(6), 725–735 (2006)
7. Fukuda, T., Komata, Y., Arakawa, T.: Stabilization control of biped locomotion robot based learning with gas having self-adaptive mutation and recurrent neural networks. In: Proceedings of IEEE International Conference on Robotics & Automation, pp. 217–222 (1997)
8. Kajita, S., Kanehiro, F., Kaneko, K., Fujiwara, K., Harada, K., Yokoi, K., Hirukawa, H.: Biped walking pattern generation by using preview control of zero-moment point. In: Proceedings of IEEE International Conference on Robotics & Automation, pp. 1620–1626 (2003)
9. Vukobratovic, M., Juricic, D.: Contribution to the synthesis of biped gait. IEEE Trans. Bio-Med. Eng. **16**, 1–6 (1969)
10. Zhou, C., Meng, Q.: Reinforcement learning with fuzzy evaluative feedback for a biped robot. In: Proceedings of IEEE International Conference on Robotics and Automation, San Francisco, pp. 3829–3834 (2000)
11. Nishiwaki, K., Kagami, S., Kuniyoshi, Y., Inaba, M., Inoue, H.: Online generation of humanoid walking motion based on a fast generation method of motion pattern that follows desired ZMP. In: Proceedings of IEEE/RSJ International Conference on Intelligent Robots and System, vol. 3, pp. 2684–2689 (2002)
12. Djoudi, D., Chevallereau, C.: Stability analysis of a walk of a biped with control of the ZMP. In: Proceedings of IEEE/RSJ International Conference on Intelligent Robots and Systems, pp. 2461–2467 (2005)
13. Katić, D., Vukobratović, M.: Survey of intelligent control techniques for humanoid robots. J. Intell. Rob. Syst. **37**, 117–141 (2003). https://doi.org/10.1023/A:1024172417914
14. Sugihara, T., Nakamura, Y.: Whole-body cooperative balancing of humanoid robot using COG Jacobian. In: Proceedings of IEEE/RSJ International Conference on Intelligent Robots and Systems, pp. 2575–2580 (2002)
15. Zhong, Q.B., Chen, F.: Trajectory planning for biped robot walking on uneven terrain taking stepping as an example. CAAI Trans. Intell. Technol. **1**(3), 197e209 (2016)
16. Samant, R., Nair, S., Kazi, F.: Development of autonomous humanoid robot control for competitive environment using fuzzy logic and heuristic search. IFAC Papers Online; **49**(1), 373e8 (2016)
17. Wang, L.Y., Liu, Z., Zeng, X.J., Zhang, Y.: Gait control of humanoid robots via fuzzy logic and iterative optimization. In: The 30th Chinese Control Conference (CCC); p. 3931e6 (2011)
18. Cheng, C.T., Chen, H.C., Hu, Y.Y., Wong, C.C.: Fuzzy balancing control of a small-size humanoid robot based on accelerometer. Int. J. Fuzzy Syst. **11**(3), 146–153 (2009)
19. Cuevas, E., Zaldivar, D., Rojas, R.: Incremental fuzzy control for a biped robot balance. In: Proceedings of IASTED International Conference on Robotics and Applications, pp. 99–104 (2005)
20. Park, S., Han, Y., Hahn, H.: Fuzzy controller based biped robot balance control using 3D image. Int. J. Fuzzy Syst. **11**(3), 202–212 (2009)
21. Huang, Q., Yu, Z., Zhang, W., Duan, X., Huang, Y., Li, K.: Generation of humanoid walking pattern based on human walking measurement. In: 8th IEEE-RAS International Conference on Humanoid Robots, Humanoids, pp. 99–104. https://doi.org/10.1109/ICHR.2008.4755938 (2008)
22. Chao, C.H., Hsueh, B.Y., Hsiao, M.Y., Tsai, S.H., Li, T.H.S.: Fuzzy target tracking and obstacle avoidance of mobile robots with a stereo vision system. Int. J. Fuzzy Syst. **11**(3), 183–191 (2009)

23. Singh, M.K., Parhi, D.R.: Path optimization of a mobile robot using an artificial neural network controller. Int. J. Syst. Sci. **42**(1), 107e20 (2011)
24. Singh, M.K., Parhi, D.R., Pothal, J.K.: ANFIS approach for navigation of mobile robots. In: IEEE International Conference on Advances in Recent Technologies in Communication and Computing, p. 727e31 (2009)
25. P.K. Mohanty, D.R. Parhi, "Optimal path planning for a mobile robot using cuckoo search algorithm", J Exp Theor Artif Intell;28(1e2):35e52, 2016.
26. Mohanty, P.K., Parhi, D.R.: A new hybrid optimization algorithm for multiple mobile robots navigation based on the CS-ANFIS approach. Memet. Comput. **7**(4), 255e73 (2016)
27. Fakoor, M., Kosari, A., Jafarzadeh, M.: Humanoid robot path planning with fuzzy Markov decision processes. J. Appl. Res. Technol. **14**(5), 300e10 (2016)
28. Boukezzoula, R., Coquin, D., Nguyen, T.L., Perrin, S.: Multi-sensor information fusion: combination of fuzzy systems and evidence theory approaches in color recognition for the NAO humanoid robot. Robot Autonom. Syst. **100**, 302e16 (2018)
29. Lei, S.H., Qiang, W.A.: Research on humanoid robot soccer system based on fuzzy logic. Int. J. Comput. Netw. Inf. Secur. **2**(1), 38 (2010)
30. O'Flaherty, R., Vieira, P., Grey, M., Oh, P., Bobick, A., Egerstedt, M., Stilman, M.: Kinematics and inverse kinematics for the humanoid robot HUBO2þ. Georgia Institute of Technology (2013)
31. Pierezan, J., Freire, R.Z., Weihmann, L., Reynoso-Meza, G., dos Santos Coelho, L.: Static force capability optimization of humanoids robots based on modified self-adaptive differential evolution. Comput. Oper. Res. **84**, 205e15 (2017)
32. Pothal, J.K., Parhi, D.R.: Navigation of multiple mobile robots in a highly clutter terrains using adaptive neuro-fuzzy inference system. Robot Autonom. Syst. **72**, 48e58 (2015)
33. Eliot, E., Deepak, B.B.V.L., Parhi, D.R., Srinivas, J.: Design & kinematic analysis of an articulated robotic manipulator (2012)
34. Parhi, D.R., Mohanta, J.C.: Navigational control of several mobile robotic agents using petri-potential-fuzzy hybrid controller. Appl. Soft Comput. **11**(4), 3546e57 (2011)
35. S. Kundu, D.R. Parhi, "Navigation of Underwater Robot Based on Dynamically Adaptive Harmony Search Algorithm", Memetic Computing; 8(2):125e46, 2016.
36. Lei, J., Song, M., Li, Z.N., Chen, C.: Whole-body humanoid robot imitation with pose similarity evaluation. Signal Process **108**, 136e46 (2015)
37. Parhi, D.R.: Navigation of mobile robots using a fuzzy logic controller. J. Intell. Rob. Syst. **42**(3), 253e73 (2005)
38. Rath, A.K., et al.: Analysis and use of fuzzy intelligent technique for navigation of humanoid robot in obstacle prone zone. In: Defence Technology, vol. 14(6), 677–682 (2018)
39. Guan, Y., Yokoi, K., Tanie, K.: Can humanoid robots overcome given obstacles. In: Proceedings of IEEE International Conference on Robotics and Automation, Barcelona, Spain, pp. 1054–1059 (2005)
40. Guan, Y., Neo, E.S., Yokoi, K., Tanie, K.: Stepping over obstacles with humanoid robots. IEEE Trans. Robot. **22**(5), 958–973 (2006)
41. Stasse, O., Verrelst, B., Vanderborght, B., Yokoi, K.: Strategies for humanoid robots to dynamically walk over large obstacles. IEEE Trans. Robotics **25**(4), 960–967 (2009)
42. Park, D.H., Hoffmann, H., Pastor, P., Schaal, S.: Movement reproduction and obstacle avoidance with dynamic movement primitives and potential fields. In: Proceedings of 8th IEEE-RAS International Conference on Humanoid Robots, Daejeon, Korea, pp. 91–98 (2008)
43. Tsetserukou, D., Kawakami, N., Tachi, S.: Obstacle avoidance control of humanoid robot arm through tactile interaction. In: Proceedings of 8th IEEE-RAS International Conference on Humanoid Robots, pp. 379–384, Daejeon, Korea (2008)
44. Digioia, G., Arisumi, H., Yokoi, K.: Trajectory planner for a humanoid robot passing through a door. In: Proceedings of 9th IEEE-RAS International Conference on Humanoid Robots, Paris, France, pp. 134–141 (2009)
45. Sanada, H., Yoshida, E., Yokoi, K.: Passing under obstacles with humanoid robots. In: Proceedings of IEEE/RSJ International Conference on Intelligent Robots and Systems, San Diego, CA, pp. 4028–4034 (2007)

46. Seraji, H., Howard, A.: Behavior-based robot navigation on challenging terrain: a fuzzy logic approach. IEEE Trans. Robot. Automat. **18**, 308–321 (2002)
47. Cang, Y., Yung, N.H.C., Danwei, W.: A fuzzy controller with supervised learning assisted reinforcement learning algorithm for obstacle avoidance. IEEE Trans. Syst. Man Cybern. **33**(1), 17–27 (2003)
48. Lilly, J.H.: Evolution of a negative-rule fuzzy obstacle avoidance controller for an autonomous vehicle. IEEE Trans. Fuzzy Syst. **15**, 718–728 (2007)
49. Wong, C.C., Cheng, C.T., Huang, K.H., Yang, Y.T.: Fuzzy control of humanoid robot for obstacle avoidance. Int. J. Fuzzy Syst. **10**(1), 261–270 (2008)
50. Wong, C.C., Hwang, C.L., Huang, K.H., Hu, Y.Y., Cheng, C.T.: Design and implementation of vision-based fuzzy obstacle avoidance method on humanoid robot. Int. J. Fuzzy Syst. **13** (2011)
51. Dadios, E.P., Biliran, J.J.C., Garcia, R.R.G., Johnson, D., Valencia, A.R.B.: Humanoid robot: design and fuzzy logic control technique for its intelligent behaviors. In: Proceedings of Fuzzy Logic-Controls, Concepts, Theories and Applications (2012)
52. Pandey, A., Sonkar, R.K., Pandey, K.K., Parhi, D.R.: Path planning navigation of mobile robot with obstacles avoidance using fuzzy logic controller. In: Proceedings of IEEE 8th International Conference on Intelligent Systems and Control (ISCO), p. 39e41 (2014)
53. Pandey, A., Parhi, D.R.: MATLAB simulation for mobile robot navigation with hurdles in cluttered environment using minimum rule based fuzzy logic controller. Procedia Technol. **14**:28e34 (2014)

Humanoid Robots and Neutrosophic Sets

Theory of Single Valued Trapezoidal Neutrosophic Numbers and Their Applications to Multi Robot Systems

Irfan Deli

Abstract The chapter explains how the concept of single valued trapezoidal neutrosophic numbers (SVTN-numbers) can be applied in the field of humanoid Robotics. To explain the concept of SVTN-numbers, a multi Robot scenario is considered consisting of a central server and a group of mobile Robots patrolling a given area for surveillance application. Using the correlation coefficient of SVTN-numbers, the sensor readings were properly interpreted for proper identification of the problem faced by the Robot. To do this, the chapter first give the concept of normal SVTN-numbers which are generalization of fuzzy numbers, intuitionistic fuzzy numbers and so on. The chapter second, discuss some operations of normal SVTN-numbers. The chapter third propose a approach to find the correlation coefficient value, which is strength of relationship of two normal SVTN-numbers. Finally, the chapter propose a method and an application for a multi Robot system considered consists of a central controller and more than one patrolling Robots in a large area by using normal SVTN-number.

1 Introduction

To handle inexact and imprecise data, theory of fuzzy set by using a membership function defined by Zadeh [50] and then extended by Atanassov [2] to theory of intuitionistic fuzzy sets by using both membership function and non-membership function cope with the presence of hesitancy originating from imprecise information. Since the theory of intuitionistic fuzzy sets have some restriction such as "$0 \leq$ membership degree + non-membership degree ≤ 1", the theory were extended by Smarandache [42] to neutrosophic sets which can only handle incomplete information not the indeterminate information and inconsistent information from philo-

I. Deli (✉)
Kilis 7 Aralık University, 79000 Kilis, Turkey
e-mail: irfandeli@kilis.edu.tr

C. Kahraman and E. Bolturk (eds.), *Toward Humanoid Robots: The Role of Fuzzy Sets*, Studies in Systems, Decision and Control 344,
https://doi.org/10.1007/978-3-030-67163-1_10

sophical point of view. After Smarandache [42], by using study of Li [29], Ye [47, 48] and Deli [16, 17] proposed single valued trapezoidal neutrosophic numbers (SVTN-numbers) from different angles with their operations and properties. Also they gave different aggregation operators for multi-criteria decision making problems. Deli and Subas [15] introduced the cut sets of SVTN-numbers and then applied to single valued trapezoidal neutrosophic numbers (SVTN-numbers) and triangular neutrosophic numbers (SVTrN-numbers). Mohamed et al. [35] introduced a method to find the critical path in network diagram by using triangular neutrosophic numbers, instead of crisp numbers. Basset et al. [5] investigated a method based on Analytic Hierarchy Process (AHP) into Delphi framework in SVTN-numbers. Basset et al. [3] and Biswas et al. [8] applied the TOPSIS method to neutrosophic numbers. Pramanik and Mallick [39] developed a method based on the VIKOR method under neutrosophic environment. Karaaslan [27] defined Gaussian single-valued neutrosophic number (GSVNN) and cut sets of a GSVNN. Basset et al. [4] and Liang et al. [31] introduced single-valued trapezoidal neutrosophic preference relations for multi-criteria decision-making problems. Broumi et al. [11, 12, 14] studied the SVTN minimum spanning tree problems where the edge weights are assumed to be SVTN-numbers, neutrosphic shortest path problem in a network in which each edge weight is represented as SVTrN-numbers and a neutrosophic network problem to find the shortest path length with SVTN-numbers, respectively. Liu and Zhang [32] gave the Maclaurin symmetric mean operator under SVTN-numbers. Biswas et al. [7] defined the score and accuracy values of SVTN-numbers. Porchelvi and Umamaheswari [36] researched a Principal pivoting method to handle the Fuzzy Linear Complementarity Problems. Wu et al. [46] put forwarded a method based on SVTN-numbers for multi-criteria group decision-making problems. Liang et al. [30] initiated normalized weighted Bonferroni mean operator and developed a approach for multi-criteria group decision-making by using SVTN-numbers. Biswas et al. [6] defined expected value of SVTN-numbers to find the attribute weight. Aal et al. [1] developed a types of ranking methods under weighted value and ambiguity based on SVTrN-numbers. Biswas et al. [10] developed a ranking method for ranking SVTrN-numbers and gave some operational rules as well as cut sets of SVTrN-numbers. Also, Kahraman and Otay [25] collected some neutrosophic studies including SVTN-numbers. Then, Giri et al. [20] and Biswas et al. [9] proposed interval trapezoidal neutrosophic numbers by defining some arithmetic operations to handle multi attribute decision making problems.

Hanafy et al. [21] and Hung et al. [24] initiated a approach to find the correlation coefficient value, which is strength of relationship of two intuitionistic fuzzy set, by means of "centroid". Ruan et al. [41] introduced the centroid technique with Pearson's correlation coefficient for measuring the correlation coefficient of fuzzy data. Robinson and Amirtharaj [38] studied the correlation coefficient of trapezoidal fuzzy intuitionistic fuzzy sets and applied to multi-attribute decision making problems. Zeng and Li [51] proposed correlation and correlation coefficient of intuitionistic fuzzy sets based on the point of view of geometrical representation of intuitionistic fuzzy sets. Zhang et al. [52] defined some similarity measures which are correlation coefficients between two triangular fuzzy numbers. The methods described to measure the correlation coefficient of crisp, fuzzy and intutionistic fuzzy sets. Since neutrosophic set and especially SVTN-numbers are often involved an ill-known quantity which contain indeterminacy information, defined methods cannot be directly used to find the correlation coefficient of neutrosophic set and SVTN-numbers. Therefore, Ye [49] defined three vector similarity measures between neutrosophic sets based on Jaccard, Dice and cosine similarity measures in vector space. Hanafy et al. [23] introduce correlation coefficient of neutrosophic sets in finite and continous sets.

"A humanoid robot is widely used because of its ability to imitate human actions and the selection of navigational techniques is of prime importance because the quality of the opted technique directly affects the success of output" [28]. Therefore, some researchers have been successfully studied on humanoid robot. For example; Kashyap et al. [28] developed hybridization of the Dynamic Window Approach (DWA) and the Teaching–Learning-Based Optimization (TLBO) technique. Fakoor et al. [13] proposed decision system to investigates the conception of a real-time decision system. Fakoor et al. [19] investigated a approach according to fuzzy Markov decision processes. Lau et al. [33] studied on humanoid robot based on children with diabetes. Kahraman et al. [26] investigated a comprehensive literature review about humanoid robots that presents the recent technological developments and the theories associated with fuzzy set models. Faraj et al. [18] proposed a relatively inexpensive, open-source robot that can serve as a platform for research into emotional communication between humans and machines. Samant et al. [43] gave a new approach for interaction between humanoid robots by using fuzzy logic. Murillo et al. [34] investigated how visual constraints such as homographies and fundamental matrices can be integrated tightly into the locomotion controller of a humanoid robot to drive it from one configuration to another (pose-regulation), only by means of images. Rath et al. [37] developed a navigational controller for a humanoid by using fuzzy logic as an intelligent algorithm for avoiding the obstacles present in the environment and reach the desired target position safely.

Some humanoid robot is a robot with its overall appearance based on that of the human body and the humanoid robot uses ultrasonic and infra red sensors to detect the obstacles on its path [22]. Therefore, this chapter explains how the concept of normal SVTN-Numbers can be applied in the field of Robotics. Shinoj and John [44] said that "Robots are machines which reduces human effort. Robots can be given intelligence to perform tasks that humans can and cannot do. They can be programmed for doing a task monotonously or they can work intelligently or dynamically according to the situations around them. Some of the applications of a mobile Robot include mine detection, surveillance, bomb detection, remote surgery, welding, cleaning small pipes, window panes and glass doors of buildings using snake-like Robots etc. A Robot mainly contains: sensors, actuators and a controller. An accelerometer sensor is used for detecting shock/vibration, a temperature sensor can detect the temperature variations, an ultrasonic sensor/Infra-Red sensor/PIR sensor is used to detect obstacles, bump sensor senses a bump (collision), cliff sensor senses the presence of a cliff and so on. With the help of these sensors a Robot moves easily through its programmed path".

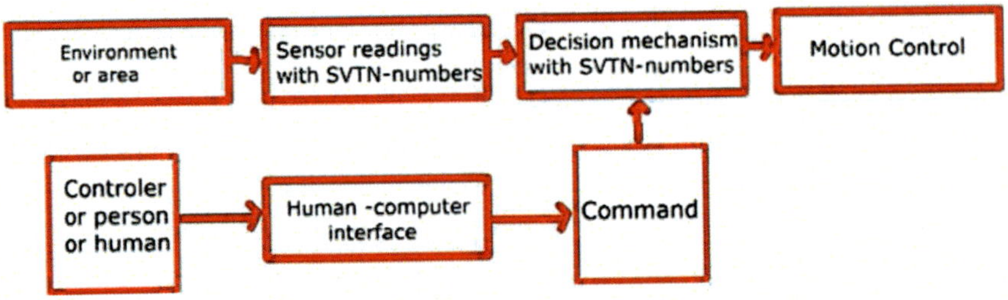

Fig. 1 Control of multi robot systems in normal SVTN-numbers

Fig. 2 Control of multi robot systems in normal SVTN-numbers

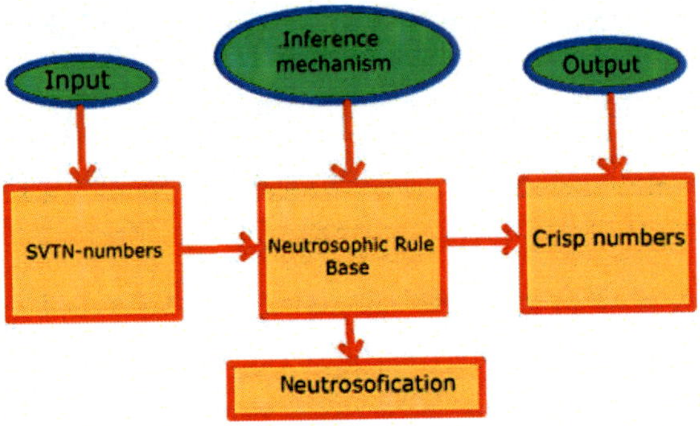

The chapter is organized as: We first gave the concept of normal SVTN-numbers which are generalization of fuzzy numbers, intuitionistic fuzzy numbers and so on. Then we discussed some operations of normal SVTN-numbers. The operations give a normal SVTN-number. Also we proposed a approach to find the correlation coefficient value, which is strength of relationship of two normal SVTN-numbers. Finally, we presented a method and an application for a multi Robot system considered consists of a central controller and more than one patrolling Robots in a large area by using normal SVTN-numbers. In application, the multi Robot system considered consists of a central controller and five patrolling Robots in a large area. The total area is divided into five equal parts and assigned to each Robot to control by the controller with sensor. The control and decision mechanism of Multi Robot Systems in normal SVTN-numbers was summed in Figs. 1 and 2, respectively.

2 Preliminary

In this section, we recall some basic notions of fuzzy sets, intuitionistic fuzzy sets, neutrosophic sets, single valued neutrosophic sets and single valued neutrosophic numbers.

Definition 1 ([50]) Let E be a universe. Then a fuzzy set X over E is defined by

$$X = \{(\mu_X(x)/x) : x \in E\}$$

where μ_X is called membership function of X and defined by $\mu_X : E \to [0.1]$. For each $x \in E$, the value $\mu_X(x)$ represents the degree of x belonging to the fuzzy set X.

Definition 2 ([53]) t-norms are associative, monotonic and commutative two valued functions t that map from $[0, 1] \times [0, 1]$ into $[0, 1]$. These properties are formulated with the following conditions:

1. $t(0, 0) = 0$ and $t(\mu_{X_1}(x), 1) = t(1, \mu_{X_1}(x)) = \mu_{X_1}(x),\ \ x \in E$
2. If $\mu_{X_1}(x) \leq \mu_{X_3}(x)$ and $\mu_{X_2}(x) \leq \mu_{X_4}(x)$, then $t(\mu_{X_1}(x), \mu_{X_2}(x)) \leq t(\mu_{X_3}x), \mu_{X_4}(x))$
3. $t(\mu_{X_1}(x), \mu_{X_2}(x)) = t(\mu_{X_2}(x), \mu_{X_1}(x))$
4. $t(\mu_{X_1}(x), t(\mu_{X_2}(x), \mu_{X_3}(x))) = t(t(\mu_{X_1}(x), \mu_{X_2})(x), \mu_{X_3}(x))$.

Also, s-norm are associative, monotonic and commutative two placed functions s which map from $[0, 1] \times [0, 1]$ into $[0, 1]$. These properties are formulated with the following conditions:

1. $s(1, 1) = 1\ and\ s(\mu_{X_1}(x), 0) = s(0, \mu_{X_1}(x)) = \mu_{X_1}(x), x \in E$
2. if $\mu_{X_1}(x) \leq \mu_{X_3}(x)$ and $\mu_{X_2}(x) \leq \mu_{X_4}(x)$, then $s(\mu_{X_1}(x), \mu_{X_2}(x)) \leq s(\mu_{X_3}(x), \mu_{X_4}(x))$
3. $s(\mu_{X_1}(x), \mu_{X_2}(x)) = s(\mu_{X_2}(x), \mu_{X_1}(x))$
4. $s(\mu_{X_1}(x), s(\mu_{X_2}(x), \mu_{X_3}(x))) = s(s(\mu_{X_1}(x), \mu_{X_2})(x), \mu_{X_3}(x))$.

t-norm and t-conorm are related in a sense of logical duality. Typical dual pairs of non parametrized t-norm and t-conorm are complied below:

1. Drastic product:

$$t_w(\mu_{X_1}(x), \mu_{X_2}(x)) = \begin{cases} min\{\mu_{X_1}(x), \mu_{X_2}(x)\}, & max\{\mu_{X_1}(x)\mu_{X_2}(x)\} = 1 \\ 0, & otherwise \end{cases} \quad (1)$$

2. Drastic sum:

$$s_w(\mu_{X_1}(x), \mu_{X_2}(x)) = \begin{cases} max\{\mu_{X_1}(x), \mu_{X_2}(x)\}, & min\{\mu_{X_1}(x)\mu_{X_2}(x)\} = 0 \\ 1, & otherwise \end{cases} \quad (2)$$

3. Bounded product:

$$t_1(\mu_{X_1}(x), \mu_{X_2}(x)) = max\{0, \mu_{X_1}(x) + \mu_{X_2}(x) - 1\} \quad (3)$$

4. Bounded sum:

$$s_1(\mu_{X_1}(x), \mu_{X_2}(x)) = min\{1, \mu_{X_1}(x) + \mu_{X_2}(x)\} \tag{4}$$

5. Einstein product:

$$t_{1.5}(\mu_{X_1}(x), \mu_{X_2}(x)) = \frac{\mu_{X_1}(x).\mu_{X_2}(x)}{2 - [\mu_{X_1}(x) + \mu_{X_2}(x) - \mu_{X_1}(x).\mu_{X_2}(x)]} \tag{5}$$

6. Einstein sum:

$$s_{1.5}(\mu_{X_1}(x), \mu_{X_2}(x)) = \frac{\mu_{X_1}(x) + \mu_{X_2}(x)}{1 + \mu_{X_1}(x).\mu_{X_2}(x)} \tag{6}$$

7. Algebraic product:

$$t_2(\mu_{X_1}(x), \mu_{X_2}(x)) = \mu_{X_1}(x).\mu_{X_2}(x) \tag{7}$$

8. Algebraic sum:

$$s_2(\mu_{X_1}(x), \mu_{X_2}(x)) = \mu_{X_1}(x) + \mu_{X_2}(x) - \mu_{X_1}(x).\mu_{X_2}(x) \tag{8}$$

9. Hamacher product:

$$t_{2.5}(\mu_{X_1}(x), \mu_{X_2}(x)) = \frac{\mu_{X_1}(x).\mu_{X_2}(x)}{\mu_{X_1}(x) + \mu_{X_2}(x) - \mu_{X_1}(x).\mu_{X_2}(x)} \tag{9}$$

10. Hamacher sum:

$$s_{2.5}(\mu_{X_1}(x), \mu_{X_2}(x)) = \frac{\mu_{X_1}(x) + \mu_{X_2}(x) - 2.\mu_{X_1}(x).\mu_{X_2}(x)}{1 - \mu_{X_1}(x).\mu_{X_2}(x)} \tag{10}$$

11. Miniumum:

$$t_3(\mu_{X_1}(x), \mu_{X_2}(x)) = min\{\mu_{X_1}(x), \mu_{X_2}(x)\} \tag{11}$$

12. Maximum:

$$s_3(\mu_{X_1}(x), \mu_{X_2}(x)) = max\{\mu_{X_1}(x), \mu_{X_2}(x)\} \tag{12}$$

Definition 3 ([45]) Let E be a universe. A single valued neutrosophic sets (SVN-sets) A over E is defined by

$$A = \{< x, (T_A(x), I_A(x), F_A(x)) >: x \in E\}.$$

where T_A, I_A and F_A are called truth-membership function, indeterminacy-membership function and falsity-membership function, respectively. They are respectively defined by

$$T_A : E \to [0, 1],\ I_A : E \to [0, 1]\ and\ F_A : E \to [0, 1]$$

such that $0 \le T_A(x) + I_A(x) + F_A(x) \le 3$ for all $x \in E$.

Definition 4 ([49]) Let $A = <x_i, (T_A(x_i), I_A(x_i), F_A(x_i))>$ and $B = <x_i, (T_B(x_i), I_B(x_i), F_B(x_i))>$ be two SVN-sets in a universe of $X = \{x_1, x_2, ..., x_n\}$. Then

1. Jaccard similarity measure between SVN-sets A and B in the vector space is defined as follows:

$$J(A, B) = \frac{1}{n}\sum_{i=1}^{n} \times \left(\frac{T_A(x_i)T_B(x_i) + I_A(x_i)I_B(x_i) + F_A(x_i)F_B(x_i)}{\begin{array}{l}[((T_A)^2(x_i) + (I_A)^2(x_i) + (F_A)^2(x_i))+ \\ ((T_B)^2(x_i) + (I_B)^2(x_i) + (F_B)^2(x_i))- \\ (T_A(x_i)T_B(x_i) + I_A(x_i)I_B(x_i) + F_A(x_i)F_B(x_i))]\end{array}} \right) \tag{13}$$

   Then, this similarity measure satisfies the following properties:

   a. $0 \le J(A, B) \le 1$;
   b. $J(A, B) = J(B, A)$;
   c. $J(A, B) = 1$ for $A = B$ i.e.$T_A(x_i) = T_B(x_i),\ I_A(x_i) = I_B(x_i),\ F_A(x_i) = F_B(x_i)(i = 1, 2..., n)\ \forall\, x_i(i = 1, 2, ..., n) \in X$.

2. Dice similarity measure between SVN-sets A and B in the vector space is defined as follows:

$$D(A, B) = \frac{1}{n}\sum_{i=1}^{n} \times \left(\frac{2(T_A(x_i)T_B(x_i) + I_A(x_i)I_B(x_i) + F_A(x_i)F_B(x_i))}{\begin{array}{c}[((T_A)^2(x_i) + (I_A)^2(x_i) + (F_A)^2(x_i))+ \\ ((T_B)^2(x_i) + (I_B)^2(x_i) + (F_B)^2(x_i))]\end{array}} \right) \tag{14}$$

   It satisfies the following properties:

   a. $0 \le D(A, B) \le 1$;
   b. $D(A, B) = D(B, A)$;
   c. $D(A, B) = 1$ for $A = B$ i.e.$T_A(x_i) = T_B(x_i),\ I_A(x_i) = I_B(x_i),\ F_A(x_i) = F_B(x_i)(i = 1, 2..., n)\ \forall\, x_i(i = 1, 2, ..., n) \in X$.

3. Cosine similarity measure between SVN-sets A and B in the vector space is defined as follows:

$$C(A,B)=\frac{1}{n}\sum_{i=1}^{n}\times\left(\frac{T_A(x_i)T_B(x_i)+I_A(x_i)I_B(x_i)+F_A(x_i)F_B(x_i)}{[\sqrt{((T_A)^2(x_i)+(I_A)^2(x_i)+(F_A)^2(x_i))}.\sqrt{((T_B)^2(x_i)+(I_B)^2(x_i)+(F_B)^2(x_i))}]}\right) \quad (15)$$

It satisfies the following properties:

a. $0 \le C(A,B) \le 1$;
b. $C(A,B)=C(B,A)$;
c. $C(A,B)=1$ for $A=B$ i.e.$T_A(x_i)=T_B(x_i)$, $I_A(x_i)=I_B(x_i)$, $F_A(x_i)=F_B(x_i)(i=1,2...,n)$ $\forall\, x_i(i=1,2,...,n)\in X$.

Definition 5 ([23]) Let $A=<x,(T_A(x),I_A(x),F_A(x))>$ and

$B=<x,(T_B(x),I_B(x),F_B(x))>$ be two SVN-sets o the universe X. Then, correlation formula of A and B is defined as;

$$C(A,B)=m(T_A)m(T_B)+m(I_A)m(I_B)+m(F_A)m(F_B)$$

where

$$m(T_A)=\begin{cases}\frac{\int_X xT_A(x)dx}{\int_X T_A(x)dx} & X\ is\ continuous\\ \frac{\sum_X xT_A(x)dx}{\sum_X T_A(x)dx} & X\ is\ finite\end{cases}$$

$$m(T_B)=\begin{cases}\frac{\int_X xT_B(x)dx}{\int_X T_B(x)dx} & X\ is\ continous\\ \frac{\sum_X xT_B(x)dx}{\sum_X T_B(x)dx} & X\ is\ finite\end{cases}$$

$$m(I_A)=\begin{cases}\frac{\int_X xI_A(x)dx}{\int_X I_A(x)dx} & X\ is\ continous\\ \frac{\sum_X xI_A(x)dx}{\sum_X I_A(x)dx} & X\ is\ finite\end{cases} \quad (16)$$

$$m(I_B)=\begin{cases}\frac{\int_X xI_B(x)dx}{\int_X I_B(x)dx} & X\ is\ continous\\ \frac{\sum_X xI_B(x)dx}{\sum_X I_B(x)dx} & X\ is\ finite\end{cases}$$

$$m(F_A)=\begin{cases}\frac{\int_X xF_A(x)dx}{\int_X F_A(x)dx} & X\ is\ continous\\ \frac{\sum_X xF_A(x)dx}{\sum_X F_A(x)dx} & X\ is\ finite\end{cases}$$

$$m(F_B)=\begin{cases}\frac{\int_X xF_B(x)dx}{\int_X F_B(x)dx} & X\ is\ continous\\ \frac{\sum_X xF_B(x)dx}{\sum_X F_B(x)dx} & X\ is\ finite\end{cases}$$

Furthermore, correlation coefficient of A and B is defined as;

$$\rho(A, B) = \frac{C(A, B)}{(C(A, A)C(B, B))^{\frac{1}{2}}} \tag{17}$$

Then, this correlation coefficient satisfies the following properties:

1. $0 \leq \rho(A, B) \leq 1$;
2. $\rho(A, B) = \rho(B, A)$;
3. $\rho(A, B) = 1$ for $A = B$

Definition 6 ([15]) A single valued trapezoidal neutrosophic number (SVTN-number)

$$\tilde{a} = \langle (a_1, b_1, c_1, d_1); w_{\tilde{a}}, u_{\tilde{a}}, y_{\tilde{a}} \rangle$$

is a special neutrosophic set on the real number set R, whose truth-membership, indeterminacy-membership, and a falsity-membership are given as follows:

$$\mu_{\tilde{a}}(x) = \begin{cases} (x - a_1)w_{\tilde{a}}/(b_1 - a_1) & (a_1 \leq x < b_1) \\ w_{\tilde{a}} & (b_1 \leq x \leq c_1) \\ (d_1 - x)w_{\tilde{a}}/(d_1 - c_1) & (c_1 < x \leq d_1) \\ 0 & otherwise, \end{cases}$$

$$\nu_{\tilde{a}}(x) = \begin{cases} (b_1 - x + u_{\tilde{a}}(x - a_1))/(b_1 - a_1) & (a_1 \leq x < b_1) \\ u_{\tilde{a}} & (b_1 \leq x \leq c_1) \\ (x - c_1 + u_{\tilde{a}}(d_1 - x))/(d_1 - c_1) & (c_1 < x \leq d_1) \\ 1 & otherwise \end{cases}$$

and

$$\lambda_{\tilde{a}}(x) = \begin{cases} (b_1 - x + y_{\tilde{a}}(x - a_1))/(b_1 - a_1) & (a_1 \leq x < b_1) \\ y_{\tilde{a}} & (b_1 \leq x \leq c_1) \\ (x - c_1 + y_{\tilde{a}}(d_1 - x))/(d_1 - c_1) & (c_1 < x \leq d_1) \\ 1 & otherwise \end{cases}$$

respectively.

Definition 7 ([15]) Let $\tilde{a} = \langle (a_1, b_1, c_1, d_1); w_{\tilde{a}}, u_{\tilde{a}}, y_{\tilde{a}} \rangle$ and $\tilde{b} = \langle (a_2, b_2, c_2, d_2); w_{\tilde{b}}, u_{\tilde{b}}, y_{\tilde{b}} \rangle$ be two SVTN-numbers and $\gamma > 0$, then

1. $\tilde{a} \oplus \tilde{b} = \langle (a_1 + a_2, b_1 + b_2, c_1 + c_2, d_1 + d_2); w_{\tilde{a}} \wedge w_{\tilde{b}}, u_{\tilde{a}} \vee u_{\tilde{b}}, y_{\tilde{a}} \vee y_{\tilde{b}} \rangle$
2. $\tilde{a} \otimes \tilde{b} = \begin{cases} \langle (a_1a_2, b_1b_2, c_1c_2, d_1d_2); w_{\tilde{a}} \wedge w_{\tilde{b}}, u_{\tilde{a}} \vee u_{\tilde{b}}, y_{\tilde{a}} \vee y_{\tilde{b}} \rangle & (d_1 > 0, d_2 > 0) \\ \langle (a_1d_2, b_1c_2, c_1b_2, d_1a_2); w_{\tilde{a}} \wedge w_{\tilde{b}}, u_{\tilde{a}} \vee u_{\tilde{b}}, y_{\tilde{a}} \vee y_{\tilde{b}} \rangle & (d_1 < 0, d_2 > 0) \\ \langle (d_1d_2, c_1c_2, b_1b_2, a_1a_2); w_{\tilde{a}} \wedge w_{\tilde{b}}, u_{\tilde{a}} \vee u_{\tilde{b}}, y_{\tilde{a}} \vee y_{\tilde{b}} \rangle & (d_1 < 0, d_2 < 0) \end{cases}$
3. $\gamma \odot \tilde{a} = \{ \langle (\gamma a_1, \gamma b_1, \gamma c_1, \gamma d_1); w_{\tilde{a}}, u_{\tilde{a}}, y_{\tilde{a}} \rangle$
4. $\tilde{a}^{\vec{\gamma}} = \{ \langle (a_1^{\gamma}, b_1^{\gamma}, c_1^{\gamma}, d_1^{\gamma}); w_{\tilde{a}}, u_{\tilde{a}}, y_{\tilde{a}} \rangle$

Example 1 Let $\tilde{a} = \langle (0.6, 0.7, 0.8, 0.9);\ 0.6, 0.7, 0, 4 \rangle$ and $\tilde{b} = \langle (0.7, 0.8, 0.9, 1.0); 0.5, 0.2, 0, 4 \rangle$ be two SVTN-numbers, then

1. $\tilde{a} \oplus \tilde{b} = \langle (1.3, 1.5, 1.7, 1.9); 0.5, 0.7, 0.4 \rangle$
2. $\tilde{a} \otimes \tilde{b} = \langle (0.42, 0.56, 0.72, 0.9); 0.5, 0.7, 0.4 \rangle$
3. $2 \odot \tilde{a} = \langle (1.2, 1.4, 1.6, 1.8); 0.6, 0.7, 0, 4 \rangle$
4. $\tilde{a}^{\overrightarrow{2}} = \langle (0.36, 0.49, 0.64, 0.81); 0.6, 0.7, 0, 4 \rangle$

Note 1 If $a, b, c, d \in [0, 1]$ then $\tilde{a} = \langle (a, b, c, d); w_{\tilde{a}}, u_{\tilde{a}}, y_{\tilde{a}} \rangle$ is normal SVTN-number. In Example 1, $\tilde{a}$ and $\tilde{b}$ are normal SVTN-numbers but $\tilde{a} + \tilde{b}$ is not normal SVTN-numbers. Therefore, based on Definition 2 and definition of [38], we update the Definition 7 as follows.

Definition 8 Let $\tilde{a} = \langle (a_1, b_1, c_1, d_1); w_{\tilde{a}}, u_{\tilde{a}}, y_{\tilde{a}} \rangle$ and $\tilde{b} = \langle (a_2, b_2, c_2, d_2); w_{\tilde{b}}, u_{\tilde{b}}, y_{\tilde{b}} \rangle$ be two normal SVTN-numbers and $\gamma > 0$, then

1. $\tilde{a} + \tilde{b} = \langle (s(a_1, a_2), s(b_1, b_2), s(c_1, c_2), s(d_1, d_2)); s(w_{\tilde{a}}, w_{\tilde{b}}), t(u_{\tilde{a}}, u_{\tilde{b}}), t(y_{\tilde{a}}, y_{\tilde{b}}) \rangle$
2. $\tilde{a} \times \tilde{b} = \begin{cases} \langle (t(a_1, a_2), t(b_1, b_2), t(c_1, c_2), t(d_1, d_2)); t(w_{\tilde{a}}, w_{\tilde{b}}), s(u_{\tilde{a}}, u_{\tilde{b}}), s(y_{\tilde{a}}, y_{\tilde{b}}) \rangle & (d_1 > 0, d_2 > 0) \\ \langle (t(a_1, d_2), t(b_1, c_2), t(c_1, b_2), t(d_1, a_2)); t(w_{\tilde{a}}, w_{\tilde{b}}), s(u_{\tilde{a}}, u_{\tilde{b}}), s(y_{\tilde{a}}, y_{\tilde{b}}) \rangle & (d_1 < 0, d_2 > 0) \\ \langle (t(d_1, d_2), t(c_1, c_2), t(b_1, b_2), t(a_1, a_2)); t(w_{\tilde{a}}, w_{\tilde{b}}), s(u_{\tilde{a}}, u_{\tilde{b}}), s(y_{\tilde{a}}, y_{\tilde{b}}) \rangle & (d_1 < 0, d_2 < 0) \end{cases}$
3. $\gamma \times \tilde{a} = \{ \langle (1 - (1 - a_1)^\gamma, 1 - (1 - b_1)^\gamma, 1 - (1 - c_1)^\gamma, 1 - (1 - d_1)^\gamma); 1 - (1 - w_{\tilde{a}})^\gamma, u_{\tilde{a}}^\gamma, y_{\tilde{a}}^\gamma \rangle$
4. $\tilde{a}^\gamma = \{ \langle (a_1^\gamma, b_1^\gamma, c_1^\gamma, d_1^\gamma); w_{\tilde{a}}^\gamma, 1 - (1 - u_{\tilde{a}})^\gamma, 1 - (1 - y_{\tilde{a}})^\gamma \rangle$

From now on, we use normal SVTN-numbers.

Example 2 Let us consider the Example 1 for Algebraic product and Algebraic sum in Eqs. 7–8 which are norms, then

1. $\tilde{a} + \tilde{b} = \langle (0.88, 0.94, 0.98, 1); 0.8, 0.14, 0.16 \rangle$
2. $\tilde{a} \times \tilde{b} = \langle (0.42, 0.56, 0.72, 0.9); 0.3, 0.76, 0.64 \rangle$
3. $2 \times \tilde{a} = \langle (0.84, 0.91, 0.96, 0.99); 0.84, 0.49, 0.16 \rangle$
4. $\tilde{a}^2 = \langle (0.49, 0.64, 0.81, 1); 0.25, , 0.36, 0.64 \rangle$

Theorem 1 *Let* $\tilde{a} = \langle (a_1, b_1, c_1, d_1); w_{\tilde{a}}, u_{\tilde{a}}, y_{\tilde{a}} \rangle$ *and* $\tilde{b} = \langle (a_2, b_2, c_2, d_2); w_{\tilde{b}}, u_{\tilde{b}}, y_{\tilde{b}} \rangle$ *be two normal SVTN-numbers and* $\gamma, \gamma_1, \gamma_2 > 0$. *Then,* $\tilde{a} + \tilde{b}$, $\tilde{a} \times \tilde{b}$, $\gamma \times \tilde{a}$ *and* $\tilde{a}^\gamma$ *are normal SVTN-numbers and the followings are valid.*

1. $\tilde{a} + \tilde{b} = \tilde{b} + \tilde{a}$
2. $\tilde{a} \times \tilde{b} = \tilde{b} \times \tilde{a}$
3. $\gamma \times (\tilde{a} + \tilde{b}) = (\gamma \times \tilde{a}) + (\gamma \times \tilde{b})$
4. $(\gamma_1 + \gamma_2) \times \tilde{a} = (\gamma_1 \times \tilde{a}) + (\gamma_2 \times \tilde{a})$
5. $\tilde{a}^{\gamma_1} \times \tilde{a}^{\gamma_2} = \tilde{a}^{(\gamma_1 + \gamma_2)}$

Proof: It is easy from Definitions 2 and 8.

Now, we give a neutrosofication method, by using centroid of area method in Definition 5, for normal SVTN-numbers based on [23, 40] as;

Definition 9 Let $\tilde{a} = \langle (a, b, c, d); w_{\tilde{a}}, u_{\tilde{a}}, y_{\tilde{a}} \rangle$ be a normal SVTN-number. Then,

1. the neutrosofication of $\tilde{a}$ with centroid of area method based on Eq. 16, is denoted by $CAM(\tilde{a})$, is defined as;

$$CAM(\tilde{a}) = (\phi(\mu_{\tilde{a}}(x)), \phi(\nu_{\tilde{a}}(x)), \phi(\lambda_{\tilde{a}}(x))) \tag{18}$$

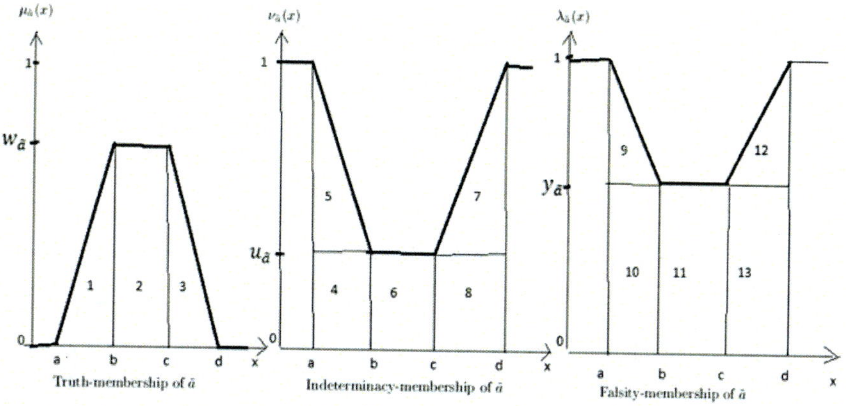

Fig. 3 Graphic of $\tilde{a}$

where

$$\phi(\mu_{\tilde{a}}(x)) = \frac{\sum_1^3 A_{i\mu_{\tilde{a}}} \times \overline{x_i}}{\sum_1^3 A_i},\ \phi(\nu_{\tilde{a}}(x)) = \frac{\sum_4^8 A_{i\nu_{\tilde{a}}} \times \overline{x_i}}{\sum_4^8 A_i},\ \text{and}\ \phi(\lambda_{\tilde{a}}(x)) = \frac{\sum_9^N A_{i\lambda_{\tilde{a}}} \times \overline{x_i}}{\sum_9^N A_i}.$$

Here "N = 13" indicates the number of sub-areas, $A_{i\mu_{\tilde{b}}}$ and $\overline{x_i}$ represents the area and centroid of area, respectively, of ith sub-area. In the aggregated normal SVTN-numbers as shown in Fig. 3, the total area is divided into 13 sub-areas and then we have to calculate the area and centroid of area of each sub-area.

Remark 1 This method provides a neutrosophic value based on the center of gravity of the normal SVTN-numbers. The total area of the truth-membership, indeterminacy-membership and falsity-membership function distribution used to represent the combined control action is divided into 3 sub-areas based on truth-membership, 5 sub-areas based on indeterminacy-membership and 5 sub-areas based on falsity-membership function. The area and the center of gravity or centroid of each sub-area is calculated and then the summation of 3 sub-areas based on truth-membership, 5 sub-areas based on indeterminacy-membership and 5 sub-areas based on falsity-membership function are taken to find the neutrosophic value for normal SVTN-numbers.

Example 3 Let $\tilde{a} = \langle(0.2, 0.4, 0.5, 0.8\,);\ 0.8, 0.3, 0.5\rangle$ be a normal SVTN-number, then 13 sub-areas and centroid of these sub-areas in Fig. 4 can be calculated as;

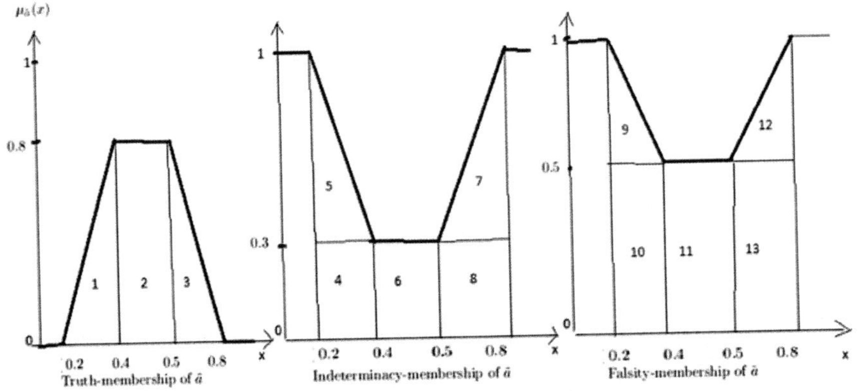

Fig. 4 Graphic of $\tilde{a}$

- The total area of the sub-area 1 is calculated as; $A_{1\mu_{\tilde{a}}} = \frac{0.4-0.2}{2} \times 0.8 = 0,080$
- The total area of the sub-area 2 is calculated as; $A_{2\mu_{\tilde{a}}} = (0.5-0.4) \times 0.8 = 0,080$
- The total area of the sub-area 3 is calculated as; $A_{3\mu_{\tilde{a}}} = \frac{0.8-0.5}{2} \times 0.8 = 0,120$
- The total area of the sub-area 4 is calculated as; $A_{4\nu_{\tilde{a}}} = (0.4-0.2) \times 0.3 = 0,060$
- The total area of the sub-area 5 is calculated as; $A_{5\nu_{\tilde{a}}} = \frac{0.4-0.2}{2} \times 0.7 = 0,070$
- The total area of the sub-area 6 is calculated as; $A_{6\nu_{\tilde{a}}} = (0.5-0.4) \times 0.3 = 0,030$
- The total area of the sub-area 7 is calculated as; $A_{7\nu_{\tilde{a}}} = \frac{0.8-0.5}{2} \times 0.7 = 0,105$
- The total area of the sub-area 8 is calculated as; $A_{8\nu_{\tilde{a}}} = (0.8-0.5) \times 0.3 = 0,090$
- The total area of the sub-area 9 is calculated as; $A_{9\lambda_{\tilde{a}}} = \frac{0.4-0.2}{2} \times 0.5 = 0,050$
- The total area of the sub-area 10 is calculated as; $A_{10\lambda_{\tilde{a}}} = (0.4-0.2) \times 0.5 = 0,100$
- The total area of the sub-area 11 is calculated as; $A_{11\lambda_{\tilde{a}}} = (0.5-0.4) \times 0.5 = 0,050$
- The total area of the sub-area 12 is calculated as; $A_{12\lambda_{\tilde{a}}} = \frac{0.8-0.5}{2} \times 0.5 = 0,075$
- The total area of the sub-area 13 is calculated as; $A_{13\lambda_{\tilde{a}}} = (0.8-0.5) \times 0.5 = 0,150$
- Centroid of sub-area 1 is calculated as; $\overline{x_1} = \frac{0.2+0.4+0.4}{3} = 0.330$
- Centroid of sub-area 2 is calculated as; $\overline{x_2} = \frac{0.4+0.5}{2} = 0.450$
- Centroid of sub-area 3 is calculated as; $\overline{x_3} = \frac{0.5+0.8+0.8}{3} = 0.600$
- Centroid of sub-area 4 is calculated as; $\overline{x_4} = \frac{0.2+0.4}{2} = 0.300$
- Centroid of sub-area 5 is calculated as; $\overline{x_5} = \frac{0.2+0.4+0.4}{3} = 0.270$
- Centroid of sub-area 6 is calculated as; $\overline{x_6} = \frac{0.4+0.5}{2} = 0.450$
- Centroid of sub-area 7 is calculated as; $\overline{x_7} = \frac{0.5+0.8+0.8}{32} = 0.700$
- Centroid of sub-area 8 is calculated as; $\overline{x_8} = \frac{0.5+0.8}{2} = 0.650$
- Centroid of sub-area 9 is calculated as; $\overline{x_9} = \frac{0.2+0.2+0.4}{3} = 0.270$
- Centroid of sub-area 10 is calculated as; $\overline{x_{10}} = \frac{0.2+0.4}{2} = 0.300$

Table 1 The $A_{i\mu_{\tilde{a}}}$, $\overline{x_i}$ and $A_{i\mu_{\tilde{a}}} \times \overline{x_i}$ for (i = 1, 2, ..., 13)

Sub-area number i, ($i = 1, 2, ..., 13$)	Area $A_{i\mu_{\tilde{a}}}$	Centroid of sub-area $\overline{x_i}$	$A_{i\mu_{\tilde{a}}} \times \overline{x_i}$
1	0.080	0.330	0.027
2	0.080	0.450	0.036
3	0.120	0.600	0.072
4	0.060	0.300	0.018
5	0.070	0.270	0.019
6	0.030	0.450	0.014
7	0.105	0.700	0.074
8	0.090	0.650	0.059
9	0.050	0.270	0.013
10	0.100	0.300	0.030
11	0.050	0.450	0.023
12	0.075	0.700	0.053
13	0.150	0.650	0.098

- Centroid of sub-area 11 is calculated as; $\overline{x_{11}} = \frac{0.5+0.4}{2} = 0.450$
- Centroid of sub-area 12 is calculated as; $\overline{x_{12}} = \frac{0.5+0.8+0.8}{3} = 0.700$
- Centroid of sub-area 13 is calculated as; $\overline{x_{13}} = \frac{0.8+0.5}{2} = 0.650.$

Now we can calculate $A_{i\mu_{\tilde{a}}} \times \overline{x_i}$ $(i = 1, 2, ..., 13)$ and is shown in Table 1.

Finally, the neutrosophic value based on Eq. 18, is calculated as;

$$CAM(\tilde{a}) = (\phi(\mu_{\tilde{a}}(x)), \phi(\nu_{\tilde{a}}(x)), \phi(\lambda_{\tilde{a}}(x))) = (0.481, 0.513, 0.508)$$

where

$$\phi(\mu_{\tilde{a}}(x)) = \frac{\sum_1^3 A_{i\mu_{\tilde{a}}} \times \overline{x_i}}{\sum_1^3 A_i},\ \phi(\nu_{\tilde{a}}(x)) = \frac{\sum_4^8 A_{i\nu_{\tilde{a}}} \times \overline{x_i}}{\sum_4^8 A_i},\ \text{and}\ \phi(\lambda_{\tilde{a}}(x)) = \frac{\sum_9^{13} A_{i\lambda_{\tilde{a}}} \times \overline{x_i}}{\sum_9^{13} A_i}.$$

3 The Correlation Coefficient of Normal SVTN-Numbers

Since the concept of correlation can be used to study the nature of the relations between the variables based on neutrosophic sets, Ye [49] and Hanafy et al. [23] developed the correlation of neutrosophic sets A and B in a finite space and in a infinite space, respectively. By using Definitions 4 and 5, in this section, we proposed correlation coefficient measurements in normal SVTN-numbers in uncountable universe.

Definition 10 Let $\tilde{a} = \langle (a_1, b_1, c_1, d_1); w_{\tilde{a}}, u_{\tilde{a}}, y_{\tilde{a}} \rangle$ and $\tilde{b} = \langle (a_2, b_2, c_2, d_2); w_{\tilde{b}}, u_{\tilde{b}}, y_{\tilde{b}} \rangle$ be two normal SVTN-numbers. Then,

1. correlation coefficient of $\tilde{a}$ and $\tilde{b}$ based on centroid point , is denoted by $\Phi_1(\tilde{a}, \tilde{b})$, is defined as;

$$\Phi_1(\tilde{a}, \tilde{b}) = \frac{C(\tilde{a}, \tilde{b})}{C(\tilde{a}, \tilde{a}) + C(\tilde{b}, \tilde{b}) - C(\tilde{a}, \tilde{b})} \tag{19}$$

2. correlation coefficient of $\tilde{a}$ and $\tilde{b}$ based on centroid point, is denoted by $\Phi_2(\tilde{a}, \tilde{b})$, is defined as;

$$\Phi_2(\tilde{a}, \tilde{b}) = \frac{2C(\tilde{a}, \tilde{b})}{C(\tilde{a}, \tilde{a}) + C(\tilde{b}, \tilde{b})} \tag{20}$$

3. correlation coefficient of $\tilde{a}$ and $\tilde{b}$ based on centroid point, is denoted by $\Phi_3(\tilde{a}, \tilde{b})$, is defined as;

$$\Phi_3(\tilde{a}, \tilde{b}) = \frac{C(\tilde{a}, \tilde{b})}{(C(\tilde{a}, \tilde{a})C(\tilde{b}, \tilde{b}))^{\frac{1}{2}}} \tag{21}$$

where

$$\begin{aligned}
C(\tilde{a}, \tilde{b}) &= \phi(\mu_{\tilde{a}}(x)).\phi(\mu_{\tilde{b}}(x)) + \phi(\nu_{\tilde{a}}(x)).\phi(\nu_{\tilde{b}}(x)) + \phi(\lambda_{\tilde{a}}(x)).\phi(\lambda_{\tilde{b}}(x)) \\
C(\tilde{a}, \tilde{a}) &= \phi(\mu_{\tilde{a}}(x)).\phi(\mu_{\tilde{a}}(x)) + \phi(\nu_{\tilde{a}}(x)).\phi(\nu_{\tilde{a}}(x)) + \phi(\lambda_{\tilde{b}}(x)).\phi(\lambda_{\tilde{b}}(x)) \\
C(\tilde{b}, \tilde{b}) &= \phi(\mu_{\tilde{b}}(x)).\phi(\mu_{\tilde{b}}(x)) + \phi(\nu_{\tilde{b}}(x)).\phi(\nu_{\tilde{b}}(x)) + \phi(\lambda_{\tilde{b}}(x)).\phi(\lambda_{\tilde{b}}(x))
\end{aligned}$$

and where

$$\phi(\mu_{\tilde{a}}(x)) = \frac{\sum_1^3 A_{i\mu_{\tilde{a}}} \times \overline{x_i}}{\sum_1^3 A_i}, \ \phi(\nu_{\tilde{a}}(x)) = \frac{\sum_4^8 A_{i\nu_{\tilde{a}}} \times \overline{x_i}}{\sum_4^8 A_i}, \ \phi(\lambda_{\tilde{a}}(x)) = \frac{\sum_9^{13} A_{i\lambda_{\tilde{a}}} \times \overline{x_i}}{\sum_9^{13} A_i}.$$
$$\phi(\mu_{\tilde{b}}(x)) = \frac{\sum_1^3 A_{i\mu_{\tilde{b}}} \times \overline{x_i}}{\sum_1^3 A_i}, \ \phi(\nu_{\tilde{b}}(x)) = \frac{\sum_4^8 A_{i\nu_{\tilde{b}}} \times \overline{x_i}}{\sum_4^8 A_i}, \ \phi(\lambda_{\tilde{b}}(x)) = \frac{\sum_9^{13} A_{i\lambda_{\tilde{b}}} \times \overline{x_i}}{\sum_9^{13} A_i}.$$

Theorem 2 *Let* $\tilde{a} = \langle (a_1, b_1, c_1, d_1); w_{\tilde{a}}, u_{\tilde{a}}, y_{\tilde{a}} \rangle$ *and* $\tilde{b} = \langle (a_2, b_2, c_2, d_2); w_{\tilde{b}}, u_{\tilde{b}}, y_{\tilde{b}} \rangle$ *be two normal SVTN-numbers. Then, for* $i \in \{1, 2, 3\}$ *the followings are valid.*

1. $\Phi_i(\tilde{a}, \tilde{b}) = \Phi_i(\tilde{b}, \tilde{a})$
2. $\tilde{a} = \tilde{b} \Rightarrow \Phi_i(\tilde{a}, \tilde{b}) = 1$
3. $0 \leq \Phi_i(\tilde{b}, \tilde{a}) \leq 1$

Proof: It is easy from Definition 10.

Example 4 $\tilde{a} = \langle (0.2, 0.4, 0.5, 0.8); 0.8, 0.3, 0, 5 \rangle$ and $\tilde{b} = \langle (0.1, 0.3, 0.6, 0.7); 0.4, 0.5, 0, 6 \rangle$ be two single valued trapezoidal neutrosophic numbers. Based on Eq. 18, the neutrosophic values of $\tilde{a}$ and $\tilde{b}$ are calculated as;

$$CAM(\tilde{a}) = (\phi(\mu_{\tilde{a}}(x)), \phi(\nu_{\tilde{a}}(x)), \phi(\lambda_{\tilde{a}}(x))) = (0.481, 0.513, 0.508)$$

$$CAM(\tilde{b}) = (\phi(\mu_{\tilde{b}}(x)), \phi(\nu_{\tilde{b}}(x)), \phi(\lambda_{\tilde{b}}(x))) = (0.422, 0.387, 0.390)$$

respectively.

Also, $C(\tilde{a}, \tilde{b})$ $C(\tilde{a}, \tilde{b})$ and $C(\tilde{a}, \tilde{b})$ is given as;

$$C(\tilde{a}, \tilde{a}) = 0.481 \times 0.422 + 0.387 \times 0.513 + 0.508 \times 0.390 = 0.753$$
$$C(\tilde{b}, \tilde{b}) = 0.481 \times 0.481 + 0.513 \times 0.513 + 0.508 \times 0.508 = 0.480$$
$$C(\tilde{a}, \tilde{b}) = 0.422 \times 0.422 + 0.387 \times 0.387 + 0.390 \times 0.390 = 0.600$$

Finally,

1. 1. correlation coefficient $\Phi_1(\tilde{a}, \tilde{b})$ in Eq. 19, is calculated as;

$$\Phi_1(\tilde{a}, \tilde{b}) = \frac{0.600}{0.753 + 0.480 - 0.600} = 0.948$$

2. 2. correlation coefficient $\Phi_2(\tilde{a}, \tilde{b})$ in Eq. 20, is calculated as;

$$\Phi_2(\tilde{a}, \tilde{b}) = \frac{2 \times 0.600}{0.753 + 0.480} = 0.487$$

3. 3. correlation coefficient $\Phi_3(\tilde{a}, \tilde{b})$ in Eq. 21, is calculated as;

$$\Phi_3(\tilde{a}, \tilde{b}) = \frac{0.600}{(0.480 \times 0.753)^{\frac{1}{2}}} = 0.998$$

4 Multi Robot System Based on Normal SVTN-Numbers: An Application

Let $R = \{R_p : p = 1, 2, ..., k\}$ be a set of Robots, $S = \{s_i : i = 1, 2, ..., m\}$ be a set of sensors deployed on each Robot and $C = \{c_j : j = 1, 2, ..., n\}$ be a set of situations. If $\tilde{A}_{ij} = \langle (a_{ij}, b_{ij}, c_{ij}, d_{ij}); w_{ij}, u_{ij}, y_{ij} \rangle$ and $\tilde{B}_{pi} = \langle (a_{pi}, b_{pi}, c_{pi}, d_{pi}); w_{pi}, u_{pi}, y_{pi} \rangle$ be two normal SVTN-numbers, then each sensor reading value based on sensors and situations and sensor readings monitored for one minutes, one reading one minute based on Robots and the sensor is described by normal SVTN-numbers as Table 2 and Table 4, respectively.

Example 5 Let us consider the multi Robot system adapted from [44]. Assume that $R = \{R_1, R_2, R_3, R_4, R_5\}$ be a set of five Robots, $C = \{c_1 = Fire, c_2 = Obstacle, c_3 = Bump, c_4 = Cliff, c_5 = Vibration\}$ be a set of situations and $S = \{s_1 = Temperature\ sensor, s_2 = Ultrasonic\ sensor, s_3 = Bump\ sensor, s_4 = Cliff\ sensor, s_5 = Accelerometer\ sensor\}$ be a set of sensors deployed on each Robot. A single Robot can be assigned different normal SVTN-number for the five different sensor readings. Whether from a single reading can we conclude what

Table 2 The corresponding normal SVTN-number values based on sensors and situations

	c_1	c_2	$\cdots$	c_n
s_1	A_{11}	A_{12}	$\cdots$	A_{1n}
s_2	A_{11}	A_{21}	$\cdots$	A_{2n}
$\vdots$	$\ddots$	$\ddots$	$\ddots$	$\vdots$
s_m	A_{m1}	A_{m2}	$\cdots$	A_{mn}

Table 3 Algorithm based on decision mechanism

Decision mechanism	
Step 1	Describe the corresponding normal SVTN-number values based on sensors and situations as a table based on Table 2,
Step 2	Insert the sensor readings of each robot monitored for one minutes, one reading as table based on Table 4 (R_p : (p = 1,2,...,k) Robots and s_i, (i = 1, 2,...,m) sensors deployed on each Robot)
Step 3	**a.** Calculate the means 1. correlation coefficient D^1_{pj} ((p = 1,2,...,k and j=1,2,...,n)) of each Robot to the situation based on Eq. 19 considered as; $D^1_{pj} = \frac{1}{m}\sum_{i=1}^{m} \frac{C(A_{ij},B_{pi})}{C(A_{ij},Aij)+C(B_{pi},B_{pi})-C(A_{ij},Aij)}$, for p=1,2,...,k, and j = 1, 2,...,n. or **b.** Calculate the means 2. correlation coefficient D^2_{pj} ((p = 1, 2,...,k and j = 1, 2,...,n)) of each Robotto the situation Eq, 20 considered as; $D^2_{pj} = \frac{1}{m}\sum_{i=1}^{m} \frac{C(A_{ij},B_{pi})}{C(A_{ij},Aij)+C(B_{pi},B_{pi})}$, for $p = 1, 2, ..., k,\ and\ j = 1, 2, ..., n.$ or **c.** Calculate the means 3. correlation coefficient D^3_{pj} ((p = 1,2,...,k and j = 1, 2,...,n)) of each Robot to the situation Eq. 21 considered as; $D^3_{pj} = \frac{1}{m}\sum_{i=1}^{m} \frac{C(A_{ij},B_{pi})}{(C(A_{ij},Aij)\times C(B_{pi},B_{pi}))^{\frac{1}{2}}}$, for $p = 1, 2, ..., k,\ and\ j = 1, 2, ..., n.$
Step 4	Determine the correct situation of each Robot using the correlation coefficient of normal SVTN-numbers

Table 4 The sensor readings of each robot

	s_1	s_2	$\cdots$	s_m
R_1	B_{11}	B_{12}	$\cdots$	B_{1m}
R_2	B_{11}	B_{21}	$\cdots$	B_{2m}
$\vdots$	$\ddots$	$\ddots$	$\ddots$	$\vdots$
R_k	B_{k1}	B_{k2}	$\cdots$	B_{km}

Table 5 The means correlation coefficient of each robot based on the situation

	c_1	c_2	$\cdots$	c_n
R_1	D_{11}	D_{12}	$\cdots$	D_{1n}
R_2	D_{11}	D_{21}	$\cdots$	D_{2n}
$\vdots$	$\ddots$	$\ddots$	$\ddots$	$\vdots$
R_k	D_{k1}	D_{k2}	$\cdots$	D_{kn}

Table 6 The corresponding normal SVTN-number values based on sensors and situations

sensor	Fire	Obstacle
Temperature	<(0.2, 0.4, 0.51.0); 0.3, 0.7, 0.5>	<(0.1, 0.3, 0.6, 0.9); 0.4, 0.1, 0.6>
Ultrasonic	<(0.5, 0.6, 0.7, 0.8); 0.5, 0.9, 0.1>	<(0.2, 0.4, 0.5, 0.9); 0.9, 0.7, 0.9>
Bump	<(0.1, 0.5, 0.5, 0.7); 0.9, 0.4, 0.3>	<(0.8, 0.9, 0.91.0); 1.0, 0.6, 0.8>
Cliff	<(0.5, 0.6, 0.81.0); 0.4, 0.8, 0.3>	<(0.2, 0.5, 0.5, 0.8); 0.9, 0.7, 0.5>
Accelerometer	<(0.1, 0.2, 0.6, 0.7); 0.7, 0.5, 0.6>	<(0.5, 0.6, 0.71.0); 0.1, 1.0, 0.8>
Sensor	Bump	Cliff
Temperature	<(0.2, 0.4, 0.5, 0.9); 0.5, 0.2, 0.1>	<(0.1, 0.3, 0.6, 0.7); 0.8, 0.7, 0.6>
Ultrasonic	<(0.5, 0.6, 0.7, 0.8); 0.2, 0.8, 0.6>	<(0.3, 0.4, 0.5, 0.9); 0.9, 0.4, 0.7>
Bump	<(0.2, 0.5, 0.51.0); 0.6, 0.1, 0.7>	<(0.8, 0.9, 0.9, 1.0); 0.4, 0.5, 0.8>
Cliff	<(0.5, 0.7, 0.81.0); 0.1, 0.7, 0.3>	<(0.3, 0.4, 0.51.0); 0.3, 0.21.0>
Accelerometer	<(0.2, 0.5, 0.5, 0.7); 0.8, 0.4, 0.9>	<(0.8, 0.9, 0.91.0); 0.6, 0.6, 0.8>
Sensor	Shock/Vibration	
Temperature	<(0.2, 0.4, 0.81.0); 0.5, 0.7, 0.4>	
Ultrasonic	<(0.2, 0.5, 0.5, 0.9); 1.0, 0.4, 0.5>	
Bump	<(0.1, 0.2, 0.6, 0.7); 0.7, 0.3, 0.1>	
Cliff	<(0.5, 0.6, 0.7, 0.8); 0.6, 0.5, 0.1>	
Accelerometer	<(0.2, 0.4, 0.5, 0.8); 0.1, 0.7, 0.5>	

are the situations faced by the Robots? The sensor readings from the Robots have to be monitored for a particular time, say for one minutes. For example, if the ultrasonic sensor in R_1 experiences a vibration, it sends a message to the controller so that the corrective measure could be taken. The controller has to make sure whether the R_1 is really faced with a vibration or not. For that purpose, the controller monitors the ultrasonic sensor reading for one minutes.

Now, we solution the problem according to algorithm based on Decision mechanism in Table 3

Step 1: We described the Robots and the corresponding normal SVTN-number values to the sensor values in Tables 5 and 6;

Step 2: We inserted the sensor readings monitored for one minutes, one reading one minute in Table 7;

Table 7 The sensor readings of each robot

	Temperature sensor	Ultrasonic sensor
R_1	<(0.1, 0.3, 0.6, 1.0); 0.5, 0.2, 1.0>	<(0.5, 0.6, 0.7, 0.8); 1.0, 0.7, 0.1>
R_2	<(0.3, 0.4, 0.5, 0.7); 0.7, 0.2, 0.9>	<(0.1, 0.5, 0.5, 0.7); 0.9, 0.1, 0.9>
R_3	<(0.8, 0.9, 0.9, 1.0); 1.0, 0.2, 0.8>	<(0.2, 0.4, 0.5, 0.7); 0.3, 0.6, 1.0>
R_4	<(0.1, 0.5, 0.5, 0.9); 0.7, 0.7, 1.0>	<(0.2, 0.9, 0.9, 1.0); 1.0, 0.6, 0.2>
R_5	<(0.5, 0.4, 0.8, 1.0); 0.4, 0.8, 0.4>	<(0.2, 0.5, 0.5, 0.8); 0.9, 0.3, 0.3>
	Bump sensor	Cliff sensor
R_1	<(0.1, 0.5, 0.6, 0.9); 0.4, 0.6, 0.6>	<(0.5, 0.6, 0.7, 1.0); 0.9, 0.9, 0.4>
R_2	<(0.2, 0.4, 0.5, 0.8); 0.8, 0.3, 0.5>	<(0.2, 0.3, 0.6, 0.7); 0.4, 0.5, 0.6>
R_3	<(0.5, 0.6, 0.7, 0.8); 0.4, 0.9, 0.1>	<(0.3, 0.4, 0.5, 0.7); 0.2, 0.2, 0.9>
R_4	<(0.1, 0.5, 0.5, 0.7); 0.9, 0.4, 0.4>	<(0.8, 0.9, 0.9, 1.0); 0.5, 0.6, 0.2>
R_5	<(0.2, 0.6, 0.7, 0.8); 0.1, 0.2, 0.7>	<(0.2, 0.4, 0.5, 0.8); 1.0, 0.7, 0.7>
	Accelerometer sensor	
R_1	<(0.1, 0.3, 0.6, 0.7); 0.4, 0.5, 0.6>	
R_2	<(0.5, 0.6, 0.7, 0.8); 0.7, 0.9, 0.1>	
R_3	<(0.3, 0.4, 0.5, 0.9); 1.0, 0.1, 0.4>	
R_4	<(0.1, 0.5, 0.6, 0.7); 0.4, 0.7, 0.6>	
R_5	<(0.5, 0.6, 0.7, 1.0); 0.7, 1.0, 0.4>	

Table 8 1. Correlation coefficient D^1_{pj} of each Robot to the situation

D^1_{pj}	Fire	Obstacle	Bump	Cliff	Shock/Vibration
R_1	0.955	0.842	0.909	0.819	**0.968**
R_2	0.923	0.883	**0.967**	0.857	0.927
R_3	0.931	0.928	**0.953**	0.927	0.903
R_4	0.902	0.790	0.870	0.751	**0.959**
R_5	0.914	**0.954**	0.943	0.937	0.933

Step 3: **a**. We calculated the means 1. correlation coefficient D^1_{pj} ((p = 1, 2,...,k and j = 1, 2,...,n)) of each Robot to the situation based on Eq. 19 considered in Table 8

b. We calculated the means 2. correlation coefficient D^2_{pj} ((p = 1, 2,...,k and j = 1, 2,...,n)) of each Robot to the situation based on Eq. 20 considered in Table 9

c. We calculated the means 3. correlation coefficient D^3_{pj} ((p = 1, 2,...,k and j = 1, 2,...,n)) of each Robot to the situation based on Eq. 21 considered in Table 10

Step 4: We determined the correct situation of each Robot using the 1., 2. and 3. correlation coefficient of normal SVTN-numbers as;

In Table 8 the highest values point gives the accuracy of the Robot. Robot R_1 experiences a vibration, R_2 is bumped, R_3 is bumped, R_4 experiences a vibration and R_5 near an obstacle.

Table 9 2. Correlation coefficient D^2_{pj} of each robot to the situation

D^2_{pj}	Fire	Obstacle	Bump	Cliff	Shock/Vibration
R_1	0.977	0.910	0.951	0.894	**0.984**
R_2	0.959	0.933	**0.983**	0.919	0.962
R_3	0.963	0.961	**0.975**	0.960	0.948
R_4	0.946	0.876	0.929	0.853	**0.979**
R_5	0.952	**0.976**	0.971	0.967	0.965

Table 10 3. Correlation coefficient D^3_{pj} of each robot to the situation

D^3_{pj}	Fire	Obstacle	Bump	Cliff	Shock/Vibration
R_1	**1.000**	0.999	0.999	0.998	0.999
R_2	0.998	**0.999**	**0.999**	**0.999**	**0.999**
R_3	0.998	**0.999**	**0.999**	0.998	**0.999**
R_4	**0.997**	**0.997**	**0.997**	0.996	**0.997**
R_5	0.998	0.998	**0.999**	0.996	0.998

Similarly, Tables 9 and 10 can be interpreted.

Finally, the normal SVTN-number is able to make out the correct situation of each Robot.

5 Conclusion

In the chapter, we first gave the concept of normal SVTN-numbers which are generalization of fuzzy numbers, intuitionistic fuzzy numbers and so on. We second, discussed some operations of normal SVTN-numbers. The operations gave a normal SVTN-number. We third propose a approach to find the correlation coefficient value, which is strength of relationship of two normal SVTN-number. Finally, we proposed a method and an application for a multi Robot system considered consists of a central controller and more than one patrolling Robots in a large area by using normal SVTN-numbers. In future work, we will applied this concept to intelligent robots, automatic navigation of robots, and so on.

References

1. Aal, S.I.A., Ellatif, M.M.A.A., Hassan, M.M.: Two ranking methods of single valued triangular neutrosophic numbers to rank and evaluate information systems quality. Neutros Sets Syst **19**, 132–141 (2018)
2. Atanassov, K.T.: Intuit. Fuzzy Sets. Pysica-Verlag, Springer, New York (1999)
3. Basset, M.A., Atef, A., Smarandache, F.: A hybrid neutrosophic multiple criteria group decision making approach for project selection. Cogn. Syst. Res. (2018). https://doi.org/10.1016/j.cogsys.2018.10.023
4. Basset, M.A., Mohamed, M., Hussien, A.N., Sangaiah, A.K.: A novel group decision-making model based on triangular neutrosophic numbers. Soft Comput. **22**, 6629–6643 (2018)
5. Basset, A.B., Mohamed, M., Sangaiah, A.K.: Neutrosophic AHP-Delphi Group decision making model based on trapezoidal neutrosophic numbers. J. Ambient Intell. Hum. Comput. **9**, 1427–1443 (2018)
6. Biswas, P., Pramanik, S., Giri, B.C.: Cosine similarity measure based multi-attribute decision-making with trapezoidal fuzzy neutrosophic numbers. Neutros. Sets Syst. **8**, 46–56 (2014)
7. Biswas, P., Pramanik, S., Giri, B.C.: Aggregation of triangular fuzzy neutrosophic set information and its application to multi-attribute decision making. Neutros. Sets Syst. **12**, 20–40 (2016)
8. Biswas, P., Pramanik, S., Giri, B.C.: TOPSIS strategy for MADM with trapezoidal neutrosophic numbers. Neutros. Sets Syst. **19**, 29–39 (2018)
9. Biswas, P., Pramanik, S., Giri, B.C.: T distance measure MADM strategy with interval trapezoidal neutrosophic numbers. Neutros. Sets Syst. **19**, 40–46 (2018)
10. Biswas, P., Pramanik, S., Giri, B.C.: Value and ambiguity index based ranking method of single-valued trapezoidal neutrosophic numbers and its application to multi-attribute decision making. Neutros. Sets Syst. **12**, 127–138 (2016)
11. Broumi, S., Talea, M., Bakali, A., Smarandache, F., Patro, S.K.: Minimum spanning tree problem with single- valued trapezoidal neutrosophic numbers, studies in fuzziness and soft computing. Soft Comput. **22** (Springer)
12. Broumi, S., Talea, M., Bakali, A., Smarandache, F.: Shortest path problem under triangular fuzzy neutrosophic information. In: 10th International Conference on Software. Knowledge, Information Management and Applications (SKIMA) (2016)
13. Boukezzoula, R., Coquin, D., Nguyen, T.L.: Multi-sensor information fusion: combination of fuzzy systems and evidence theory approaches in color recognition for the NAO humanoid robot. Robot. Auton. Syst. **100**, 302–316 (2018)
14. Broumi, S., Talea, M., Bakali, A., Smarandache, F.: Computation of shortest path problem in a network with SV-trapezoidal neutrosophic numbers. In: Proceedings of the 2016 International Conference on Advanced Mechatronic Systems. Melbourne, Australia (2016)
15. Deli, I., Subas, Y.: A ranking method of single valued neutrosophic numbers and its applications to multi-attribute decision making problems. Int. J. Mach. Learn. Cybern. **8**(4), 1309–1322 (2017)
16. Deli, I., Şubaş, Y.: Some weighted geometric operators with SVTrN-numbers and their application to multi-criteria decision making problems. J. Intell. Fuzzy Syst. **32**(1), 291–301 (2017)
17. Deli, I.: Operators on single valued trapezoidal neutrosophic numbers and SVTN-Group decision making. Neutros. Sets Syst. **22**, 130–150 (2018)
18. Faraj, Z., Selamet, M., Morales, C., Torres, P., Hossain, M., Lipson, H.: Facially expressive humanoid robotic face. HardwareX (2020). https://doi.org/10.1016/j.ohx.2020.e00117
19. Fakoor, M., Kosari, A., Jafarzadeh, M.: Original humanoid robot path planning with fuzzy Markov decision processes. J. Appl. Res. Technol. **14**, 300–310 (2016)
20. Giri, B.C., Molla, M.U., Biswas, P.: TOPSIS method for MADM based on interval trapezoidal neutrosophic number. Neutros. Sets Syst. **22**, 151–167 (2018)
21. Hanafy, I.M., Salama, A.A., Mahfouz, K.M.A.: Correlation coefficients of generalized intuitionistic fuzzy sets by centroid method. J. Mech. Civil Eng. **3**(5), 11–14 (2012)

22. Hirai, K.: Current and future perspective of Honda humanoid robot. In: Proceedings of IEEE/RSJ International Conference on Intelligent Robots and Systems (IROS), pp. 500–508 (1997)
23. Hanafy, I.M., Salama, A.A., Mahfouz, K.M.: Correlation coefficients of neutrosophic sets by centroid method. Int. J Prob. Stat. **2**(1), 9–12 (2013)
24. Hung, W.L., Wu, J.W.: Correlation of intuitionistic fuzzy sets by centroid method. Inf. Sci. **144**, 219–225 (2002)
25. Kahraman, C., Otay, I. (eds.). Fuzzy Multi Criteria Decision Making Using Neutrosophic Sets, Studies in Fuzziness and Soft Computing, vol. 369. Springer
26. Kahraman, C., Deveci, M., Boltürk, E., Türk, S.: Fuzzy controlled humanoid robots: a literature review. Robot, Auton. Syst. **134**, 1–12 (2020)
27. Karaaslan, F.: Gaussian single-valued neutrosophic numbers and its application in multi-attribute decision making. Neutros. Sets Syst. **22**, 101–117 (2018)
28. Kashyap, A.K., Parhi, D.R., Muni, M.K., Pandey, K.K.: A hybrid technique for path planning of humanoid robot NAO in static and dynamic terrains. Appl. Soft Comput. J. **96**, 1–18 (2020)
29. Li, D.F.: Decision and game theory in management with intuitionistic fuzzy sets. In: Studies in Fuzziness and Soft Computing, vol. 308. Springer
30. Liang, R.X., Wang, J.Q., Li, L.: Multi-criteria group decision-making method based on interdependent inputs of single-valued trapezoidal neutrosophic information. Neural Comput. Appl. **30**, 241–260 (2018)
31. Liang, R.X., Wang, J.Q., Zhang, H.Y.: A multi-criteria decision-making method based on single-valued trapezoidal neutrosophic preference relations with complete weight information. Neural Comput. Appl. **30**, 3383–3398 (2018)
32. Liu, P., Zhang, X.H.: Some Maclaurin symmetric mean operators for single-valued trapezoidal neutrosophic numbers and their applications to group decision making. Int. J. Fuzzy Syst. **20**(1), 45–61 (2018)
33. Lau, Y., Chee, D.G.H., Chow, X.P., Wong, S.H., Cheng, L.J., Lau, S.T.: Humanoid robot-assisted interventions among children with diabetes: a systematic scoping review. Int. J. Nurs. Stud. **111**, 1–10 (2020)
34. Murillo, N.G.A., Sandoval, L., Hayet, J.B., Esteves, C., Becerra, H.M.: Coupling humanoid walking pattern generation and visual constraint feedback for pose-regulation and visual path-following. Robot. Auton. Syst. **128**, 1–23 (2020)
35. Mohamed, M., Zhou, Y.Q., Baset, M.A., Smarandache, F.: A critical path problem using triangular neutrosophic number. In: Neutrosophic Operational Research, vol. I, Section X. Pons Brussels (2017)
36. Porchelvi, R.S., Umamaheswari, M.M.: A study on intuitionistic fuzzy multi objective LPP into LCP with neutrosophic triangular numbers approach. J. Appl. Sci. Comput. **5**(9), 570–576 (2018)
37. Rath, A.K., Parhi, D.R., Das, H.C., Muni, M.K., Kumar, P.B.: Analysis and use of fuzzy intelligent technique for navigation of humanoid robot in obstacle prone zone. Defence Technol. **14**, 677–682 (2018)
38. Robinson, J., Amirtharaj, H.: MADM problems with correlation coefficient of trapezoidal fuzzy intuitionistic fuzzy sets. Adv. Decis. Sci. Article ID 159126 (2014)
39. Pramanik, S., Mallick, R.: VIKOR based MAGDM strategy with trapezoidal neutrosophic numbers. Neutros. Sets Syst. **22**, 118–129 (2018)
40. Paksoy, T., Pehlivan, N.Y., Ozceylan, E.: Fuzzy Set Theory (In Turkish). Nobel Publisher, Ankara (2013)
41. Ruan, J., Chan, F.T.S., Zhu, F., Shi, Y., Wang, Y.: A centroid based correlation coefficient of fuzzy numbers. In: Proceedings of the 3rd World Congress on Mechanical, Chemical, and Material Engineering (MCM'17) Rome, Italy (2017)
42. Smarandache, F.: A unifying field in logics. In: Neutrosophy: Neutrosophic Probability, Set and Logic. American Research Press, Rehoboth (1998)
43. Samant, R., Nair, S., Kazi, F.: Development of autonomous humanoid robot control for competitive environment using fuzzy logic and heuristic search. IFAC-PapersOnLine **49–1**, 373–378 (2016)

44. Shinoj, T.K., John, S.J.: Accuracy in collaborative robotics: i–13 an intuitionistic fuzzy multiset approach. Glob. J. Sci. Front. Res. Math. Decis. Sci. **13**(10), 21–28 (2013)
45. Wang, H., Smarandache, F., Zhang, Q., Sunderraman, R.: Single valued neutrosophic sets. Multispace Multistruct. **4**, 410–413 (2010)
46. Wu, X., Qian, J., Peng, J.J., Xue, C.C.: A multi-criteria group decision-making method with possibility degree and power aggregation operators of single trapezoidal neutrosophic numbers. Symmetry **10**(590), 1–21 (2018). https://doi.org/10.3390/sym10110590
47. Ye, J.: Some weighted aggregation operators of trapezoidal neutrosophic numbers and their multiple attribute decision making method. Informatica **28**(2), 387–402 (2017)
48. Ye, J.: Trapezoidal neutrosophic set and its application to multiple attribute decision making. Neural Comput. Appl. **26**(5), 1157–1166 (2015)
49. Ye, J.: Vector similarity measures of simplified neutrosophic sets and their application in multicriteria decision making. Int. J. Fuzzy Syst. **16**(2), 204–211 (2014)
50. Zadeh, L.A.: Fuzzy sets. Inf. Control **8**, 338–353 (1965)
51. Zeng, W., Li, H.X.: Correlation coefficient of intuitionistic fuzzy sets. J. Ind. Eng. Int. Islamic Azad Univ. South Tehran Branch **3**(5), 33–40 (2007)
52. Zhang, L., Xu, X.H., Tao, L.: Some similarity measures for triangular fuzzy number and their applications in multiple criteria group decision-making. J. Appl. Math. Article ID 538261 (2013). https://doi.org/10.1155/2013/538261
53. Zimmermann, H.-J.: Fuzzy Set Theory and Its Applications. Kluwer Academic Publishers (1993)

Trends on Extension and Applications of Neutrosophic Graphs to Robots

Said Broumi, Kifayat Ullah, Tahir Mahmood, Mohamed Talea, Assia Bakali, Florentin Smarandache, D. Nagarajan, and M. Lathamaheswari

Abstract Neutrosophic set (NS) theory has a diverse nature in dealing with impreciseness of real life events and has a wider range of applications in logic, algebra, topology, operation research, pattern recognition, artificial intelligence, robotics, neural networks and several other fields. After the initiation of NS theory, some quality research has been done in this field and as a result a new dimension in graph theory is evolved known as neutrosophic graph (NG) theory (combination of graph

S. Broumi (✉) · M. Talea
Laboratory of Information Processing, Faculty of Science Ben M'Sik, University Hassan II, Sidi Othman, B.P 7955 Casablanca, Morocco
e-mail: broumisaid78@gmail.com

M. Talea
e-mail: taleamohamed@yahoo.fr

K. Ullah
Department of Mathematics, Riphah International University Lahore, 13km, Raiwand Road, Lahore 54000, Pakistan
e-mail: kifayat.ullah@riphah.edu.pk

T. Mahmood
Department of Mathematics and Statistics, International Islamic University Islamabad, Islamabad, Pakistan
e-mail: tahirbakhat@iiu.edu.pk

A. Bakali
Ecole Royale Navale, Boulevard Sour Jdid, B.P 16303 Casablanca, Morocco
e-mail: assiabakali@yahoo.fr

F. Smarandache
Department of Mathematics, University of New Mexico, 705 Gurley Avenue, Gallup, NM 87301, USA
e-mail: fsmarandache@gmail.com

D. Nagarajan · M. Lathamaheswari
Department of Mathematics, Hindustan Institute of Technology and Science, Chennai 603 103, India
e-mail: dnrmsu2002@yahoo.com

M. Lathamaheswari
e-mail: lathamax@gmail.com

C. Kahraman and E. Bolturk (eds.), *Toward Humanoid Robots: The Role of Fuzzy Sets*, Studies in Systems, Decision and Control 344,
https://doi.org/10.1007/978-3-030-67163-1_11

theory and NS theory). The aim of this chapter is to highlight the recent developments in the field of NG theory and their generalizations including neutrosophic hypergraph, interval NGs, bipolar NGs etc. Almost all the work based on the development of NGs and their applications have been discussed thoroughly.

Keywords Neutrosophic graphs · Generalizations of neutrosophic graphs · Applications

1 Introduction

A new logic known as neutrosophic logic and based on neutrosophic logic he developed a concept in fuzzy sets (FS) known as NS [1–3]. A NS is composed of three functions that represent the membership, neutral and non-membership grades of an event or object. A NS is a generalized form of FS [4] and Atanassov's IFS [5]. After the development of this new idea, several new concepts of NSs and their generalizations have been proposed and applied to many problems of real-life such as decision making, pattern recognition, artificial intelligence, neural networks, shortest path problems, telephone networks, clustering and many areas [6–8].

Among several dimensions of NSs, one is NG theory which have been extensively worked on and applied to many problems of real-life. The concept of single valued NG (SVNG) has been introduced by Broumi et al. [9] while the framework of interval valued neutrosophic graph (IVNG) is developed by Broumi et al. [9]. A NG, is basically, an enhanced form of FG [10, 11] and IFG by Parvathi and Karunambigai [12] by handling uncertain information with the help of three functions. Akram and Shahzadi [13] developed some operations on SVNGs and some other types of NGs have been introduced by Dhavaseelan et al. [14]. Some developments in NG theory have been discussed by Kandasamy et al. [15] including applications of NGs. The idea of similarity measure in NGs have been proposed by Fathi et al. [16] and the theory of intuitionistic NGs have been proposed by Akram and Shahzadi [17]. A study of isomorphisms of NGs have been established by Shah [18] while isomorphisms of single and interval valued neutrosophic hypergraphs have been studied by Malik et al. [19]. Hassan et al. [20] developed the concepts of regular and irregular hypergraphs. Broumi et al. [21, 22] proposed the concepts of isolated SVNGs as well as isolated IVNGs. Malik et al. [23] studied the isomorphisms of bipolar single valued neutrosophic hypergraphs and Shah and Broumi [21] contributed to the theory of irregular NGs. Malik et al. [23] discussed some theory of regular bipolar single valued neutrosophic hypergraphs. A new novel theory of soft neutrosophic graphs have been proposed by Shah and Hussain [24] which were further utilized by Akram and Shahzadi [13] in some real-life applications. The terms order, degree and size for SVNGs have been defined by Broumi et al. [25]. A detailed study on bipolar SVNGs and their related terms and A study of planer SVNGs and planer bipolar NGs have been established by Broumi et al. [22, 26]. Broumi et al. [27, 28] developed a study of strong IVNGs and further contributed in developing some operations on IVNGs.

Some further contribution in the development of NG theory that is briefly discussed [29–75].

NG, as an advanced form FG and IFG, has the ability to cope with uncertain and imprecise information using three membership functions. Therefore, modeling an uncertain event using NG will give us better results than other fuzzy structures. Since the evolution of NG theory, NGs have been applied to several real-life applications. Broumi et al. [25] proposed a novel approach towards decision-making using the idea of NG. Broumi et al. [21, 22, 26, 27] successfully computed the shortest path using Dijkstra algorithm in the environment of SVNGs, SV trapezoidal neutrosophic numbers, IVNGs and triangular neutrosophic numbers. For some other work on neutrosophic shortest path, one is referred to Broumi et al. [65, 66, 76, 77, 78, 79]. Sahin [80] introduced a technique to solve multi- attribute decision making (MADM) problems using NG and applied the theory of NGs to time travel problem. Akram and Luqman [81–83] uses the framework of NGs to discuss some social and collaboration networks, marketing and biological problems and decision-making problems. Akram and Sarwar [84, 85] are also based on MADM techniques using the approach of bipolar NGs and Hamidi and Saeid [86] uses the theory of NGs in sensor networks. Akram and Siddique [87] developed neutrosophic competition graphs and applied them in solving ecosystem and job competition problems. Gulistan et al. [88] is based on the application of NG in industry and Akram et al. [84] solved decision making problems in the environment of NGs. Broumi et al. [89] provided a dynamic algorithm to save the number of incoming and outgoing calls based on SVNGS.

The aim of this chapter is to provide a detailed study of the developments of NGs and its generalizations along with their applications to several real-life phenomena. This chapter is organized as section one provides a brief introduction of basic work on NSs, NGs and its developments. In section two, a detailed study of NGs have been carried out summarizing the work in the environment of NGs since its evolution. In section three, applications of neutrosophic graphs and humanoid robots have been presented. In section four, integration of neutrosophic graph and humanoid robots is given. In section five, we provide a comparative study while section four is based on advantages of using NGs instead of FGs as well as IFGs. The chapter ended with a conclusion.

2 Developments in Neutrosophic Graphs

In this section, the developments in NG theory have been comprehensively discussed along with their application to some real-life problems. Starting from the pioneering work on NGs, we will discuss some great achievements of NGs one by one.

2.1 Single-Valued Neutrosophic Graph [9]

In this chapter, the concept of SVNG is proposed and supported with examples. The generalization of SVNGs over FG and IFG has be proved. Some related graphical terms and notions like subgraph of SVNG, complement of SVNG, strong SVNG etc. are developed and demonstrated with the help of examples.

Now will discuss some other chapters that contributed to SVNGs.

2.1.1 Operations Single-Valued Neutrosophic Graph [90]

This chapter is based on operations on SVNGs. The authors pointed out some flaws in definitions from [9] and developed some new operations. Some other characteristics of SVNGs have also been investigated using the concept of level graphs.

2.1.2 Single-Valued Neutrosophic Hypergraphs (SVNHGs) [52]

This chapter introduced four major notions in the theory of SVNGs which are:

- SVNHG
- Dual SVNHG
- Line graph of SVNHG
- Transversal SVNHG

These notions are supported with examples and their properties are elaborated.

2.1.3 A Neutrosophic Graph Similarity Measures

This chapter proposed a novel similarity measures for NGs. This new tool measures the similarity of two NGs using their nodes. The new tool is based on Hausdorff distance and it is proved that proposed distance measure satisfies the metric distance. This type of tool could be very helpful in finding two NGs of same type.

2.1.4 Some Studies in Neutrosophic Graphs [18]

The purpose of this chapter is to develop a study of isomorphisms. The authors defined weak isomorphism, co-weak isomorphism and isomorphisms between two NGs.

2.1.5 Isomorphism of Single-Valued Neutrosophic Hypergraphs (SVNHGs) [91]

This chapter discussed the theory of isomorphisms of SVNHGs. The main topics of this article include order of SVNHGs, degree of SVNHGs, size of SVNHGs, isomorphism of two SVNHGs, weak and co-weak isomorphism of two SVNHGs. The defined terms are supported with the help of examples.

2.1.6 Single Valued Neutrosophic Graphs: Degree, Order and Size [22]

This chapter is based on some certain types of sizes, order and degrees of SVNGs. Further, the concept of regular SVNGs is also proposed. Each proposed concept is supported with the help of examples.

2.1.7 Irregular Neutrosophic Graphs [92]

This chapter proposed some irregularities in NGs. The authors described neighborly irregular and totally neighborly irregular NGs. Further the concept of highly irregular and totally highly irregular NGs is also demonstrated. The criteria for all types of NGs is also considered.

2.1.8 Isolated SVNGs [21]

This chapter is based on isolated and complete SVNGs. A condition under which a SVNG becomes an isolated SVNG is stated and demonstrated with examples.

2.1.9 Single Valued Neutrosophic Planer Graphs [93]

This chapter aims to apply the theory of SVNGs to planer graphs, multi-graphs and dual graphs which results in the development of single valued neutrosophic (planer graphs, multi-graphs and dual graphs) respectively. With the help of some examples, these concepts are supported.

Regularity of Graphs in Single Valued Neutrosophic Environment [31]

This chapter discussed some regularity in SVNGs as the concept of edge regular SVNG, full edge regular SVNG and partially edge regular SVNG are proposed. Further, strongly regular SVNG and bi-regular SVNG are defined. The authors investigated properties of these concepts are investigated and some examples are provided in their support. The application of these notions in decision-making problem is also studied.

Intuitionistic Single Valued Neutrosophic Hypergraphs [94]

In this chapter, a new framework of intuitionistic SVNHGs is proposed along with some other graphical structures such as Dual intuitionistic SVNHGs, (η, ϕ, ψ)—level hypergraphs, tempered intuitionistic SVNHGs and transversal intuitionistic SVNHGs. These concepts are utilized in a clustering algorithm.

Single Valued Neutrosophic Graph Structure [33]

This chapter is based on a novel framework of SVNG structure. A graph structure is basically an extension of undirected graphs. A SVNG structure is developed by applying the concept of SVNSs to graph structure. Some related work is studied and supported with examples.

Some Operations on Neutrosophic Fuzzy Graphs [34]

This chapter is simply based on complement of NGs and some operations on NGs like union of NGs and intersection of NGs.

Accessible SVNGs [30]

This chapter aims to reduce a SVNHG to a SVNG via strong equivalence relation. Another important result that is discussed in this paper is that every weak SVNG is a derived SVNG and every linear weak single valued neutrosophic tree can be extended to linear single valued neutrosophic tree.

Certain Competition Graphs Based on Intuitionistic Neutrosophic Environment [38]

This chapter proposed the concept of intuitionistic neutrosophic competition graphs (NCGs). Extending the idea, P-competition intuitionistic NGs and M-step intuitionistic NCGs. The usefulness of these ideas lies in the fact that these concepts can cope with ecosystem and career competition problems. Such problems are addressed in the paper using proposed work.

Neutrosophic Competition Graphs with Applications [87]

This chapter merged the notion of SVNG with competition graphs to introduce the notion of single valued neutrosophic competition graphs (SVNCG). The characteristics of new work is studied. Further, an algorithm is proposed to deal with decision making problem and a problem related to ecosystem and job competition is solved using proposed algorithm.

Single Valued Neutrosophic Trees [95]

In this chapter, the importance of edge connectivity of SVNGs is discussed. The authors developed some special types of bridges, cut-vertices, cycles and trees in SVNGs. The properties of these concepts are also proposed.

On Antipodal SVNGs [40]

This chapter simply developed the concept of antipodal SVNGs. Further the authors established the isomorphisms between antipodal SVNGs.

Self-Centered SVNGs [41]

This chapter contributed to the theory of SVNGs by introducing the concept of self-centered SVNG. The authors also proposed some new graph theoretic notions such as length of SVNG, eccentricity of SVNG, distance of SVNG, radius of SVNG, diameter of SVNG, status of SVNG, total status of SVNG, central vertex of SVNG and median of SVNG. The characteristics of self-centered SVNG is explained and the new notions are demonstrated with the help some examples.

Single Valued Neutrosophic Signed Graphs (SVNSGs) [42]

This chapter is based on a new notion of SVNSG. The author generalized the concept of intuitionistic signed graphs [96] to propose SVNSG. The authors investigated the properties of new proposed model and examples are provided for their support.

Neutrosophic Incidence Graphs with Application [83]

This chapter is based on single-valued neutrosophic incidence graphs (SVNIGs). We developed some concepts including bridges of SVNIG, cut vertex of SVNIG and blocks in SVNIGs. Some properties of SVNIGs are also studied. The edge-connectivity of SVNIG, vertex-connectivity of SVNIG and pair-connectivity in

SVNIGs are also demonstrated. The situation of illegal migration from Pakistan to Europe is studied as an application of SVNIGs.

Uniform SVNGs [46]

This chapter introduced a new approach of uniform SVNGs and an algorithm for finding the complement of SVNGs as well as uniform SVNGs. The algorithm is supported with an example.

Certain Notions of Energy in SVNGs [97]

This chapter certainly contributed to the theory of SVNGs as the authors defined the energy of SVNGs, Laplacian energy of SVNGs as well as signless Laplacian energy of SVNGs. Further, the relationship among these notions is established along with their properties are elaborated.

Properties of SVNGs [49]

This chapter developed the theory of SVNGs by introducing the notions of homomorphisms and isomorphisms. Further, several types of products are defined for SVNGS including arithmetic cartesian product, arithmetic lexicographic product as well as direct product and supported these ideas with examples. The notion of union an intersection is also introduced along with their properties. Also, two kinds of addition for SVNGs are developed based on summation and union operations. A MADM problem is also discussed using SVNGs.

Perfect and Status in SVNGs [50]

This chapter introduced the notion of perfect SVNGs, complete perfect SVNGs and complete perfect vertices and edges. The concept of status in complete perfect SVVNG is also discussed and exemplified.

New Concepts in Neutrosophic Graphs with Applications [52]

This chapter contributed to the frame of SVNGs. Some edge irregular SVNGs are developed and supported with examples. An algorithm for decision making is develop and demonstrated with the help of a numerical example.

Strong Degrees in SVNGs [68]

The goal of this chapter is to develop various types of strong degrees in SVNGs and introduce some new concepts of vertex membership of truth-values, vertex membership of indeterminate-values and vertex membership of false-values, which are sequence of SVNG with proofs and numerical demonstration.

Now we will summarize the work done in the frame of IVNGs along with its several applications.

Neutrosophic Logic Used in Robotics [72, 73]

An affiliate theory for designing and implementation of decision support system is neutrosophy. Neutrosophic logic includes truth, indeterminacy and falsity values (indeterminacy expressing unknown parameters). There is no restriction for the sum of these components.

The neutrosophic robot is unique by being indeterminacy-permissive in performance compared to its fuzzy analogues. Controllers will act according to the problem to be solved. Neutrosophic system has the capacity of utilizing the knowledge collected from human operators. For conventional systems, it is very difficult to design a mathematical structure to simulate the behavior of the system. In addition, the data collected by the system cannot be complete and determinate, since the human behavior is uncertain by nature.

Constitutive non-linearity, time-varying nature, environmental disturbances that are highly unpredictable, and difficulties faced to obtain actual and decisive measurements may cause indeterminacy and incompleteness in the data. In all these situations, humans can take knowledgeable decisions, but their linguistic expression often contains imprecise linguistic terms. This type of linguistic description would be clearly explained by the operator with relevant content.

This type of neutralities is missing in fuzzy logic, which considers only membership and non-membership values of a specified element. By contrary, the neutrosophic logic is very efficient to deal with incomplete and indeterminate situations. The conversion from a normal operating model based on fuzzy logic to neutrosophic case can be done by applying the set of formulae in the Dezert-Smarandache Theory (DSmT).

Every input value maps into an element of a class described to deal with the multiple dimensions as properties, where to form and express coherently the tuple is needed. There are four processes of neutrosophic logic that can be applied in robotics, namely: neutrosophication, neutrosophic rule base, neutrosophic engine, and de-neutrosophication.

Hence, applying neutrosophic logic in robotics is an efficient and reliable endeavor.

2.2 *Interval Valued Neutrosophic Graph [25]*

An IVFG enhanced the frame of SVNG by representing the grades of three functions in terms of sub-intervals of unit interval. In this chapter, along with the pioneering concept of IVNGs, some of graphical terms are also demonstrated with the help of examples. Using some restriction over IVNG, it can be transformed to SVNG, IFG as well as FG.

2.2.1 Concepts of IVNGs [98]

In this chapter, Akram and Maryam proposed a new definition for IVNG by challenging the definition of Broumi et al. [25]. Further, in view of new definition, some operations are developed.

2.2.2 Regular and Totally Regular Interval Valued Neutrosophic Graphs [20]

This chapter contributed to the theory of IVNHGs by proposing the concepts of regular IVNHG and totally regular IVNHG. The authors defined the order and size along with properties of the totally regular IVNHGs. The work is further generalizing to the complete IVNHGs.

2.2.3 Isomorphisms of IVNHGs [19]

This chapter introduced the notion of homomorphism and further isomorphism for IVNHGs. The idea of weak isomorphism is also illustrated. The properties of degree of vertices, order and size in context of isomorphisms are explored. It is also proved that isomorphism of IVNHG is an equivalence relation while weak isomorphism of IVNHG is a partial order. These ideas are demonstrated with the help of examples.

2.2.4 An Isolated IVNG [25]

This chapter introduced the necessary and sufficient condition for an IVNG to be isolated IVNG. This idea of isolation is further elaborated with the help of examples.

2.2.5 Operations on IVNGs [99]

This chapter developed the basic operations of intersection, union, Cartesian product and join of two IVNGs. For illustration and validity of these operations, some examples and results are discussed.

2.2.6 On Strong IVNGs [28]

This chapter is based on a special case of IVNG known as strong IVNG. The definition of strong IVNG is proposed and supported by examples. Some operations of strong IVNG such as Cartesian product, union, composition and intersection of strong IVNGs are proposed. Few results have also been proposed based on new operations.

2.2.7 Representation of Graph Structures Based on IVNSs [32]

This chapter introduced the notion of interval valued neutrosophic graph structures (IVNGSs) upon applying the notion of IVNS to graph structures. The authors proposed some certain operations on IVNGSs and described their properties. Some open problems are also raised in the context of IVNGSs.

2.2.8 Interval Valued Neutrosophic Competition Graphs (IVNCGs) [36]

This chapter aimed to introduce the concepts of IVNCGs. Some other special cases which are discussed in this article are:

- K-competition IVNGs
- P-competition IVNGs
- M-step IVNCGs
- M-step interval valued neutrosophic neighborhood graphs.

2.2.9 IVNG Representation of Concept of Lattice and Its (α, β, γ)-Decomposition [39]

In this chapter, the concept of IVNS, IVNG and lattice is used to formulate two methods. In first method, some undefined concepts are discovered and their super-sub-concept hierarchical reflection in the concept is lattice. In second method, some (α, β, γ)—cut of three membership grades of IVNG are defined. Each method is supported by examples.

IVNGSs [100]

This chapter proposed some certain notions of IVNGSs which are:

- $\varphi - complement$ of IVNGS.
- Self-complementary IVNGSs
- Totally self-complementary IVNGSs.

Some properties of these defined notions are studied along with few examples.

Interval Valued Neutrosophic Signed Graphs (IVNSG) [53]

This chapter is based on the concept of IVNSG. The authors proposed some examples in the support of IVNSGs and described its various properties.

Self-Centered Interval Valued Signed Neutrosophic Graph (SCIVSNG) [58]

In this chapter, the authors discussed the notion of SCIVSNG and studied its elated terms. Further, several properties of SCIVSNGs are examined along with their proofs and some examples.

2.3 *Bipolar Neutrosophic Graphs (BNGs) [35]*

In this section, we aim to summarize the work related to bi-polar neutrosophic graphs and discussed several developments that are achieved in this regard.

2.3.1 On Bi-Polar Single Valued Neutrosophic Graph (BSVNG) [26]

This chapter is based on development of BSVNG. A BSVNG is a combination of both bi-polar FG as well as SVNGs. This type of graphs can cope with an uncertain event where each membership grades have bi-polarity. In this article, the concept of strong BSVNG, complete BSVNG, regular BSVNG and irregular BSVNGs have been developed and supported with examples.

2.3.2 An Introduction to Bi-Polar SVNG Theory [27]

This chapter is based on the concepts of bi-polar FG, IFG, bi-polar IFG, SVNG and bi-polar SVNG. The novelty and generalization of bi-polar SVNGs is described over the existing ideas.

2.3.3 A Bi-Polar Single Valued Neutrosophic Isolated Graph (BSVNIG) [48]

In this chapter, the model of bi-polar SVNG is proposed which is a generalized form of SVNG. The novelty of proposed new model is described and the concept of BSVNIG and complete BSVNIG are proposed. Some results are proved based on new concepts.

2.3.4 Isomorphism of Bi-Polar Single Valued Neutrosophic Hypergraphs (BSVNHG) [19]

This chapter is based on the notion of isomorphism for BSVNHGs. The idea of weak isomorphism is also illustrated. The properties of degree of vertices, order and size in context of isomorphisms are explored. It is also proved that isomorphism of BSVNHG is an equivalence relation while weak isomorphism of BSVNHG is a partial order. These ideas are demonstrated with the help of examples.

2.3.5 Regular Bi-Polar SVNHGs [101]

This chapter proposed the concepts of regular and totally regular BSVNHGs along with order and degree of BSVNHGs. The authors also proposed the definition of complete regular BSVNHG and irregular BSVNHG.

2.3.6 Bi-Polar Neutrosophic Graph Structure (BNGS) [35]

In this chapter, the concept of BNGS is developed and few methods for the construction of BNGS are proposed. The characteristics of BNGS are also investigated.

2.3.7 Special Types of BSVNGs [43]

A BSVNG is a generalization of SVNG. In this chapter, some contributions to the theory of BSVNG have been made. The authors defined middle BSVNG, subdivision BSVNG, total BSVNG and supported these concepts with examples. The concept of bi-polar single valued neutrosophic line graph is also introduced and exemplified. The isomorphism of all the defined structures has also been discussed.

2.3.8 Bi-Polar Neutrosophic Planer Graphs (BNPGs) [102]

This chapter is a contribution to the theory of BNG as the notions of BNPG, bi-polar neutrosophic multi-graph and dual BNG are proposed. Each of these is supported with examples and their properties are studied. An application of proposed models has also been discussed in road networks.

2.4 *Neutrosophic Soft Graph (NSG) and Complex Neutrosophic Graph (CNG)*

In this section, the developments achieved in the frame of NCGs and CNGs. A NSG is a complex graphical approach having the characteristics of soft graph as well as NGs whereas the approach of CNG is based on the notion of complex neutrosophic set (CNS).

2.4.1 Neutrosophic Soft Graphs (NSGs) with Applications [90]

This chapter is based on a novel approach of NSG i.e. the graph of neutrosophic soft set is developed. This type of framework is useful in handing neutrosophic soft information. The manuscript also contained the ideas of strong NSG, complete NSG and some examples to support each concept. An application of NSG in decision making is also introduced.

2.4.2 Representation of Graphs Using Intuitionistic Neutrosophic Soft Sets (INSSs) [17]

In this chapter, the authors introduced the notion of intuitionistic neutrosophic soft graph (INSG) based on a comprehensive concept of INSS. The authors applied this novel concept of INSG to MADM problem and provided and algorithm of proposed method.

2.4.3 Generalized Interval Valued Neutrosophic Graphs of First Type 1 [44]

This chapter is based on a graphical approach for the newly developed concept of CNS known as CNGs. A CNG extends the approaches of SVNG, IFG as well as FG. A matrix representation for CNG is also proposed. Some properties of CNG are explained with examples and some useful results are described.

2.4.4 On Regular Complex Neutrosophic Graphs [37]

This chapter contributes to the framework of CNG by introducing the notion of regular CNG. The author also defined the notion of degree of CNG and studied some properties. It is also proved that the concept of regular CNG is an extended form of complex neutrosophic fuzzy graph of type 1.

2.4.5 Bi-Polar Neutrosophic Graphs (BNGs) of Type 1 [55]

This chapter is based on generalization of SVNGs of type 1. A new concept of BNGs of type 1 is developed in this manuscript along with a study of its matrix representation. The novelty of this new approach is studied through a comparative study.

2.4.6 Generalized Bi-Polar Neutrosophic Graphs of Type 1 [45]

This chapter is based on generalized SVNGs of type 1. A new model of generalized bi-polar NGs of type 1 is proposed in this manuscript along with a study of its matrix representation. The generalization of this new approach is also proved.

2.4.7 Interval Complex Neutrosophic Graph of Type 1 (ICNGT1) [103]

This chapter extended the concept of NGs of type 1 to ICNGT1. The idea of adjacency matrix is discussed in the environment of ICNGT1. The novelty and applicability of proposed idea is studied and demonstrated with examples.

2.4.8 Generalized Interval Valued Neutrosophic Graph of Type 1 (GIVNGT1) [79]

This chapter presented the concept of GIVNGT1 inspired by the concept of generalized single valued neutrosophic graph of type 1. The matrix representation of proposed new concept is studied and the novelty of GIVNGT1 is shown.

2.4.9 Characterizations of Strong and Balanced Neutrosophic Complex Graphs [69]

In this chapter, the concepts of neutrosophic complex graph, complete neutrosophic complex graph, strong neutrosophic complex graph, balanced neutrosophic complex graph and strictly balanced neutrosophic complex graph are introduced. Some of the interesting properties and related examples are established.

Now we discussed an article based on neutrosophic approach to finite groups.

2.5 Neutrosophic Graphs of Finite Groups [104]

This chapter presented the concept of NGs in context of finite groups. The authors discussed some related ideas and supported these ideas with examples.

3 Applications of Neutrosophic Graphs

The goal of this section is to discuss the achievements of NG theory in modeling real life problems. We will summarize the applications of NG theory in dealing with some real-life problems.

3.1 Applications in Solving Shortest Path Problems (SPP)

In this subsection, we summarized the work that is being done for computation of shortest path in different neutrosophic networks.

3.1.1 Applying Dijkstra Algorithm for Solving Neutrosophic SPP [21]

This chapter aims to apply the famous Dijkstra algorithm for computing shortest path in the environment of SVNGs. Applying Dijkstra algorithm in neutrosophic environment provides a flexibility of dealing with uncertainty with three types of membership grades. The neutrosophic score and accuracy functions are also used in this phenomenon. The process is explained numerically and studied comparatively.

3.1.2 Computation of SPP in a Network with Single Valued-Trapezoidal Neutrosophic Numbers (SV-TNN) [22]

This chapter enables us to cope with SPP in the environment of SV-TNN. The method involves the score function for computation of convenient path form starting node to final node. At each path the weights are in the form of SV-TNN showing the significance of that particular path. The method is explored with the help of a numerical problem.

3.1.3 Application of Dijkstra Algorithm for Solving Interval Valued Neutrosophic SPP [26]

This chapter is a generalization of [61]. Again, it uses the score function to compute the shortest distance between two nodes. The difference is that here the weights of path are in the form of IVNNs. An IVNN modeled the imprecise phenomena in a better way than a SVNN. This process is explained numerically and studied comparatively with existing techniques.

3.1.4 SPP Under Triangular Fuzzy Neutrosophic (TFN) Information [27]

The work in this chapter is obtained from [61, 63]. The authors used ranking function to compute the shortest path from one node to another. The difference between this and other SPP is the weights of path are based on TFN information. The method proposed is illustrated with an example.

3.1.5 Operations on SVNGs with Applications [105]

This chapter is a contribution to the theory of SVNGs. The authors developed some operations on SVNG including Cartesian product, direct product, strong and semi-strong product, union, ring sum, join and lexicographic product. Each of these ideas is numerically demonstrated. An application of new proposed work is described in computing SPP.

3.1.6 SPP on SVNGs [65]

In the theory of NGs, the SPP has been solved for the first time in the environment of SVNGs in this chapter. The authors used score function of SVNS to compute the shortest path where weight of each path is a SVNN. This novel approach is demonstrated numerically along with some advantages of new method are studied.

3.1.7 SPP Under Bi-Polar Neutrosophic (BN) Settings [66]

This chapter introduced a method of finding convenient path (shortest path) in the environment of BN graphs. The method involves a ranking function to compute the convenient path form one node to other. In this type of problems, each of the path have weights in the form of BN numbers.

3.1.8 SPP by Minimal Spanning Tree (MST) Algorithm Using BN Numbers [76]

This chapter developed a technique of finding minimum distant route in a network where the paths have uncertainty and inconsistency. The imprecision of paths is denoted by BN numbers. The algorithm involves a matrix algorithm for computing minimum distant path.

3.1.9 BN Minimum Spanning Tree [78]

This chapter basically improved the work done by [76] using a matrix algorithm for computing SPP. The environment considered in this article is based on undirected graphs where weights are in the form of BN numbers. The authors provided a flowchart of proposed algorithm and established a comparison of new approach with existing technique. The work presented in this manuscript can be performed using a computer program and is therefore considered best.

3.1.10 Improved Similarity Measure in Neutrosophic Environment and its Application in Finding MST [70]

MST finds its vast applicability in designing of networks, approximation algorithms for NP-hard problems and clustering etc. Many research works have been done to find MST due to its various applications. But, till date very few research works are available in computing MST in neutrosophic environment. This chapter contributes significantly by defining the weight of each network edge using single valued neutrosophic set (SVNS) and introduce a new approach using similarity measure to find MST in neutrosophic environment. Use of SVNS makes the problem realistic as it can describe the uncertainty, indeterminacy and hesitancy of the real world in a better way. We introduce two new and simple similarity measures to overcome some disadvantages of existing Jaccard, Dice and Cosine similarity measures of SVNSs for ranking the alternatives. Further from the similarity measures we have developed two formulas for the entropy measure proving a fundamental relation between similarity measure and entropy measure. The new entropy measures define the uncertainty more explicitly in comparison to other entropy measure existing in the literature which has been established using an example.

3.1.11 A New Algorithm for Finding MSTs with Undirected NGs [59]

A MST problem of an undirected weighted connected NG in which a SVNN, instead of a real number, is assigned to each arc as its arc length. The chapter proposed this type of MST as neutrosophic MST. The utility of neutrosophic numbers as arc lengths and its application in different real-world MST problems are described. Further, a novel algorithm of MST problem in neutrosophic environment is introduced. The new algorithm cope with uncertainty using Kruskal algorithm for designing MST using neutrosophic number as arc length. A neutrosophic score function is also utilized in the algorithm. A comparison of this weight of the neutrosophic MST with that of an equivalent classical MST with real numbers as arc lengths is also established. Finally, the new algorithm is demonstrated by an example.

3.1.12 Applying Floyd's Algorithm for Solving Neutrosophic Shortest Path Problems [56]

For solving SPP, so far Dijkstra algorithm has been followed. In this article, the authors proposed Floyd's algorithm for solving neutrosophic SPP. The algorithm is supported with numerical examples.

3.1.13 SSP Under Interval Neutrosophic Settings [77]

This chapter provides an algorithm for computing shortest path from starting node to destination node in a network where each path length is described in terms of IVNNs. The main component is ranking function which is used to evaluate the shortest path.

3.1.14 Spanning Tree Problem with Neutrosophic Edge Weights [79]

This chapter introduces a new algorithm for finding minimum spanning tree (MST) in an environment where path lengths are represented by SVNNs. The algorithm is based on matrix approach followed by a numerical example and a comparative study is developed with existing methods.

3.2 Applications in Decision-Making (DM)

This subsection is based on contribution of NGs in DM problems. DM problems have been greatly discussed in fuzzy environment but in graph theory still there is huge gap for work in this area. Now we present some neutrosophic articles based on DM.

3.2.1 Some Results on the Graph Theory of CNGs [54]

This chapter introduced the notion of CNGs of type 1 (CNGT1) as a generalization of GNGT1 and GSVNGT1. The matrix representation of new concept is defined and the approach of CNGT1 is used in a MADM problem where the problem of software selection is carried out.

3.2.2 Bi-Polar Neutrosophic Hypergraphs with Applications [82]

This chapter developed the theory of bi-polar neutrosophic hypergraphs (BNHGs) keeping in view the importance of hypergraphs. The proposed several graph theoretic results based on BNHGs. There are two really important issues are discussed using proposed approach. The authors described a marketing problem using BNHGs and further they applied the concept of BNHGs to genes mutation.

3.2.3 Notions of Rough Neutrosophic Diagraphs (RNDGs) [51]

This chapter introduces the theory of roughness to NGs and developed the idea of RNDGs. The authors described several events having roughness as motivation for their work. Further, they proposed several types of products for RNDGs and studied their properties. At the end, the authors applied RNDGs to a MADM problem and discussed its applicability.

3.2.4 Entropy Based Single Valued Neutrosophic Diagraphs (SVNDGs) and Its Applications [71]

This chapter is based on the basic notions of SVNDGs. The authors defined the concept of SVNDGs and proposed some related theory. This new concept of SVNDGs is used to solve MADM problem.

3.2.5 Novel Applications of SVNG Structures in Decision Making [85]

This chapter developed the theory of SVNG structures and defined several operations on SVNG structures. Each concept is supported with examples. The authors utilized the notion of SVNG structures in DM problem and proposed an algorithm for this method.

3.2.6 Novel Multi-Criteria DM Methods Based on BNSs and BNGs [84]

This chapter is based on domination theory of BNGs and few algorithms for computation of domination in BNGs. This concept of domination in BNGs is further utilized in DM problems.

3.2.7 A New DM Method Based on Bi-Polar Neutrosophic Directed Graphs (BNDHGs) [81]

In this chapter, the viability and flexibility of BNDHGs are studied. The authors contributed to this theory by developing the concepts of regular BNDHGs, homomorphisms and isomorphisms of BNDHGs and demonstrated these ideas with examples. The isomorphic properties of strong BNDHGs are also studied. Finally, an algorithm is proposed for solving DM problem using BNDHGs.

3.2.8 An Approach to NG Theory with Applications [80]

This chapter contributed to the theory of SVNGs by discussing some of its pioneer concepts and demonstrated those ideas with examples. Then a DM technique is developed using SVNGs and two problems are solved based on proposed technique. The viability and advantages of new technique are also studied.

3.2.9 DM Method Based on the IVNGs [25]

These chapters discussed the MADM problems in the environment of SVNGs and support the approach with some numerical examples.

3.3 Application of NGs in Networks

In this subsection, some applications of NGs in different types of Networks are studied. We present some prominent work as follows:

3.3.1 Achievable SVNGs in Wireless Sensor Networks [86]

In this chapter, the authors derived the concept of SVNGs from SVNHGs via equivalence relation and applied this concept to wireless sensor networks. The concepts equivalence relation and SVNGs are used for creating sensor cluster and accessing cluster heads. Further, the SVNG is considered as energy of clustering of wireless sensor networks.

3.3.2 Certain Networks Models Using Single Valued Neutrosophic Directed Hypergraphs (SVNDHGs) [81]

This chapter is based on SVNDHGs by applying the concept of SVNGs to directed hypergraphs. The authors further proposed dual SVNDHGs and single valued neutrosophic directed line hypergraphs. These concepts are then utilized in some networks including social networks, collaboration networks as well as production networks. Some algorithms are also developed for dealing with such types of networks problems.

Now we present the application of NGs in handling a dynamic situation of telephone networks.

3.4 Novel System and Method for Telephone Network Planning Based on Neutrosophic Graph [89]

Literature shows no research of handling the dynamic situation of incoming and outgoing phone calls based on any fuzzy structure. This chapter presents a new algorithm that can handle some uncertain phenomena of incoming and outgoing calls using SVNGs. This new approach is capable of handling huge number of incoming and outgoing calls which may contained received, rejected and dropped calls also. The advantages of proposed work are demonstrated, and the proposed method is illustrated with an example.

Finally, we present an application of NGs in industry in the following article.

3.5 A Study on Neutrosophic Cubic Graphs (NCGs) with Real Life Applications in Industries [88]

This chapter developed a new model of NG known as NCGs and proposed several related terms and notions. An important property is discussed that states that R-union of two NCGs may not be a NCGs. The concept of join, union, Cartesian product and composition of two NCGs are developed and supported with examples. To check the validity of new approach, and application of proposed work is studied in industry.

3.6 Humanoid Robotics [106]

This chapter explained the developments of humanoid robots before the development of modern robotics engineers, philosophers and artists were attentive machines similar to humans.

3.7 Humanoid Robot "TRON" [107]

This chapter discussed the voice recognized humanoid robot.

3.8 Humanoid Robotics: A UCD Review [108]

This chapter contributes an innovation based overview of the fifty main humanoid robots which are used in the rebellious robotics field. Also this study presents a comparison of the principal aesthetic and interaction characteristics related to the environment and users in such a way that each humanoid robot was designed.

3.9 Impression Change in Humanoid Robot [109]

This chapter examined about an impression change on nonverbal non-humanoid robot by interaction with humanoid robot.

3.10 Dynamics of Humanoid Robots [110]

This chapter reviewed the mathematical foundation to define the agitation of a humanoid mechanism. It begins with the kinematics of an anthropomorphic mechanism, carried out by the equation of motion of the system and the contact mechanics that follows with the motions. Also summarized few compact representations of both the robot dynamics and the contact mechanics. Further shown some of the techniques to incorporate the intended motion into the joint influence torques based on limitations of contact forces.

3.11 Humanoid Robots from the Past to the Present [111]

This chapter reviewed the mathematical foundation to define the agitation of a humanoid mechanism. It begins with the kinematics of an anthropomorphic mechanism, carried out by the equation of motion of the system and the contact mechanics that follows with the motions. Also summarized few compact representations of both the robot dynamics and the contact mechanics. Further shown some of the techniques to incorporate the intended motion into the joint influence torques based on limitations of contact forces.

3.12 Humanoid Robots from the Past to the Present [111]

This chapter provides technical details of the development of the humanoid robots from 1970s to present day. It also presents the design, method and approach for the development of the humanoid robots. Compare to conventional robots, recent robots are doing well as they are sharp, stable and having human like capabilities such as climbing stairs, running and carrying heavy loads. Hence this chapter illustrates the technological progress of humanoid robots for the mentioned period.

3.13 Humanoid Robot in Smart Restaurant [112]

This chapter provides the role of humanoid robots in smart restaurant.

4 Integration of Neutrosophic Graphs and Humanoid Robots

Humanoid robotics is a developing and challenging research field which has been receiving significant attention during the past years. Humanoid robotics has the capacity of understanding human-like information processing and latent mechanisms of the human brain in handling with the real world. A robot which looks like the human body is a humanoid robot. The purpose of this design is for interacting with the human tools and environments and for experimental purposes. Humanoid robots have a body, a head, two arms and two legs. But there may be some of the humanoids have only part of the body i.e., a part of the body between the ribs and hips. Also some of the humanoid robots have heads modelled to imitate human facial features such as eyes and mouths. For example androids which epithetically resemble humans. As mentioned in the previous section there are many types of graphs have been introduced by the researchers under neutrosophic environment. In the dynamics of robotic system, a common feature of conventional neural architecture treats the system as a single component. For example, in general, while designing robotic arm, the input will be a vector containing all the angles and angular velocities of our arm. Hence this can be done with a special neural network which operates graph data. Here, the nodes and edges have been used in each layer of the system where each edge is updated using the node from which it begins and that in which it ends. Therefore we need to use the concepts of the graphs. Since humanoid robots are human like robots which are assisting humans in the real world it needs to handle all kinds of data which are uncertain and indeterminate. Since fuzzy and intuitionistic fuzzy and their extensions dealt only with membership and nonmembership we use neutrosophic logic as it deals with membership, nonmembership and also indeterminacy, this logic will be the optimized one for dealing with the real world problems. Hence integration of neutrosophic graphs and humanoid robots may be a good system.

5 Comparison

This chapter developed a new model of NG known as NCGs and proposed several related terms and notions. An important property is discussed that states that R-union of two NCGs may not be a NCGs. The concept of join, union, Cartesian product and composition of two NCGs are developed and supported with examples. To check the validity of new approach, and application of proposed work is studied in industry.

Crisp Graphs: In crisp graph theory, A crisp graph is based on vertices and edges. An edge between two vertices exists only if they are related to each other however it does not describe the strength of their relation For two vertices v_i and v_j, an edge $(e_{i,j})$ between v_i and v_j if exist then $e_{i,j} = 1$ if does not exist then $e_{i,j} = 0$ In the graph edges between every two vertices exist except v_j and v_k The main disadvantage is there is no information about the strength or affinity of edge relations. It is unable to represent the uncertainty in data set	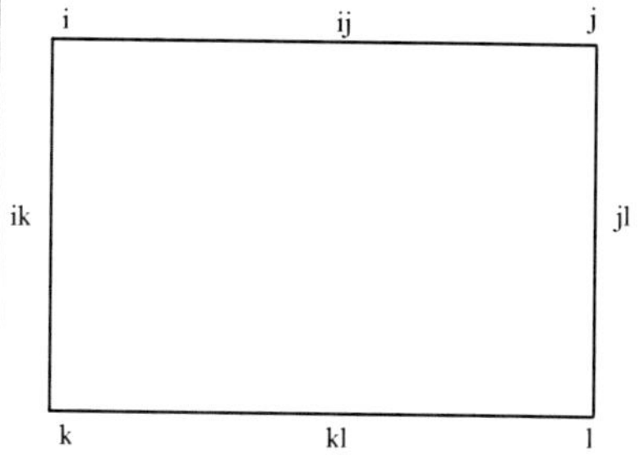

(continued)

(continued)

Fuzzy Graph: In fuzzy graph theory, edges and vertices are described in terms of membership grades from [0, 1]. In this case edge between two vertices exist if $e_{i,j} \neq 0$ and $e_{i,j} \in [0, 1]$ The graph clearly indicates the strength of each edge relation. Moreover, edge between vertex v_i and v_l does not exist as $e_{i,l} = 0$. This graph is more informative than that of crisp graph Crisp graph only discusses the concept of good or bad but fuzzy graphs can provide the degree of goodness and badness Fuzzy graphs have also a disadvantage that is fuzzy graphs only deals with membership grades of vertices and edges and provides no information about the non-membership grades as well as indeterminacy membership-values	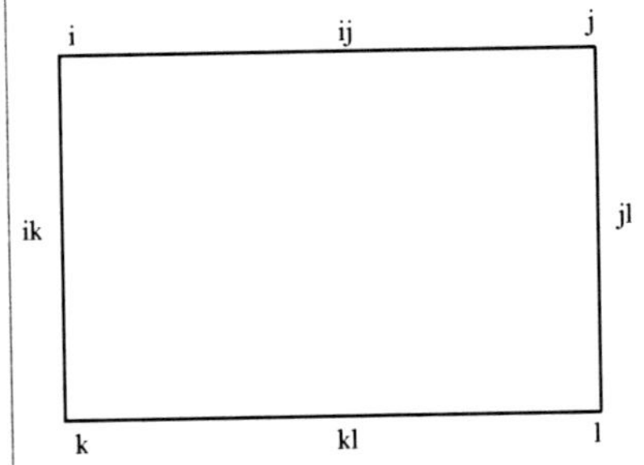 Vertex Set $P = \{\langle i, (0.5)\rangle, \langle j, (0.6)\rangle, \langle k, (0.2)\rangle, \langle l, (0.4)\rangle\}$ Edge Set $= Q = \{\langle ij, (0.4)\rangle, \langle jl, (0.3)\rangle, \langle kl, (0.2)\rangle, \langle ik, (0.1)\rangle\}$
Intuitionistic Fuzzy Graph: In intuitionistic fuzzy graph, vertices and edges are in the form of intuitionistic fuzzy number which is generalized form of fuzzy number describing non-membership grade along with membership grade. So, the main advantage of Intuitionistic fuzzy graphs over fuzzy graphs is its diverse structure The only drawback of IFGs is its limited structure. Let α, β be the membership and non-membership grades of an IFN. Then due to condition $0 \leq \alpha + \beta \leq 1$ we are unable to assign values to α and β freely	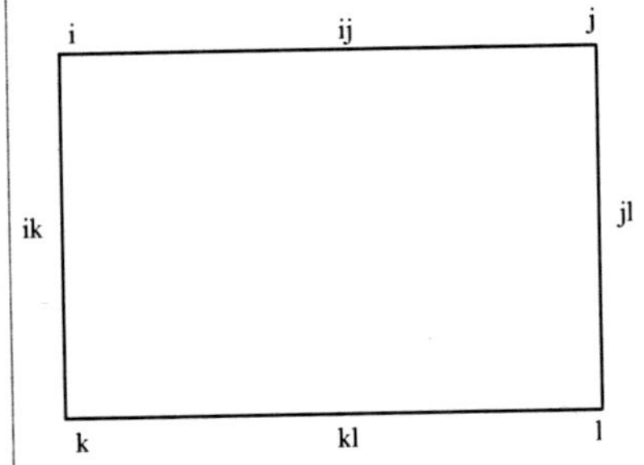 Vertex Set $P = \{\langle i, (0.5, 0.3)\rangle, \langle j, (0.6, 0.4)\rangle, \langle k, (0.2, 0.3)\rangle, \langle l, (0.4, 0.2)\rangle\}$ Edge Set $= Q = \{\langle ij, (0.4, 0.6)\rangle, \langle jl, (0.3, 0.3)\rangle, \langle kl, (0.2, 0.4)\rangle, \langle ik, (0.1, 0.4)\rangle\}$

(continued)

(continued)

<table>
<tr><td>Vague Graph:
In vague graph vertices and edges in format of vague set. In this graph the vertex and edges represented by vague set. One of the main advantages of this set is that it works on evidence to accept or reject the particular events. It is useful while analysis of opinion in a democratic country. The drawback is that the truth and false membership-values are not independent to each other</td><td>
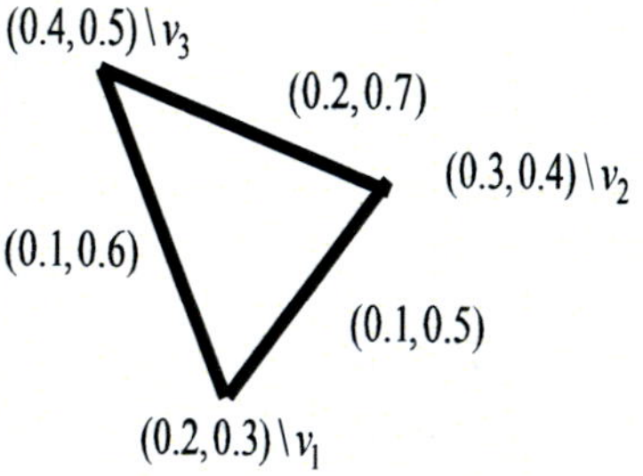

</td></tr>
<tr><td>Neutrosophic Graph (NG):
In NG, vertices and edges are represented by neutrosophic set (NS). The three components of NS namely, truth, indeterminacy and false membership functions are independent to each other and its values lie between 0 and 1. The main advantage of this graph is that it can deal with indeterminacy of the vertices and edges. Also when the terminal points and edges cannot be predicted for a network then NG can be used</td><td>Vertex Set = $P = \{\langle i, (0.5, 0.1, 0.4)\rangle, \langle j, (0.6, 0.3, 0.2)\rangle, \langle k, (0.2, 0.3, 0.4)\rangle, \langle l, (0.4, 0.2, 0.5)\rangle\}$
Edge Set = $Q = \{\langle ij, (0.4, 0.3, 0.5)\rangle, \langle jk, (0.2, 0.4, 0.5)\rangle, \langle kl, (0.2, 0.4, 0.6)\rangle, \langle il, (0.3, 0.4, 0.6)\rangle\}$
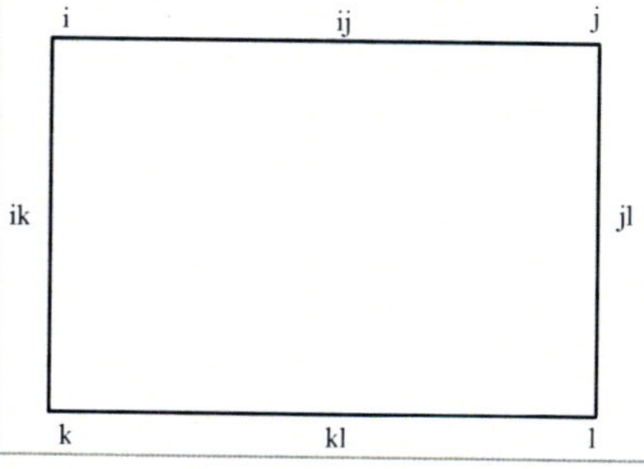

</td></tr>
</table>

6 Advantages of Working in Environment of NGs [44]

Many works have been done under crisp, fuzzy and intuitionistic fuzzy graph environments by considering vertex and edge sets by crisp/ fuzzy and intuitionistic fuzzy sets. But if the relationship of the nodes/edges are indeterminate then the above mentioned graphs are failed. NGs can be applied. Working with neutrosophic environment has the advantages such as adaptability, accuracy and harmony in relationship of vertices and edges.

In general, the relationship among many individuals can be represented by graph theory. This theory has many applications in different fields such as theory of database, recent trends in sciences and technology, neural network, intelligent systems, capturing images, grouping analysis and control theory. But all these applications and real world problems have uncertain in nature with indeterminacy of the data since human behavior is always uncertain. Hence the difficulties faced by the mentioned fields due to indeterminacy can be dealt by neutrosophic graphs.

7 Conclusion

In this chapter, several results on neutrosophic graphs and humanoid robots have been referred. Also integration of neutrosophic graphs and humanoid robots has been presented to show the effectivenss of the neutrosophic concept in robotics field especially in humanoid robots. Further, comparative analysis has been done on various types of graphs. Hence, this chapter will be a compendium for the researchers to work in the field of neutrosophic graph theory and robotics. In future, the present chapter may be extended to the integration of plithogenic graph and humanoid robots.

References

1. Smarandache, F.: Neutrosophic logic and set. mss 105 (1995)
2. Smarandache, F.: A unifying field in logics: Neutrosophic logic, in Philosophy. American Research Press, pp. 1–141 (1999)
3. Smarandache, F.: Neutrosophic set-a generalization of the intuitionistic fuzzy set. Int. J. Pure Appl. Math. **24**(3), 287 (2005)
4. Zadeh, L.A.: Fuzzy Sets. Information and Control **8**(3), 338–353 (1965)
5. Atanassov, K.T.: Intuitionistic fuzzy sets. Fuzzy Sets Syst. **20**(1), 87–96 (1986)
6. Haibin W., et al.: Single valued neutrosophic sets. Infinite Study (2010)
7. Zhang, H.J., Wang and Chen X, : An outranking approach for multi-criteria decision-making problems with interval-valued neutrosophic sets. Neural Comput. Appl. **27**(3), 615–627 (2016)
8. Kandasamy, I.: Double-valued neutrosophic sets, their minimum spanning trees, and clustering algorithm. J. Intell. Syst. **27**(2), 163–182 (2018)
9. Broumi, S., et al.: Single valued neutrosophic graphs, p. 187. Florentin Smarandache, Surapati Pramanik (2015)
10. Kaufmann, A.: Introduction à la théorie des sous-ensembles flous à l'usage des ingénieurs: Éléments théoriques de base. vol. 1, Masson (1973)
11. Rosenfeld, A.: Fuzzy graphs, in Fuzzy sets and their applications to cognitive and decision processes, Elsevier pp. 77–95 (1975)
12. Parvathi, R., Karunambigai, M.: Intuitionistic fuzzy graphs, in Computational Intelligence, Theory and Applications. Springer , pp. 139–150 (2006)
13. Akram, M., Shahzadi, S.: Neutrosophic soft graphs with application. J. Intell. Fuzzy Syst. **32**(1), 841–858 (2017)
14. Dhavaseelan, R., Vikramaprasad, R., Krishnaraj, V.: Certain types of neutrosophic graphs. Int. J. Math. Sci. Appl. **5**(2), 333–339 (2015)
15. Kandasamy, V., Ilanthenral, K., Smarandache, F.: Neutrosophic graphs: A new dimension to graph theory. Infinite Study
16. Fathi, S., Elchawalby, H., Salama, A.: A neutrosophic graph similarity measures. In: New Trends in Neutrosophic Theory and Applications-Florentin Smarandache and Surpati Pramanik (Editors), pp. 223–230 (2016)
17. Akram, M., Shahzadi, S.: Representation of graphs using intuitionistic neutrosophic soft sets. Infinite Study (2016)
18. Shah, N.: Some studies in neutrosophic graphs. Neutrosophic Sets and Systems **12**, 54–64 (2016)
19. Malik, M.A., et al.: Isomorphism of single valued neutrosophic hypergraphs. a Publication of Society for Mathematics of Uncertainty **13**, 19 (2016)

20. Hassan, A., Malik, M.A., Smarandache, F.: Regular and totally regular interval valued neutrosophic hypergraphs. Infinite Study (2016)
21. Broumi, S., et al.: An Isolated Interval Valued Neutrosophic Graphs. Critical Review **13**, 67–80 (2016)
22. Broumi, S., et al.: Isolated single valued neutrosophic graphs. Infinite Study (2016)
23. Malik, M.A., et al.: Isomorphism of bipolar single valued neutrosophic hypergraphs. a Publication of Society for Mathematics of Uncertainty **13**, 79 (2016)
24. Shah, N., Hussain, A.: Neutrosophic soft graphs. Neutrosophic Sets and Systems **11**, 31–44 (2016)
25. Broumi, S., et al.: Interval valued neutrosophic graphs. Critical Review, XII **2016**, 5–33 (2016)
26. Broumi, S., et al.: Single valued neutrosophic graphs: degree, order and size. In: Fuzzy Systems (FUZZ-IEEE), 2016 IEEE International Conference on. 2016. IEEE (2016)
27. Broumi, S., et al.: On bipolar single valued neutrosophic graphs. J. New Theory **11**, 84–102 (2016)
28. Broumi, S. et al.: An introduction to bipolar single valued neutrosophic graph theory. In: Applied Mechanics and Materials. 2016. Trans Tech Publication (2016)
29. Broumi, S., Bakali, A., Talea, M., Smarandache, F., and Singh, P. K.: Properties of Interval-Valued Neutrosophic Graphs. in :C. Kahraman and ˙ I. Otay (eds.), Fuzzy Multicriteria Decision Making Using Neutrosophic Sets, Studies in Fuzziness and Soft Computing 369, 173-202 (2019) https://doi.org/10.1007/978-3-030-00045-5_8
30. Hamidi, M., Saeid, A.B.: Accessible single-valued neutrosophic graphs. J. Appl. Math. Comput. **2017**, 1–26 (2017)
31. Ashraf, et al.: Regularity of graphs in single valued neutrosophic environment. J. Intelli. Fuzzy Syst. **33**(1), 529–542 (2017)
32. Akram, M., Sitara, M.: Representation of graph Structure based on IV neutrosophic sets. Int. J. Algebra Stat. **6**, 56–80 (2017)
33. Akram, M., Sitara, M.: Single-Valued Neutrosophic Graph Structures. Appl. Math. E-Notes **17**, 277–296 (2017)
34. Divya, G., Malarvizhi, J.: Some operations on neutrosophic fuzzy graphs. Infinite Study (2017)
35. Akram, M., Sitara, M.: Bipolar neutrosophic graph structures. Infinite Study (2017)
36. Akram, M., Nasir, M.: Interval-valued neutrosophic competition graphs. Infinite Study (2017)
37. Thirunavukarasu, P., Suresh, R.: On regular complex neutrosophic graphs. Infinite Study (2017)
38. Akram, M., Nasir, M.: Certain competition graphs based on intuitionistic neutrosophic environment. Information **8**(4), 132 (2017)
39. Singh, P.K.: Interval-Valued Neutrosophic Graph Representation of Concept Lattice and Its ($$\alpha,\beta,\gamma $$)-Decomposition. Arabian Journal for Science and Engineering **43**(2), 723–740 (2018)
40. Malarvizhi, J., Divya, G.: On antipodal single valued neutrosophic graph. Ann. Pure Appl. Math. **15**(2), 235–242 (2017)
41. Krishnaraj, V., Vikramaprasad, R., Dhavaseelan, R.: Self-centered single valued neutrosophic graphs. Infinite Study (2017)
42. Mehra, S., Singh, M.: Single valued neutrosophic signed graphs. Infinite Study (2017)
43. Hassan, A., et al.: Special types of bipolar single valued neutrosophic graphs. Infinite Study (2017)
44. Broumi, S., et al.: Generalized single valued neutrosophic graphs of first type. In: Proceedings of the 2017 IEEE International Conference on Innovations in Intelligent Systems and Applications (INISTA), Gdynia Maritime University, Gdynia, Poland (2017)
45. Broumi, S., et al.: Generalized Bipolar Neutrosophic Graphs of Type 1. Infinite Study (2017)
46. Broumi, S., Dey, A., Bakali, A., Talea, M., Smarandache, F., Son, L.H., Koley, D.: Uniform Single Valued Neutrosophic Graphs. Neutrosophic Sets and Systems, Vol. **17**, 42–49 (2017)
47. Broumi, S., et al.: Generalized Interval valued Neutrosophic graphs of first type. In: 2017 IEEE International Conference on Innovations in Intelligent SysTems and Applications (INISTA), Gdynia Maritime University, Gdynia, Poland (2017)

48. Broumi, S., et. al.: A bipolar single valued neutrosophic isolated graphs: Revisited. Infinite Study (2017)
49. Karaaslan, F., Davvaz, B.: Properties of single-valued neutrosophic graphs. J. Intell. Fuzzy Syst. **34**(1), 57–79 (2018)
50. Dhavaseelan, R., Krishnaraj, V., Vikrama, P.: Perfect and status in single valued neutrosophic graphs. International J. Mech. Eng. Techonol. **9**(1), 1124–1132 (2018)
51. Ishfaq, N., et al.: Notions of rough neutrosophic digraphs. Mathematics **6**(2), 18 (2018)
52. Akram, M., Shahzadi, S., Saeid, A.B.: Single-valued neutrosophic hypergraphs. TWMS J. Appl. Eng. Math. **8**(1), 122 (2018)
53. Sudhakar, V.J., Mohamed Ali, A., Vinoth, D.: Interval valued signed neutrosophic graph. Int. J. Math. Archive EISSN 2229–5046 **9**(9) (2018)
54. Quek, S., Broumi, S., Selvachandran, G., Bakali, A., Talea, M., Smarandache, F.: Some results on the graph theory for complex neutrosophic sets. Symmetry **10**(6), 190 (2018)
55. Broumi, S., et al.: Bipolar complex neutrosophic graphs of Type 1. Infinite Study (2018)
56. Jeyanthi, V., Radhika, : Applying Floyd's algorithm for solving neutrosophic shortest path problems. Int. J. Math. Trends Technol. **61**(1), 58–63 (2018)
57. Akram, M., Sayed, S., Smarandache, F.: Neutrosophic incidence graphs with application. Axioms **7**(3), 47 (2018)
58. Ali, et al.: Self-centered interval valued signed neutrosophic graph. Int. J. Math. Archive **9**(9):44–50 (2018)
59. Dey, A., et al.: A new algorithm for finding minimum spanning trees with undirected neutrosophic graphs. Granular Computing 1–7 (2018)
60. Broumi, S., et al.: Decision-making method based on the interval valued neutrosophic graph. In: IEEE conference Future technologies conference (FTC) (2016)
61. Broumi, S., et al.: Applying Dijkstra algorithm for solving neutrosophic shortest path problem. In: IEEE International Conference Advanced Mechatronic Systems (ICAMechS) (2016)
62. Broumi, S., et al.: Computation of shortest path problem in a network with SV-trapezoidal neutrosophic numbers. In: IEEE International Conference on Advanced Mechatronic Systems (ICAMechS) (2016)
63. Broumi, S., et al.: Application of Dijkstra algorithm for solving interval valued neutrosophic shortest path problem. In: 2016 IEEE Symposium Series on Computational Intelligence (SSCI) (2016)
64. Broumi, S., et al.: Shortest path problem under triangular fuzzy neutrosophic information. In: 2016 10th International Conference on Software, Knowledge, Information Management & Applications (SKIMA), IEEE (2016)
65. Broumi, S., et al.: Shortest path problem on single valued neutrosophic graphs. In: 2017 International Symposium on Networks, Computers and Communications (ISNCC) IEEE (2017)
66. Broumi, S., et al.: Shortest path problem under bipolar neutrosphic setting. Appl. Mech. Mater. **859** (2017)
67. Akram, M., Siddique, S., Davvaz, B.: New concepts in neutrosophic graphs with application. J. Appl. Math. Comput. **57**(1–2), 279–302 (2018)
68. Broumi, S., Smarandache, F., Bakali, A., Mehra, S., Talea, M., Singh, M.: Strong degrees in single valued neutrosophic graphs. In: Arai K., Kapoor S., Bhatia R. (eds) Advances in Information and Communication Networks. FICC 2018. Adv. Intell. Syst. Comput. **886**. Springer, Cham, 221–238(2019)
69. Devi, R.N., et al.: Characterizations of Strong and Balanced Neutrosophic Complex Graphs. Int. J. Eng. Technol. **7**(4.10):593–597 (2018)
70. Mandal, K., Basu, K.: Improved similarity measure in neutrosophic environment and its application in finding minimum spanning tree. J. Intell. Fuzzy Syst. **31**(3), 1721–1730 (2016)
71. Kalyan, S., Pinaki, : Entropy based Single Valued Neutrosophic Digraph and its applications. Neutrosophic Sets Syst. **19**, 119–126 (2018)
72. Vladareanu, V., et al.: Neutrosophic application for decision logic in robot Intelligent control systems. In: 2017 IEEE International Conference on INnovations in Intelligent SysTem and Applications, Gdynia, Maritime University, Poland 420–425 (2017)

73. Vladareanu, V., et al.: Robot System Identification using 3D Simulation Component Applied on VIPRO Platform. In: Proceedings of the 2016 th International Conference on Advanced Mechatronics Systems, Melbourne, Australia, 2016 406–411 (2016)
74. Hashimoto, K.: Mechanics of humanoid robot. Advanced Robotics (2020). https://doi.org/10.1080/01691864.2020.1813624
75. Martin, N., Smarandache, F., Broumi,S.: PROMTHEE Plithogenic Pythagorean Hypergraphic Approach in Smart Materials Selection, Int. J. Neutrosophic Sci. **13**, Issue1, 52–60 (2021)
76. Mullai, M., Broumi, S., Stephen, A.: Shortest path problem by minimal spanning tree algorithm using bipolar neutrosophic numbers. Infinite Study (2017)
77. Broumi, S., Bakali, A., Talea, M., Smarandache, F., Krishnan Kishore, K.P., Şahin, R.: Shortest Path Problem under Interval Valued Neutrosophic Setting. International J. Adv. Trends Comput. Sci. Eng. **8**(1.1), 216–222 (2019)
78. Broumi, S., et al.: Bipolar Neutrosophic Minimum Spanning Tree (2018)
79. Broumi, S., Bakali, A., Talea, M., Smarandache, F., Dey, A.and Son, Le H.: Spanning Tree Problem with Neutrosophic Edge Weights. 2018 Proc. Comput. Sci. **127**, 190–199 (2018)
80. Sahin, R.: An approach to neutrosophic graph theory with applications. Soft. Comput. **2017**, 1–13 (2017)
81. Akram, M., Luqman, A.: Certain networks models using single-valued neutrosophic directed hypergraphs. J. Intell. Fuzzy Syst. **33**(1), 575–588 (2017)
82. Akram, M., Luqman, A.: Bipolar neutrosophic hypergraphs with applications. J. Intell. Fuzzy Syst. **33**(3), 1699–1713 (2017)
83. Akram, M., Luqman, A.: A new decision-making method based on bipolar neutrosophic directed hypergraphs. J. Appl. Math. Comput. **57**(1–2), 547–575 (2018)
84. Akram, M., Sarwar, M.: Novel multiple criteria decision making methods based on bipolar neutrosophic sets and bipolar neutrosophic graphs. Infinite Study (2016)
85. Akram, M., Sitara, : Novel applications of single-valued neutrosophic graph structures in decision-making. J. Appl. Math. Comput. **56**(1–2), 501–532 (2018)
86. Hamidi, M., Borumand Saeid, A.: Achievable Single-Valued Neutrosophic Graphs in Wireless Sensor Networks. New Math. Nat. Comput. **14**(02), 157–185 (2018)
87. Akram, M., Siddique, S.: Neutrosophic competition graphs with applications. J. Intell. Fuzzy Syst. **33**(2), 921–935 (2017)
88. Gulistan, M., et al.: A study on neutrosophic cubic graphs with real life applications in industries. Symmetry **10**(6), 203 (2018)
89. Broumi, S., et al.: Novel system and method for telephone network planing based on neutrosophic graph. Glob. J. Comput. Sci. Tech. (2018)
90. Akram, M., Shahzadi, G.: Operations on Single-Valued Neutrosophic Graphs. J. Uncertain Syst. **11**(1), 1–26 (2017)
91. Malik, M.A., et al.: Regular bipolar single valued neutrosophic hypergraphs. Infinite Study (2016)
92. Shah, N., Broumi, S.: Irregular neutrosophic graphs. Neutrosophic Sets Syst. **13**, 47–55 (2016)
93. Akram, M.: Single-valued neutrosophic planar graphs. Infinite Study (2016)
94. Akram, M., Luqman, A.: Intuitionistic single-valued neutrosophic hypergraphs. OPSEARCH **54**(4), 799–815 (2017)
95. Hassan, A., Malik, M.: Single valued neutrosophic trees. J. Appl. Eng. Math. **8**(1a), 255–266 (2018)
96. Mishra, S., Pal, : Intuitionistic Fuzzy Signed Graphs. Int. J. Pure Appl. Math. **106**(6), 113–122 (2016)
97. Naz, S., Akram, M., Smarandache, F.: Certain notions of energy in single-valued neutrosophic graphs. Axioms **7**(3), 50 (2018)
98. Akram, M., Nasir, M.: Concepts of interval-valued neutrosophic graphs. Infinite Study (2017)
99. Broumi, S., et al.: On strong interval valued neutrosophic graphs. Critical Review **12**, 49–71 (2016)
100. Akram, M., Sitara, M.: Interval-valued neutrosophic graph structures. Infinite Study (2017)

101. Malik, M.A., et al.: Isomorphism of interval valued neutrosophic hypergraphs. a Publication of Society for Mathematics of Uncertainty **13**, 41 (2016)
102. Akram, M., Shum, K.: Bipolar neutrosophic planar graphs. Infinite Study (2017)
103. Broumi, S, et al.: Interval Complex Neutrosophic Graph of Type 1. Neutrosophic Operational Research **3**: p. https://vixra.org/pdf/1807.0114v1.pdf (2018)
104. Chalapathi, T., Kumar, R.V.V.K.: Neutrosophic graphs of finite groups. Neutrosophic Sets Syst. **2017**, 22 (2017)
105. Naz, S., Rashmanlou, H., Malik, M.A.: Operations on single valued neutrosophic graphs with application. J. Intell. Fuzzy Syst. **32**(3), 2137–2151 (2017)
106. Miheli, M., et al.: Humanoid Robotics. Robotics, In Book https://doi.org/10.1007/978-3-319-72911-4_14 (2019)
107. Shekhar, et al.: Humanoid robot "TRON". In: International Conference on Emerging Trends in Engineering, Management & Sciences (ICETEMS), Graphic Era, India (2020)
108. Casiddu, N., et al.: Humanoid Robotics: A UCD Review. In: Advances in Human Factors in Robotics, Drones and Unmanned Systems doi: https://doi.org/10.1007/978-3-030-51758-8_12 (2021)
109. Ueno, A., Hayashi, K., Mizuuchi I: Impression Change on Nonverbal Non-Humanoid Robot by Interaction with Humanoid Robot. In: 2019 28th IEEE International Conference on Robot and Human Interactive Communication (ROMAN) doi: https://doi.org/10.1109/RO-MAN46459.2019.8956240 (2019)
110. Sigihara, T., Morisawa, M.: A Survey: dynamics of humanoid robots. Advanced Robotics (2020). https://doi.org/10.1080/01691864.2020.1778524
111. Bogue, R.: Humanoid robots from the past to the present. Industrial Ahead-Of-Print (2020). https://doi.org/10.1108/IR-05-2020-0088
112. Gupta, S.: Smart Restaurant and Humanoid Robot. International Journal of Research in Applied Science and Engineering Technology (2020). https://doi.org/10.22214/ijraset.2020.5186

Index

C. Kahraman and E. Bolturk (eds.), *Toward Humanoid Robots: The Role of Fuzzy Sets*, Studies in Systems, Decision and Control 344,
https://doi.org/10.1007/978-3-030-67163-1